ELECTRIC POWER DISTRIBUTION, AUTOMATION, PROTECTION, AND CONTROL

ELECTRIC POWER DISTRIBUTION, AUTOMATION, PROTECTION, AND CONTROL

James A. Momoh
Howard University, Washington DC, USA

CRC Press
Taylor & Francis Group
Boca Raton London New York

CRC Press is an imprint of the
Taylor & Francis Group, an **informa** business

CRC Press
Taylor & Francis Group
6000 Broken Sound Parkway NW, Suite 300
Boca Raton, FL 33487-2742

Printed in the United States of America on acid-free paper
10 9 8 7 6 5 4 3 2 1

International Standard Book Number-13: 978-0-8493-6835-6 (Hardcover)

Library of Congress Cataloging-in-Publication Data

Momoh, James A., 1950-
Electric power distribution, automation, protection, and control / James A. Momoh.
p. cm.
Includes bibliographical references and index.
ISBN 978-0-8493-6835-6 (alk. paper)

1. Electric power distribution--Automation. 2. Electric power distribution--Automatic control. I. Title.
TK3001.M63 2008 621.319--dc22
2007022016

Visit the Taylor & Francis Web site at
http://www.taylorandfrancis.com

and the CRC Press Web site at
http://www.crcpress.com

Contents

Preface

This book is intended to introduce distribution engineering as a growing area suitable for studying new trends in computation, automation, and control techniques. The idea is to present the basic concepts for assessment, design, formulation, and analysis of distribution performance. This is timely, given the growing research interest, the desire for automation, and the commitment to build an efficient and cost-effective distribution system in a competitive utility environment.

The textbook is intended as a resource for electrical engineering students, as well as professional engineers, who are interested in learning the fundamentals of distribution engineering analysis. The book presents computation and automation techniques in a simple, easy-to-follow treatment. Background requirements include a basic concept of electric circuits and a working knowledge of foundation mathematics. The text is arranged from basic distribution principles through renewable energy resources, computation tools and techniques, reliability and maintenance, distribution automation, and telecommunications. The topics are covered with illustrative examples and some case studies to illuminate the topic as needed. Overall, the book provides both analytical basics and practical intuition for the future design of distribution systems.

Chapters 1 and 2 treat the foundation of distribution automation by summarizing distribution topology, modeling, and different computation techniques.

Chapter 3 introduces distribution protection and control schemes for self-defense of distribution systems under different fault types; different relay-protection schemes are also introduced, and some illustrative examples for coordination and relay settings are given.

Chapter 4 discusses distribution reliability, computation techniques, and maintenance concepts. These topics are helpful in evaluating the performance of distribution systems to guide the distribution operator, planner, and maintenance engineer in choosing among the tools available to enhance practical "rule of thumb" judgment.

Chapter 5 is dedicated to distribution automation and control functions. Here, we deal with the different automation functions and review various modeling, analytical, and computational methods using a background in optimization techniques. Here, only analytical functions and statements of outstanding work done by researchers and the author are given as working examples.

Chapter 6 deals with the extension of distribution automation functions and computation using intelligent systems (IS). This is an important topic, given the ample engineering rules and new trends in computational intelligence that can be used in the design of future distributed systems.

Chapter 7 is concerned with renewable energy sources; its models, characteristics, benefits, drawbacks, and possible areas of application are treated.

Chapter 8 presents new advances in communication technology for data acquisition, monitoring, control, load management, billing, and metering of distribution systems.

Chapter 9 provides a foundation of telecommunications from basic theory to practice, including modulation, networking, frame relay, standards, and security strategy. Communication concepts have become critical to power system distribution automation and control in today's competitive environment, which demands ever-greater reliability and efficiency.

It is hoped that the introduction of new trends in IT (information technology) and artificial intelligence (AI) will enhance future performance of distribution and that the reader will continue to engage in the developmental work done by researchers. The goal of the book will be achieved if distribution engineers will adapt and build future generations of distribution systems using the technology discussed.

The author is indebted to outstanding research by colleagues, sponsored conferences, workshops, popular text in related material, and sponsored research in distribution automation of which I have had personal involvement. These involve research and development efforts supported by NSF, DOE Oakridge National Laboratory, NREL, NASA, and LADWP in the development and testing of various algorithms for the distribution automation and reliability study of optimization for power management and distribution applicable to both utility and navy ship systems.

I remain indebted to my colleagues who offered encouragement and critical reviews of the book during the preparation stage. I wish to thank my graduate student, Garfield Boswell, who kept the hope alive, as well as other graduate and undergraduate students who came in at the last minute to help get this book done!

Finally, I offer my deepest personal gratitude to my family, who always showed encouragement for me to get this book done.

Author

James A. Momoh is a professor and former chair (1990–2000) of the Department of Electrical Engineering as well as the director of the Center for Energy Systems and Controls at Howard University, Washington, D.C. Additionally, he served as the program director of the Electrical and Communication Systems division at the National Science Foundation (2001–2004), where he was responsible for the development of the interdisciplinary program, Electric Power Network Efficiency and Security. He was also a principal consultant at Booneville Power Administration, Portland, OR, as well as the affiliate staff scientist at Pacific Northwest Laboratory, Seattle, WA. Dr. Momoh has authored *Electric Power System Applications of Optimization* and coauthored *Electric Systems, Dynamics and Stability with Artificial Intelligence Applications* (Marcel Dekker, Inc.). He has over 200 technical publications and reports in the field of power engineering. He is an associate editor of the journals *Power Letters* and *Journal of Electric Machines and Power Systems*.

Dr. Momoh received the B.S.E.E. degree (1975) with honors from Howard University, Washington, D.C.; the M.S.E.E. degree (1976) in electrical engineering from Carnegie Mellon University, Pittsburgh, PA; the M.S.S.E degree (1980) in systems engineering from the University of Pennsylvania, Philadelphia; and the Ph.D. degree (1983) in electrical engineering from Howard University, Washington, D.C. In addition, he received an M.A. degree (1991) in theology from the School of Divinity at Howard University, Washington, D.C. A recipient of the 1987 National Science Foundation U.S. Presidential Investigator Award, he is a Fellow of the Institute of Electrical and Electronics Engineers (IEEE), a member of the National Academy of Engineering (Nigeria), and also holds membership in numerous other professional and honor societies.

1

Introduction to Distribution Automation Systems

1.1 Historical Background

Power system utilities consist of generation, transmission, and distribution functions. Several advances have been made to improve the performance, efficiency, reliability, and security of power systems. The initial design of the electricity industry by Edison in 1881, with AC generation, has changed with several modifications. This design, with its modifications, has led to the development of today's power system utilities. The design of large-scale electric production has produced AC power at high voltage and current levels. The growth of the industry has led to many innovations, including economy of scale from large hydro, fossil fuel and, recently, small independent power producers (IPP), in what is called distributed generation. The designs of distributed generation have been based on criteria to improve its reliability, load management, and system performance in response to various disturbances. Over the last decade, protection schemes to detect abnormalities, control schemes to stabilize the system, and economic principles to ensure optimal allocation and bidding have all been implemented to ensure a network's competitiveness in the electric market.

The generated power is transmitted over long distances from city to city or across country boundaries. The transmission lines can be rated to operate as either DC or AC systems at low, medium, or extra-high voltage levels of 230 kV, 750 kV, or 1130 kV, respectively. Efficiency and reliability at an affordable cost is the ultimate aim of the transmission planners and operators. The line must withstand and tolerate dynamic changes in load and contingency without unreasonable impact on the continuity of service. To ensure that the system meets the expected performance, reliability, and quality of supply, some standards are preferred following the occurrence of a contingency. Simulation tools and advanced technology such as load flow, optimal power flow, state estimation, stability estimation, reliability estimation, market stimulation tools, and flexible AC transmission devices (FACTs) have been developed to ensure the reliability and security of the transmission/distribution system. The transferred power is ultimately delivered to residential, commercial, and industrial customers at local but lower voltage levels. The voltage level for industrial customers ranges from 4.0 kV to 34.4 kV. Residential customers are supplied with voltage levels at 120/240 V, while the typical voltage level for commercial customers is 440 V. The distribution reliability and the quality of utility services are easily measured by all stakeholders at the customer end. With this in mind, the progressive utility must provide adequate planning and operation, as well as reliability-centered maintenance to the system, to minimize downtime of service from the distribution level up.

1.2 Distribution System Topology and Structure

Distribution system topology can take a variety of forms. The topology is typically radial or ring, mesh, or radial mesh, depending on the configuration, quality of service, and cost. The cost of operation and maintenance (or lack of it) is usually huge, so appropriate techniques used in communication technology and automation are desirable in achieving a distribution system of high quality. For example, distribution automation functions have recently been designed to support trouble call analysis, which will reduce repair crew time and ensure timely payment of bills. Distribution automation also enhances integration to system reconfiguration and restoration, thereby minimizing losses and voltage deviation, especially during an emergency. Several optimization and intelligent-system techniques are used in the design of distribution automation schemes.

Prototype work is being carried out using optimization and intelligent-system techniques to address some of the common day-to-day problems that can affect the quality of service. Furthermore, the penetration of electronic devices such as power converters and flexible AC transmission (FACT) devices can be utilized to improve the system power quality. The future distribution network will also incorporate distributed generation, such as photovoltaic (PV), wind power, biomass, and microturbines. This has improved the capability of distributed systems to meet the ever-changing load demands at a reduced cost for capital equipment.

The transmission and distribution of electrical power is commonly based on single- and three-phase transmission using aluminum conductors from point to point or to many other points. The challenge of routing power within its capacity limits at minimum cost and minimum losses is part of the overall design problem.

Power systems (in an unbalanced state) in the new competitive environment also have to meet some regulatory requirements to ensure safety and security. The important functions and regulatory requirements that must be met are as follows:

1. Generation, transmission, and distribution must be able to meet anticipated demand with sufficient reserve margins, which could be met by demand-side management schemes or storage schemes for the distribution business units.
2. The power system, including distribution subsystems, must be cost effective with the overall goal of meeting technical, economic, environmental, and public-perception constraints.
3. The reliability and quality of power transmission and distribution must be able to meet minimum standards.
4. Appropriate cost-benefit analysis should be done to ensure priority of project execution, which will improve the performance and quality of service.

With this in mind, modern tools must be developed to support the distribution options that have traditionally been tracked as nonrigorous, simple, and error-analysis strategies.

The distribution system's main features are shown in Figure 1.1. The sample diagram in Figure 1.1 consists of fuses, reclosers, relays, a circuit breaker, transformers, regulators (voltage), and dispersed generator/storage. We describe each of them here briefly:

Relay: a device designed to protect against overvoltage, -frequency, or -current. It relays abnormal voltage or current to the circuit breaker to open (close) a circuit from further deterioration due to fault signals.

Reclosers: devices serving as special purpose, light-duty circuit breakers for interrupting overloads but not faults. It allows temporary faults to clear and then restores service quickly, but disconnects a permanent fault.

Circuit breaker: a high-current device that automatically disconnects faulted equipment. It facilitates protection of equipment from further damage or people from injury, and it is typically rated in terms of voltage and fault current. Circuit breakers come in different forms due to the arcing phenomena caused during contact (opening/closing) at high voltage. Typical models are air-blast circuit breakers, vacuum circuit breakers, oil circuit breakers, and sulfur hexafluoride circuit breakers, which use SFL gas media for extra-high voltage, which are applications above 345 kV.

Fuses: These are devices that melt when overload current passes through it. They come in different forms of low- or high-voltage fuses made from zinc, copper, silver, cadmium, or tin materials. They are rated in terms of BIL, voltage, continuous current, and interrupt-capacity fuse coordination (time it takes for the fuse to blow).

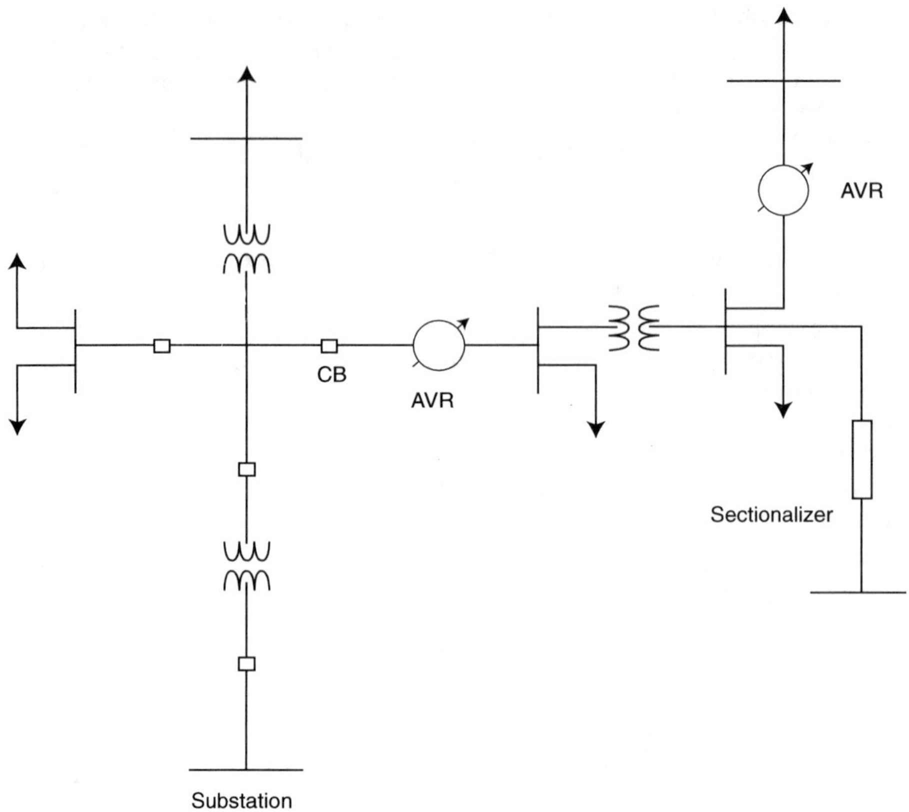

FIGURE 1.1
Distribution system.

Sectionalizer: a device that is used to automatically isolate a fault on a line segment from a disturbance. It senses any current above its activating current followed by a line and then de-energizes using a recloser.

Renewable energy/storage: referred to as IPP, an independent power producer at the customer side. It is called distributed power resulting from a renewable energy source such as photovoltaic, biomass, microturbine, or wind power.

A complete distribution subsystem includes other pieces of equipment, such as batteries, sensors, and computer application software. Overall, the additional equipment or apparati provide functionality to ensure real-time monitoring and control of the power system distribution. It is a creative art of ingenious engineering and has served the industry for years. However, as communication and intelligent-system technology advances, distribution systems can be enhanced. The potential of this automation is a fundamental concern of the text.

1.3 Distribution Automation (DA) and Control

The term "distribution automation" is used to define the application of communication, optimization, and intelligent systems to improve the performance and functions of distribution systems during normal and abnormal operation. DA facilitates system efficiency, quality of service, and the security of the power system. These abilities are classified as DA function options as follows:

Efficiency: DA function option that controls (minimizes) losses through network reconfiguration and restoration by appropriate relocation of fuses, circuit breakers, and loads for optimum performance during an overload.

Reliability and quality: To guarantee that the system is reliable at an acceptable value of risk (given the history of recorded failures and duration), an index to quality-acceptable customer-interruption service preference is proposed. Actions to manage unreliability through maintenance or demand-side management (DSM) are planned using distribution automation. New data-gathering tools such as power management unit (PMU) and frequency recorders are used for reliability assessment.

Security: The security of distribution is enhanced using integration of dispersed energy storage, distributed generators (DGs), or FACT devices. The aim here is to reduce voltage sag and eliminate harmonics that could cause low power quality and to dampen instability caused by penetration of DGs.

The integration of these DAs will provide a platform for building a future, highly competitive, and efficient autonomous distribution system that will be able to respond to different situations and be self-aware, self-organizing, and self-reconfigurable.

We present here an overview of DAs for distribution systems. The overall structure indicated in Figure 1.2 utilizes a combination of optimization and intelligent systems to develop effective DA functions. For example, the intelligent system (IS) will be based on fuzzy logic for demand-side management and restoration. Expert systems will be used for classification and ranking of control options, and artificial neural networks (ANN) will be used for fault detection and restoration as well as power quality assessment and control.

Optimization schemes based on linear and mixed integer programming and next-generation optimization techniques, evolutionary programming, adaptive dynamic programming (ADP), Tabu search, and annealing methods will be used to enhance the development of distribution automation functions.

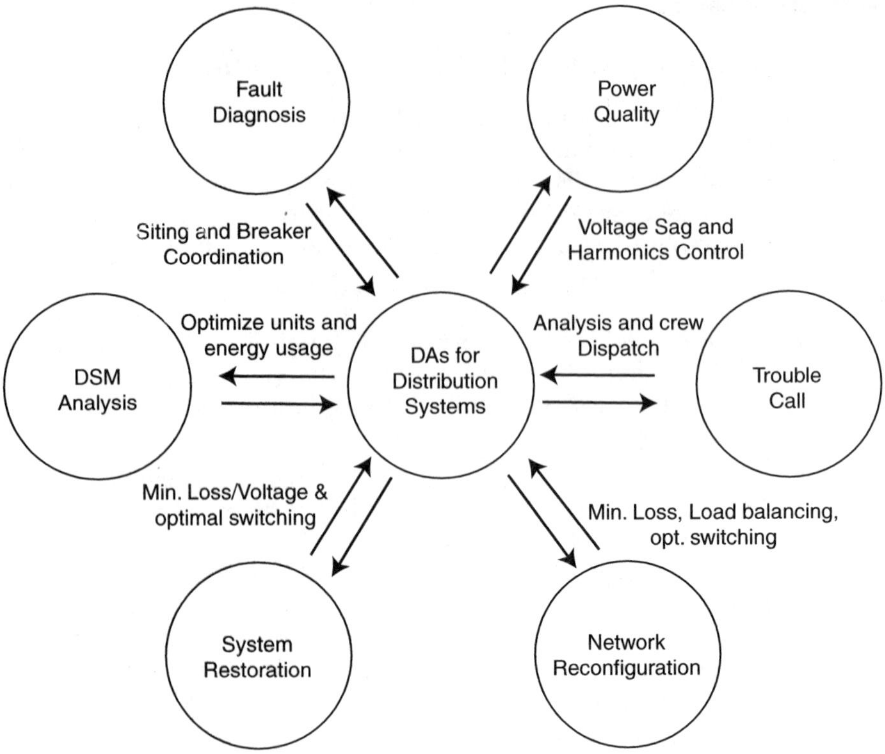

FIGURE 1.2
DA functions and structure.

Finally, this book introduces the reader to the fascinating new trend of integrating intelligent systems (IS) and telecommunication applications to power system distribution automation and control.

1.4 Summary

This chapter explained the major concepts involved in the distribution system modeling of various components in a typical distribution system. First, the concept of distribution configuration, different aspects of distribution study, and advances for automation and control of distribution system are described. Finally, a summary of open questions and new advances to be discussed in the text are given.

Problem Set 1

1. List the major advantages and disadvantages of ring, mesh, and combined distribution system topology.
2. What are the control strategies needed to improve the performance of each type of topology?
3. Discuss the advantages and disadvantages of overhead and underground distribution.
4. What are the tools for estimating distribution automation functions?
5. Consider the management and functions of a typical distribution power system.
   a. What are the important functions and regulatory requirements for power systems operating in an unbalanced state in the new competitive environment?
   b. What are the main features of the distribution system?
6. Define the following terms as they pertain to distribution automation and control:
   a. Efficiency
   b. Reliability and quality
   c. Security
7. Define the roles and operations of the following distribution system protection devices:
   a. Relay
   b. Re-closer
   c. Fuse
   d. Sectionalizer
8. Construct a diagram representing the Distribution Automation structure showing what each part of the structure entails.

2

Computational Techniques for Distribution Systems

2.1 Introduction

Distribution system networks consisting of transmission lines, generators, loads, fuses, capacitors, and reclosers have been presented in the introductory chapter. To fully analyze the system under steady-state or transient conditions requires some fundamental concepts of computational techniques. This chapter therefore presents a review of basic computational fundamentals, terminology, and notation used for the analysis of single-phase or multiphase distribution systems.

The review presented here covers such topics as instantaneous and complex power, power factor, loss calculations, and management in distribution systems, as well as single- and three-phase load-flow analysis techniques used in distribution networks. This background will provide a basis for the computational tools needed for subsequent operational and planning studies in distribution systems. These tools are applicable in fault studies, distribution reliability assessments, and distribution automation function computations. In addition, the fundamental tools used give us a greater appreciation for the use of communication and software tools designed specifically for distribution planning, protection, and control.

2.2 Complex Power Concepts

The computation of power in a circuit is generally found using the instantaneous current injection and the potential difference across the circuit elements. Consider a simple load circuit given as a generalized load $\mathbf{Z}_{\text{load}}$ connected to a voltage source $v(t)$, as shown in Figure 2.1. The sinusoidal representation of the source voltage is given as

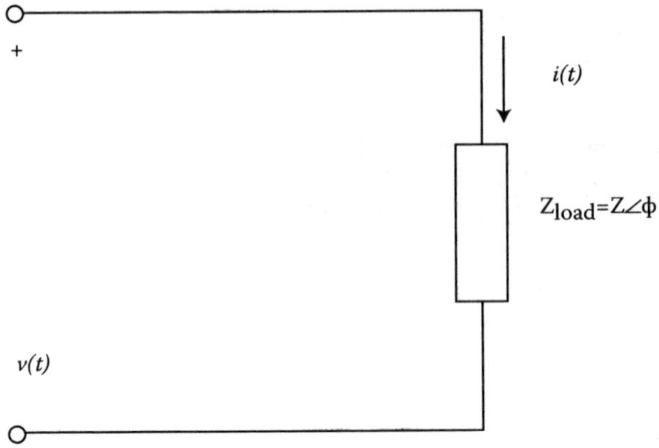

FIGURE 2.1
Load circuit.

$$v(t) = V_m \cos(\omega t) \tag{2.1}$$

and the corresponding current is

$$i(t) = I_m \cos(\omega t - \phi) \tag{2.2}$$

where

V_m = amplitude of the source voltage (in volts)

ω = angular frequency (in rad·sec^{-1})

ϕ = phase shift of the voltage waveform with respect to the current (in rads)

V_{rms} = root-mean-square (rms) value of voltage computed as $V_{rms} = \frac{V_m}{\sqrt{2}}$

$I_m = \frac{V_m}{|Z|}$ = current magnitude

Now, the complex power or apparent power is defined as the total orthogonal components in a vector space given by

$$\begin{aligned} S = P + jQ &= VI^* \\ &= VI \cos\phi + jVI \sin\phi \\ &= VI (\cos\phi + j \sin\phi) \end{aligned} \tag{2.3}$$

Using Euler's identity,

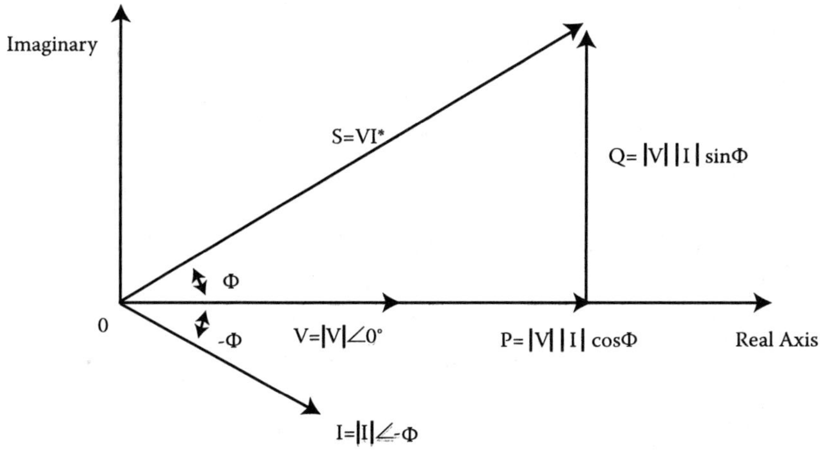

FIGURE 2.2
Phasor representation of complex power relationships.

$$S = VIe^{j\phi} = VI\angle\phi \tag{2.4}$$

and defining $I^* = |I|\angle\phi$, we can write $S = VI^*$ as an equivalent apparent power.

We also note that $V = ZI$ and $I = YV$ yield alternate forms of $S = (VI)I^*$ or $S = V(YV)^* = VV^*Y^*$. These relationships for power using the phase relationship can be denoted in graphical form, as shown in Figure 2.2.

2.2.1 Power Equations

For a typical power network consisting of R (resistive), L (inductive), and C (capacitive) elements, the following subsections summarize power equations in terms of the voltages and currents for the power dissipated or developed across these elements.

2.2.1.1 Resistive Element

In the case of a purely resistive network, we develop the following power relations:

$$P_R(t) = \frac{1}{T}\int_0^T v_R(t)\, i_R(t)\, dt \tag{2.5}$$

$$P_{\text{average}} = V_m I_m \cos^2 \omega t = \frac{V_m I_m}{2}\left[1 + \cos 2\omega t\right] \tag{2.6}$$

Using $V_{\text{rms}} = \frac{V_{\text{m}}}{\sqrt{2}}$ and $I_{\text{rms}} = \frac{I_{\text{m}}}{\sqrt{2}}$,

$$P_{\text{average}} = \frac{V_{\text{m}}}{\sqrt{2}} \frac{I_{\text{m}}}{\sqrt{2}} \left[1 + \cos 2\omega t\right] \tag{2.7}$$

The real power loss or power dissipated by the resistor in watts is computed using

$$P_{\text{R}} = VI_{\text{R}} = \frac{V^2}{R} = I_{\text{R}}^2 R \tag{2.8}$$

2.2.1.2 Inductive Element

Similarly, for a purely inductive network, we have the following formulation for the reactive power absorbed, $P_{\text{L}}(t) = v_{\text{L}}(t) i_{\text{L}}(t)$, such that

$$P_{\text{L}}(t) = \frac{1}{2} V_{\text{m}} I_{\text{m}} \cos(\omega t + \phi) \cos(\omega t + \phi - \pi / 2) \tag{2.9}$$

$$= V_{\text{m}} I_{\text{m}} \sin[2(\omega t + \phi)] \tag{2.10}$$

and the reactive power loss in VArs is $|V|^2/X_{\text{L}}$ where X_{L} is the reactive inductance.

2.2.1.3 Capacitive Element

For a purely capacitive element, the power developed is given as

$$P_{\text{C}}(t) = v_{\text{C}}(t) i_{\text{C}}(t) = V_{\text{m}} I_{\text{m}} \cos[2(\omega t + \theta) + \pi / 2]$$

$$= -VI \sin[2(\omega t + \theta)] \tag{2.11}$$

and the power loss in VArs is $|V|^2/X_{\text{C}}$ where X_{C} is the capacitive reactance.

Finally, depending on the configurations of the RLC elements (e.g., series, parallel, or series-parallel arrangements), the total real and reactive powers are computed according to the voltage and current distribution.

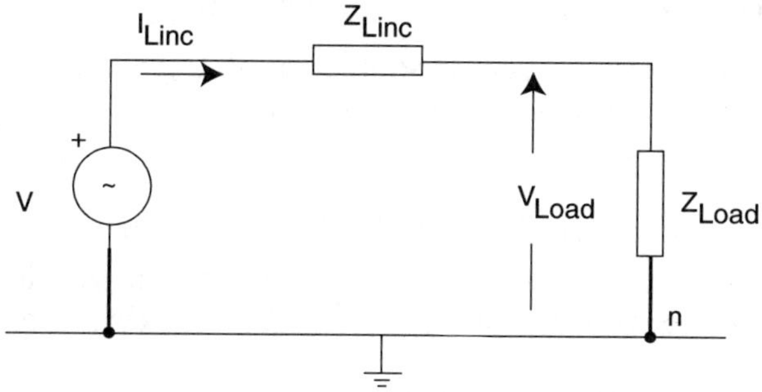

FIGURE 2.3
Single-phase generator connected to a load.

2.2.2 Single-Phase Power Formulations

Consider a single power source supplying a load with or without feeder impedance, as shown in Figure 2.3. Using the nomenclature defined above, recall that the sinusoidal representation of the source voltage is given as

$$V = V_{\mathrm{m}} \cos(\omega t) \tag{2.12}$$

and the corresponding current is

$$I_{\mathrm{Line}} = I_{\mathrm{Load}} = I_{\mathrm{m}} \cos(\omega t - \phi) \tag{2.13}$$

The power delivered to the system consisting of the network and the load is given as

$$S_{\mathrm{S}} = VI_{\mathrm{L}}^{*} = VI_{\mathrm{m}} \cos(\omega t)\cos(\omega t + \phi) \tag{2.14}$$

Similarly, the received power or power developed across the load is

$$S_{\mathrm{R}} = V_{\mathrm{Load}}I_{\mathrm{L}}^{*} = V_{\mathrm{Load}}I_{\mathrm{m}} \cos(\omega t)\cos(\omega t + \phi) \tag{2.15}$$

At the load, the real and reactive power losses are computed using $\frac{|V_{\mathrm{Load}}|^2}{R_{\mathrm{Load}}}$ and $\frac{|V_{\mathrm{Load}}|^2}{X_{\mathrm{Load}}}$, respectively. The voltage sag is computed as a function of the voltage drop across the line or feeder given by $\Delta V = V - V_{\mathrm{Load}}$.

2.2.3 Balanced Three-Phase Power Formulations

Consider a three-phase (3ϕ) power source supplying a load. For a 3ϕ, if the system (loads/generator) connected as wye (Y) or delta (Δ) is to be balanced, the following conditions must be satisfied:

1. For loads: impedances in all three phases are equal.
2. For sources: voltage magnitude and current magnitude are equal but evenly and spatially distributed by 120° for a three-phase system.

Now, let V_a, V_b, and V_c represent the voltages of phases *a*, *b*, and *c*, respectively, given as

$$V_a = V_{\mathrm{m}} \cos(\omega t) \text{ (as reference)} \tag{2.16}$$

$$V_b = V_{\mathrm{m}} \cos(\omega t - 120°) \tag{2.17}$$

$$V_c = V_{\mathrm{m}} \cos(\omega t + 120°) \tag{2.18}$$

Similarly,

$$I_a \doteq I_{\mathrm{m}} \cos(\omega t + \phi) \tag{2.19}$$

$$I_a = I_{\mathrm{m}} \cos(\omega t + \phi - 120°) \tag{2.20}$$

$$I_c = I_{\mathrm{m}} \cos(\omega t + \phi + 120°) \tag{2.21}$$

Then

$$S_{3\phi} = (V_a I_a^* + V_b I_b^* + V_c I_c^*) \tag{2.22}$$

$$P_{3\phi\mathrm{real}} = V_a I_a + V_b I_b + V_c I_c = \Re e[S_{3\phi}]$$

$$= V_{\mathrm{m}} I_{\mathrm{m}} [\cos^2 \omega t + \cos^2(\omega t - 120°) + \cos^2(\omega t + 120°) \tag{2.23}$$

But $\cos^2 \theta = \dfrac{1 + \cos 2\theta}{2}$, therefore

$$P_{3\phi} = \frac{V_{\mathrm{m}} I_{\mathrm{m}}}{2} \left[1 + \cos 2\omega t + (1 + \cos(2\omega t - 240°)) + (1 + \cos(2\omega t + 240°))\right] \tag{2.24}$$

$$P_{3\phi} = \frac{V_m I_m}{2}\left[3 + \cos\omega t - \cos(2\omega t - 240°) + \cos(\omega t + 240°)\right]$$

$$= \frac{3V_m I_m}{2} \tag{2.25}$$

And the per-phase power is therefore

$$P_\phi = \frac{V_m I_m}{2} = \frac{V_m I_m}{2} \tag{2.26}$$

2.3 Balanced Voltage to Neutral-Connected System

2.3.1 Wye- or Y-Connected System

Consider Figure 2.4.

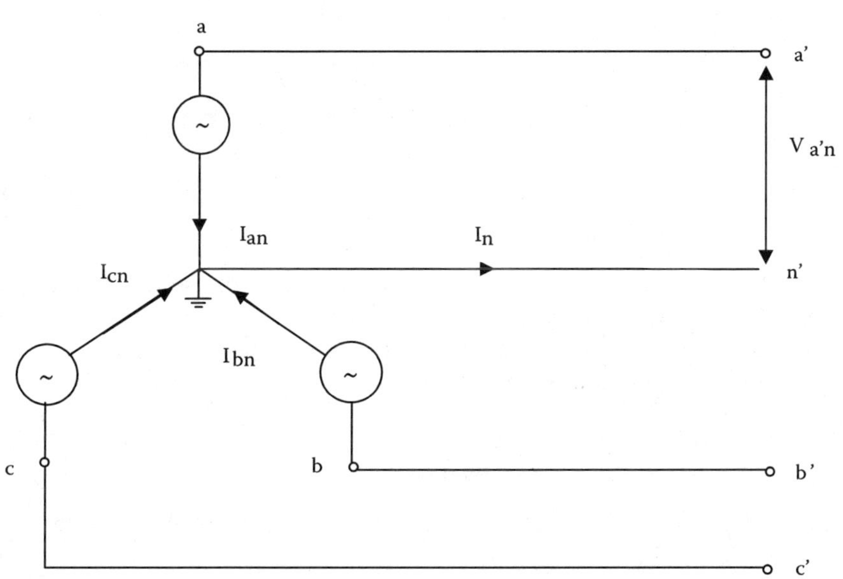

FIGURE 2.4
Wye or Y-connected system.

$$V_{ab} = V_{an} - V_{bn}$$
$$= V_m\angle 0° - V_m\angle -120° \tag{2.27}$$
$$= V_m\angle 0° - V_m\left[\frac{-1 - j\sqrt{3}}{2}\right]$$

$$V_{ab} = V_m\sqrt{3}\left[\frac{\sqrt{3}+j}{2}\right] = \sqrt{3}V_m\angle 30° \tag{2.28}$$

Similarly,

$$V_{bc} = V_{bn} - V_{cn} = \sqrt{3}\,V_m\angle 90° \tag{2.29}$$

$$V_{ca} = V_{cn} - V_{an} = \sqrt{3}\,V_m\angle 150° \tag{2.30}$$

and for a Y-connected system,

$$V_{ab} = E_{an}\sqrt{3}\angle 30° \tag{2.31}$$

$$V_{bc} = E_{bn}\sqrt{3}\angle 30° \tag{2.32}$$

$$V_{ca} = E_{cn}\sqrt{3}\angle 30° \tag{2.33}$$

Overall, the line-to-line voltages are given as a $\sqrt{3}$ phase-to-ground quantity, and $I_L = I_P$ for a Y-connected system. In the Y-connected network, if the currents are balanced in magnitude and phase, then $I_a + I_b + I_c = I_n = 0$. Otherwise, unbalanced line currents lead to $I_a + I_b + I_c = I_n \neq 0$.

2.3.2 Delta- or Δ-Connected System

Consider Figure 2.5. For the 3ϕ -connection shown, the phase voltages are equivalent to the line voltages such that

$$|V_L| = |V_P| \tag{2.34}$$

and the line current is

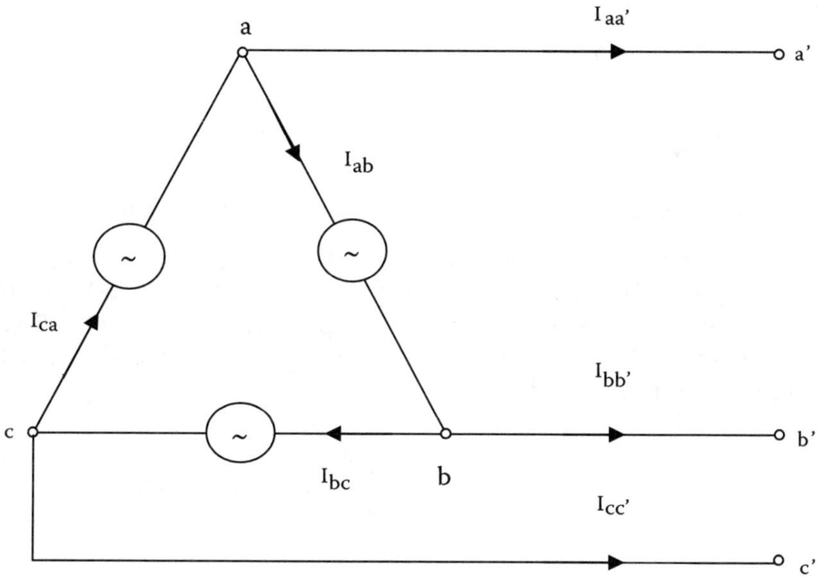

FIGURE 2.5
Delta-connected system.

$$I_L = \sqrt{3} I_P \tag{2.35}$$

The phase currents are

$$I_{ab} = I_P \angle 0° \tag{2.36}$$

$$I_{bc} = I_P \angle -120° \tag{2.37}$$

$$I_{ca} = I_P \angle 120° \tag{2.38}$$

And, using Kirchoff's current law (KCL), the line currents are

$$\therefore I_{aa'} = I_{ca} - I_{ab} \tag{2.39}$$

$$I_{aa'} = \sqrt{3}\, I_P \angle 150° \tag{2.40}$$

$$I_{bb'} = \sqrt{3}\, I_P \angle 30° \tag{2.41}$$

$$I_{cc'} = \sqrt{3}\, I_P \angle -90° \tag{2.42}$$

2.4 Power Relationship for 3ϕ Y-Δ-Connected System

For a balanced three-phase system, we write each generated voltage and current, respectively, as $V_a(t)$, $V_b(t)$, and $V_c(t)$ and $I_a(t)$, $I_b(t)$, and $I_c(t)$. From the discussion in the previous sections, the relationship between the line and phase voltage in a Y-connected system is $|V_L| = \sqrt{3}\, V_P$. Hence, the 3ϕ or total real power is computed using

$$P_{3\phi} = 3V_P I_P \cos\phi \tag{2.43}$$

or

$$P_{3\phi} = \sqrt{3}\, |V_L||I_L| \cos\phi \tag{2.44}$$

where the voltage level is specified and understood to be the line voltage. Note that the total instantaneous power is constant, having a magnitude of three times the real power per phase.

The reactive power is analogous to the summation of balanced three-phase currents, and voltages appear to cancel out mathematically but are very much alive with each phase. Reactive power is given as

$$Q_{3\phi} = \sqrt{3}\, |V_L||I_L| \sin\phi \tag{2.45}$$

Hence, the total or apparent power is

$$S_{3\phi} = P_{3\phi} + jQ_{3\phi} \tag{2.46}$$

In terms of line value, we assert that

$$S_{3\phi} = \sqrt{3}\, V_L I_L{}^{*} \tag{2.47}$$

$$P_{3\phi} = \sqrt{3}\, |V_L||I_L| \cos\phi \tag{2.48}$$

$$Q_{3\phi} = \sqrt{3}\, |V_L||I_L| \sin\phi \tag{2.49}$$

The 3ϕ circuit problem given so far can be analyzed per phase quantitatively when the system is balanced. The equivalent single phase is used with phase A as reference, and other phases are found using equivalent phase shifts of ±120° and ±240° for phases B and C, respectively. As before, the equivalent per phase circuit represents phase to neutral, with the voltage being denoted as line to neutral and the currents are line values.

2.5 Per-Unit System

The power system quantities such as voltage, current, and power are normalized for ease of computation. In power system analysis, the per unit (p.u.) or percent of specified base values is expressed for measurements in typical power systems. The advantages of using p.u. are:

1. Ease of computation and ease of comparison of results for various power systems that may have different base quantities.
2. Early detection of calculation errors, especially when device parameters fall in small ranges.
3. Elimination of ideal transformers as windings, such that voltages, currents, and external impedances and admittances expressed in p.u. do not change when they are referred from one side of the transformer to another. This leads to computational savings in a power system with hundreds of transformers and also helps to eliminate errors in calculation.

The p.u. definition allows an actual quantity to be normalized to unity and facilitates the comparison of all other measured values to that base unit value. By definition, the per-unit value of a quantity X is

$$X_{\text{p.u.}} = \frac{X_{\text{actual}}}{X_{\text{base}}} \tag{2.50}$$

The actual quantity is the value of the quantity in actual units such as watts, VArs, hertz, etc. The base value has the same unit as the actual quantity, and hence p.u. is dimensionless.

For electrical laws to be valid in the p.u. system, the following relations must be used for other bases: given $S = VI^*$, where $S\angle\phi = V\angle\alpha I\angle - \beta$, then in p.u., choose $S_{\text{base}/\phi} = Q_{\text{base}/\phi} = P_{\text{base}/\phi}$ such that

$$\frac{S\angle\phi}{S_{\text{base}}} = \frac{V\angle\alpha I\angle-\beta}{S_{\text{base}}} \tag{2.51}$$

$$S_{base} = V_{base} I_{base} \tag{2.52}$$

$$\frac{S\angle\phi}{S_{base}} = \frac{V\angle\alpha I\angle -\beta}{V_{base} I_{base}} = \frac{V}{V_{base}} \cdot \frac{I}{I_{base}} = V_{p.u.} I^{*}_{p.u.} \tag{2.53}$$

$$I_{base} = \frac{S_{base}}{V_{baseLN}} \tag{2.54}$$

$$Z_{base} = R_{base} = X_{base} = \frac{V_{baseLN}}{I_{base}} = \frac{V^{2}_{baseLN}}{S_{base/\phi}} \tag{2.55}$$

$$Y_{base} = G_{base} = B_{base} = \frac{1}{Z_{base}} \tag{2.56}$$

Note that (a) the value of $P_\phi = Q_\phi = S_\phi$ applies to the entire system and is the same for the entire circuit and (b) the ratio of voltage bases on either side of a transformer is selected to be the same as the ratio of the transformer voltage ratings. With these two rules, it is evident that the p.u. impedance is the same for transformers when referred from one side to the other.

2.5.1 Conversion of a Per Unit from a New Base of Reference

When only a component such as a transformer is considered for p.u. analysis, the nameplate ratings of the transformers involved are selected as base values. But when a different component is involved, a different nameplate rating may be given. It is necessary to convert the p.u. impedance of a device from its nameplate rating to the system base value. To convert p.u. impedance from old to new base value we use

$$Z_{p.u.\,new} = \frac{Z_{actual}}{Z_{base\,new}} = \frac{Z_{p.u.\,old} Z_{base\,old}}{Z_{base\,new}} \tag{2.57}$$

or

$$Z_{p.u.\,new} = Z_{p.u.\,old}\left(\frac{V_{base}}{V_{base\,new}}\right)^{2}\left(\frac{S_{base\,new}}{S_{base\,old}}\right) \tag{2.58}$$

2.5.2 Per-Unit Formulations for 3ϕ System

Balanced 3ϕ systems can be solved on a per-unit basis after converting Δ loads to equivalent Y impedances. Base values can be selected in single phase or in 3ϕ phase. We denote for line voltage $V_{\text{line base}}$ as the base voltage, then

$$V_{\text{LN base}} = \frac{V_{\text{base}}}{\sqrt{3}} \tag{2.59}$$

$$V_{\text{LN p.u.}} = \frac{V_{\text{LN}}}{V_{\text{LN base}}} \tag{2.60}$$

and if

$$V_{\text{LN}} = \frac{V_{\text{L}}}{\sqrt{3}}$$

then

$$V_{\text{LN p.u.}} = \frac{V_{\text{L}}/\sqrt{3}}{V_{\text{LN base}}/\sqrt{3}} = \frac{V_{\text{L}}}{V_{\text{LN base}}} = V_{\text{L p.u.}} \tag{2.61}$$

$$S_{\text{base}3\phi} = P_{\text{base}3\phi} = Q_{\text{base}3\phi}$$

Also,

$$I_{\text{base}} = \frac{S_{\text{base}\,3\phi}}{\sqrt{3}V_{\text{base LL}}} \tag{2.62}$$

$$Z_{\text{base}} = \frac{V_{\text{base LN}}}{I_{\text{base}}} = \frac{V_{\text{base}}^2}{S_{\text{base}3\phi}} \tag{2.63}$$

$$Z_{\text{base}} = \frac{V_{\text{base LN}}^2}{S_{\text{base}1\phi}} = \frac{VV_{\text{base LL}}^2}{S_{\text{base}3\phi}} \tag{2.64}$$

$$R_{\text{base}} = X_{\text{base}} = \frac{1}{Y_{\text{base}}} \tag{2.65}$$

Using single-phase bases,

$$S_{\text{base}1\phi} = \frac{S_{\text{base}3\phi}}{3} \tag{2.66}$$

$$S_{\text{base}1\phi\text{p.u.}} = \frac{S_{1\phi}}{S_{1\phi\text{base}}} = \frac{S_{3\phi}/3}{S_{\text{base}3\phi}/3} = S_{3\phi\text{p.u.}} \tag{2.67}$$

$$S_{1\phi\text{p.u.}} = S_{3\phi\text{p.u.}} \tag{2.68}$$

For a Wye-connected load,

$$Z_{\text{Y base}} = \frac{V_{\text{base LN}}^2}{S_{\text{base}1\phi}} = \frac{[V_{\text{base LL}}^2/\sqrt{3}]^2}{S_{\text{base}3\phi}/3} = \frac{V_{\text{LL}}^2}{S_{\text{base}3\phi}} \tag{2.69}$$

For a balanced 3-phase system,

$$Z_{\Delta\text{ base}} = 3Z_{\text{Y base}} \tag{2.70}$$

So that

$$Z_{\text{Y p.u.}} = Z_{\Delta\text{ p.u.}} \tag{2.71}$$

$$I_{\text{L base}} = \frac{S_{\text{base}3\phi}}{V_{\text{LN}}/\sqrt{3}} \tag{2.72}$$

2.6 Calculation of Power Losses

In power system generation, transmission, and distribution, we try to account for the efficiency of transmission and distribution. There are many sources that degrade the overall quality of power delivery. Some are due to:

- Electrical loss due to energy loss to windings and copper and iron losses
- Thermal loss due to apparati exceeding their thermal ratings, thus leading to a power loss in the delivery
- Mechanical losses due to less vibration or a deficiency in the mechanical system
- Human error in measurements
- Extreme events such as natural disasters

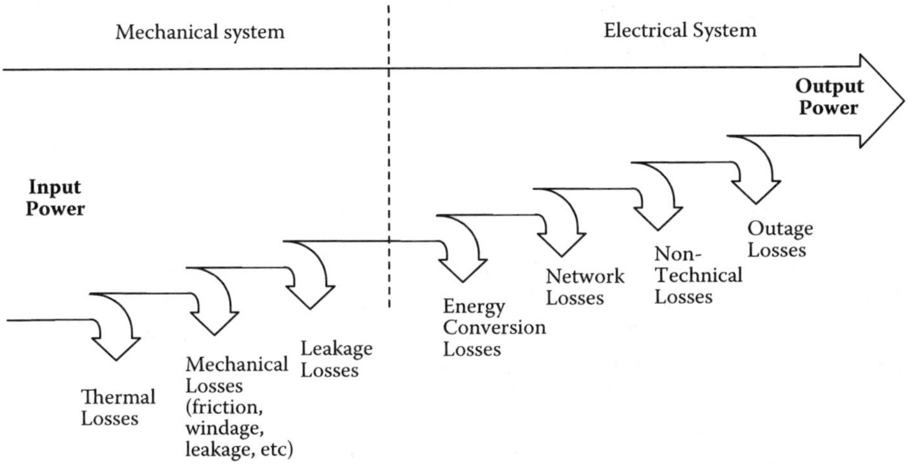

FIGURE 2.6
Loss diagram for an electrometrical energy-conversion system.

- Equipment malfunction or theft or vandalism of power system infrastructure

These losses are modeled using available parameters and physics to determine the overall efficiency of the system. Figure 2.6 illustrates some different energy loss paths that decrease power efficiency in an energy-conversion machine.

For example, electrical loss is modeled as

$$P_{\text{loss}} = I^2R \text{ (watt)} \tag{2.73}$$

$$Q_{\text{loss}} = I^2X \text{ (var)} \tag{2.74}$$

Overall, the sum of the losses is computed, and the overall efficiency is determined from

$$\eta = \text{Efficiency} = \frac{P_{\text{in}} - P_{\text{loss}}}{P_{\text{in}}} = 1 - \frac{P_{\text{loss}}}{P_{\text{in}}} \times 100\% \tag{2.75}$$

To minimize power losses, several schemes are used in distribution networks. Examples include:

- Voltage VAr control (voltage regulator)
- Network reconfiguration
- Compensation techniques/power factor correction
- Maintenance of equipment (transformer overall diagnostic, routine maintenance of system apparatus), grass trimming, etc.

These methods are discussed in appropriate sections in Chapter 5, "Distribution Automation and Control Functions."

2.7 Voltage Regulation Techniques

Given the different contingencies or faults in a distribution system, the first obvious parameter of interest to be monitored for quality and security degradation is deviation of the node voltages from the prescribed statutory standards using the following techniques.

2.7.1 Capacitor Banks for Voltage Regulation and Power Factor Correction

To boost the voltage at the customer end of service, a capacitor bank is used as a regulating device. It is connected in parallel across the line to increase voltage by reducing the inductive VArs and is generally a lagging power factor. To correct the power factor appropriate for the substation load, VAr due to computation is reduced by using a lagging power factor, while the leading power factor substation increases VAr supply, which may reduce the quality of real power supply.

Balanced capacitor banks are installed on each phase of the 3ϕ power system for effective power factor correction. Capacitors can be switchable automatically, depending on voltage degradation. The convention of the center-tapping capacitor keeps the voltage at normal values.

2.7.1.1 Shunt Capacitor Installed in Parallel to Distribution Network Model

Consider a reduced feeder length, shown in Figure 2.7. The corresponding phasor diagram is presented in Figure 2.8.

$$V_{\text{Drop}} = IR \cos \phi + IX_{\text{L}} \sin \phi \tag{2.76}$$

To reduce the V_{Drop}, we install a capacitor, as shown in Figure 2.9. The corresponding phasor diagram is shown in Figure 2.10.

$$V_{\text{Drop}} = IR \cos \phi + I(X_{\text{L}} - X_{\text{C}}) \sin \phi \tag{2.77}$$

The series-connected capacitor effectively resolved the voltage drop in the reactive element.

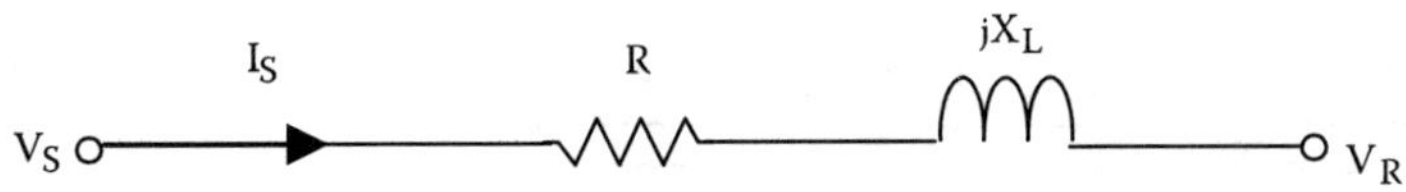

FIGURE 2.7
Reduced feeder.

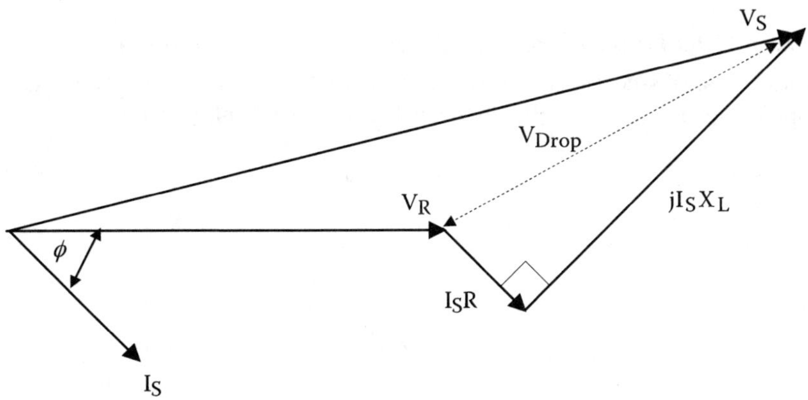

FIGURE 2.8
Phasor diagram for reduced feeder.

FIGURE 2.9
Reduced feeder with capacitor.

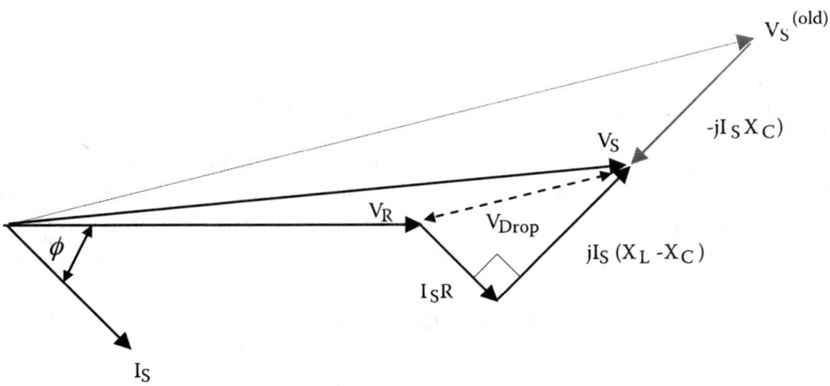

FIGURE 2.10
Phasor diagram for reduced feeder with capacitor.

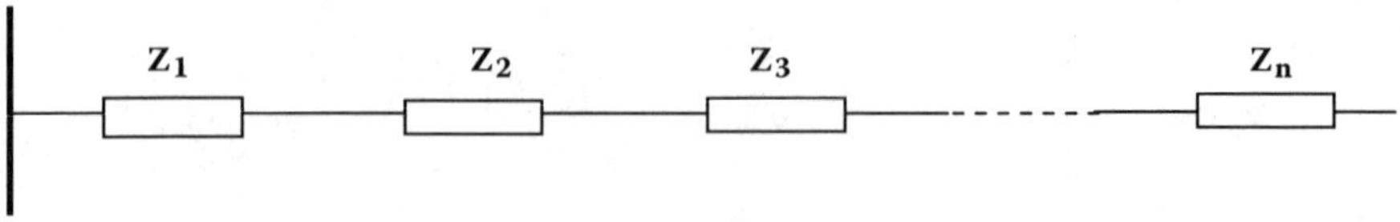

FIGURE 2.11
Series-connected radial system.

2.7.1.2 Calculation of Voltage Drop for a Distribution Feeder

Consider a series-connected (radial) system, as shown in Figure 2.11. The voltage drop per unit length for each layer (lateral) is given as

$$V = I(Z_1 + Z_2 + Z_3 + \dots + Z_n) \tag{2.78}$$

$$S = VI^* = II^*(Z_1 + Z_2 + Z_3 + \dots + Z_n) \tag{2.79}$$

$$S = \sum S_i$$

$$S_i = |I|^2 Z_i \tag{2.80}$$

If Z is given per unit length,

$$Z_{\text{total}} = z_1(l_1) + z_2(l_2) + z_3(l_3) + \dots + z_n(l_n) \tag{2.81}$$

$$S_i = |I|^2 z_i(\Delta l_i)$$

$$S_{\text{total}} = \sum S_i = \sum_{i=1}^{n} |I_i|^2 z_i(\Delta l_i) \tag{2.82}$$

2.7.2 Tap-Changing Method for Voltage Regulation

The tap-changing transformer schemes are operated manually or automatically to accommodate a variety of load types typically called ULTCs (unload tap-changing transformer). They are aimed at keeping voltages at proper levels in response to wide variations in the load and the primary voltage level.

The taps are connections on a transformer winding that change the turn ratio according to

$$\frac{N_1}{N_2} = \frac{V_1}{V_2}$$

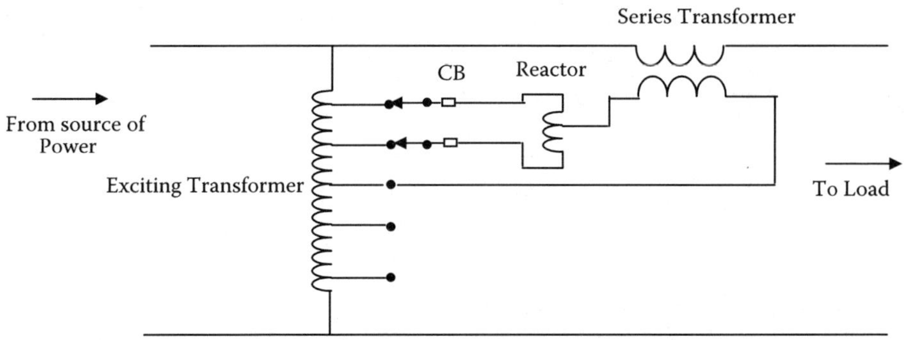

FIGURE 2.12
Tap-changing transformer.

A new voltage obtained from a particular given setting relative to the primary voltage varies from ±10% to ±2.5%. Taps are typically located on the primary side because they require less current than would be necessary on the secondary side (Figure 2.12).

2.7.3 Voltage-Regulating Transformers

We have several situations where voltage magnitude and current flows in a network need to be controlled using other devices. The automatic voltage regulation is designed to provide a boost of voltage magnitude along a line or change in phase to control flows of power between systems (Figure 2.13). In Figure 2.13, the connections and polarity of the secondary windings are used to obtain ±5% or ±10% voltage regulation. Parallel connection of the secondary windings results in a 5% voltage regulation, and series connection results in a 10% voltage regulation for this example.

Single-phase buck-boost voltage regulation can also be obtained using a voltage-to-voltage transformer configuration, as shown in Figure 2.14. This

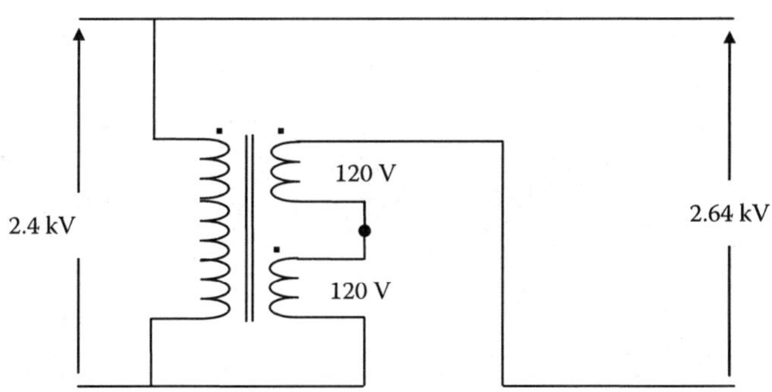

FIGURE 2.13
Single-phase booster transformer connection for 10% boost.

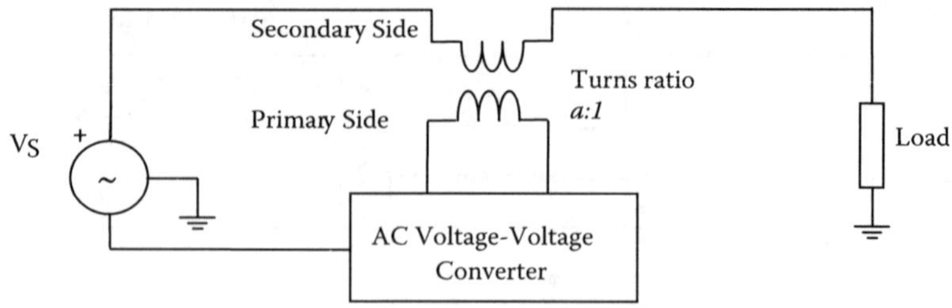

FIGURE 2.14
Voltage regulation using solid-state control.

is applicable to correcting voltage sag. The transformer is used to alter the output voltage by a converter with a duty cycle given by $V_{nominal} - V_S)(a/V_S)$, where a is the turns ratio of the buck-boost transformer, and $V_{load} = V_S + V_{secondary}$.

Figure 2.15 presents a simplified diagram of a three-phase regulating transformer. The output voltage of each phase is a fraction of the input voltage in either buck or boost operation. The tap changing is done using a ganged switch.

2.7.4 Phase Shifter or Regulating Transformer

The phase-shifter transformer is a special type of regulatory transformer that is primarily used to alter the current and voltage phase angles in a transmission or distribution system. It is used to control power flows and losses within power networks. This class of power transformer is characterized by a complex turn ratio that describes its mathematical behavior on the power

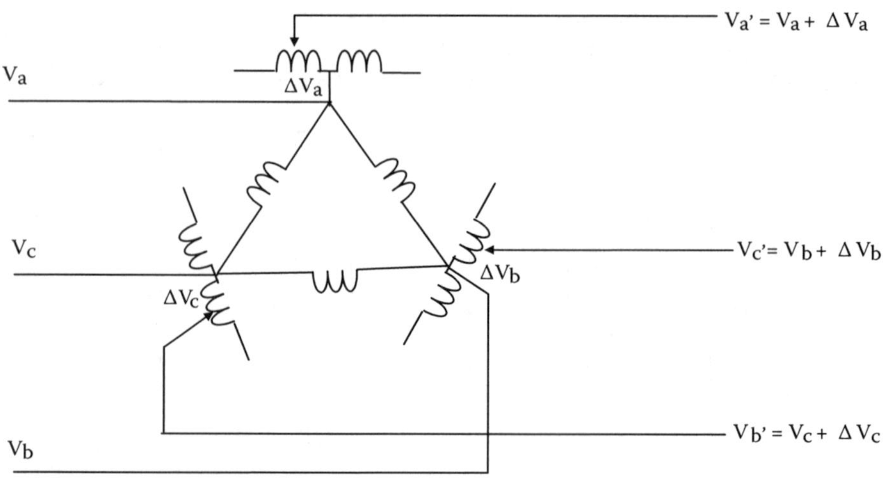

FIGURE 2.15
Three-phase regulating transformer.

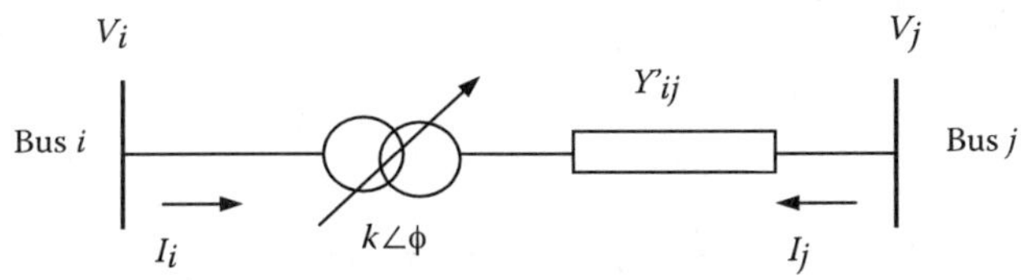

FIGURE 2.16
Network representation of the static phase-shifter transformer.

system. The network representation of the static phase-shifter transformer is shown in Figure 2.16.

The physical device consists of two sets of windings representing a booster transformer and a magnetizing transformer. The current through the magnetizing transformer uses a small voltage on the primary side of the booster transformer. This voltage is quadrature to the phase voltage. Therefore, the sending end voltage is displaced in the vector space by the presence of the booster transformer and the "quadrature" voltage. Furthermore, when the reactance of the magnetizing transformer is referred to the primary side of the booster transformer, the resultant reactance is $x_s = x_{kb} + n^2x_{km}$, as the booster transformer is at its nominal tap setting. A more exact representation is shown in Figure 2.17.

Now, by applying Kirchoff's current law to Figure 2.17,

$$I_o = -I_j = I_i - I_s \tag{2.83}$$

$$= I_i\left[1+\left(e^{j\left(\phi+\frac{\pi}{2}\right)}\right)\sin\phi\right] = k^{-1}I_s e^{j\phi} \tag{2.84}$$

where

k $= \sqrt{1+3n^2}$

ϕ $= \arctan\sqrt{3n}$

n = turns ratio of the magnetizing transformer

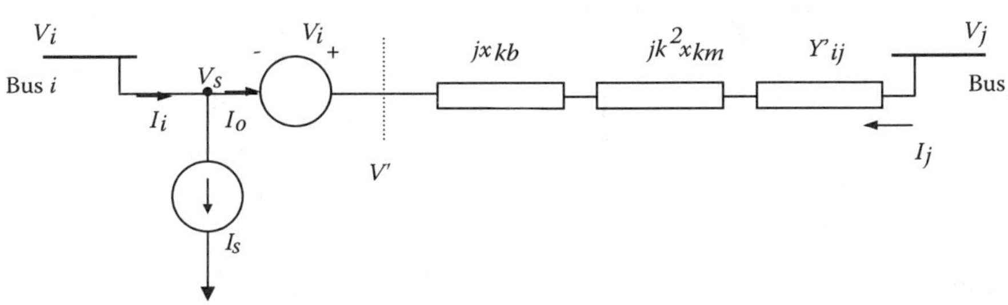

FIGURE 2.17
Equivalent circuit diagram of the phase-shifter transformer.

Further, by Kirchoff's voltage law (KVL),

$$V' = V_i + V_s \tag{2.85}$$

$$= I_i\left(1+\sqrt{3n}\,e^{j\frac{\pi}{2}}\right) = kV_s e^{j\phi} \tag{2.86}$$

Equation 2.85 and Equation 2.86 show the complex nature of the transfer characteristics of a phase-shifter transformer. The exact model will reflect this property, which is a drawback to the existing power system programming techniques and computations.

2.8 Voltage-Sag Analysis and Calculation

A voltage sag is a sudden reduction in the supply of voltage magnitude followed by a voltage recovery after a short period of time. Severe voltage sags are caused by short circuit or overloading. Methods of voltage-sag analysis include:

1. Use power factor or network analysis to determine voltage sag distribution for each supply point.
2. Locate the portion of customers with low voltage and determine the critical voltage sag.

For example, consider a distribution feeder (Figure 2.18). The voltage-sag computation for the voltage drop across the feeder that is supplying power to the downstream customers can be calculated from

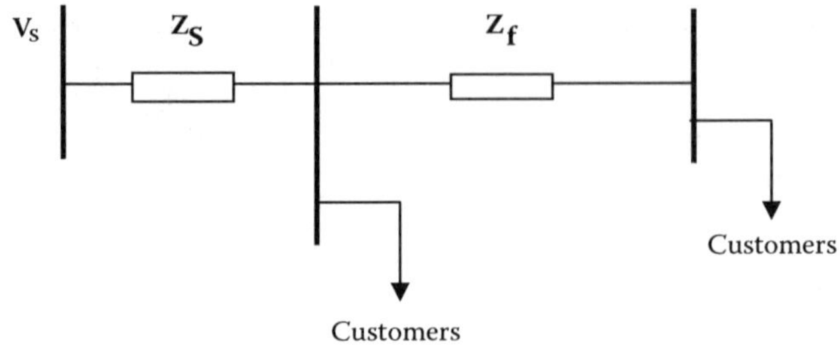

FIGURE 2.18
Distribution feeder.

$$V_{sag} = \left(\frac{Z_f}{Z_f + Z_s} \right) V_s \tag{2.87}$$

This formula gives a direct voltage-sag result when the feeder information above is given.

Voltage-sag identification is crucial for efficient and cost-effective operation of the power system. At light load, the voltage drop is noticeable, whereas at high heavy load, voltage drops as load is drawn from the source on the feeder length. If the voltage at the substation node is set at nominal voltage, the customer at the end of the line has low voltage under heavy loading. If voltage is set so that the customer at the end of the line receives the nominal voltage under heavy loading, the customer near the substation has too high a voltage, and voltage becomes high for all customers at the light-load side due to the undesirability of compromising the operating conditions: hence the different voltage regulatory schemes.

2.9 Equipment Modeling

2.9.1 Power Transformers

Power transformers are major distribution system components of importance. These are mainly devices for changing voltage and current at high/low power levels reliably and efficiently. They come in various forms. Normally the 0.1 immersed transformer type for cooling is used. It is connected from a substation to large industrial, commercial, or residential customers. Small power transformers are called the distribution transformers. Power transformers are typically sized at 1,000 kVA or 30,000 kVA up to 1,000 MVA. The operating impedance is rated in the range of 1 to 3% of 2.5 kV at 3 kVA and 4% of 15 kV at high kVA rating, and the efficiency of transformers is usually high. At full load, a transformer has an efficiency of up to 98% when of the 34-kV type.

Typical power transformers are protected using differential relays for safety, economic, and reliability reasons. Figure 2.19 shows a typical power transformer used in power networks.

2.9.2 Distribution Transformers

Distribution transformers (Figure 2.20) are used to provide electric link connections to the customer. They operate at a voltage level, providing safe usage on the customer side of the system. The voltage at the primary side is typically between 2.3 and 34.5 kV for single or three phase, depending on the customer load size. The secondary side is rated typically at 480 Y/277

FIGURE 2.19
Power transformer.

FIGURE 2.20
Distribution transformer.

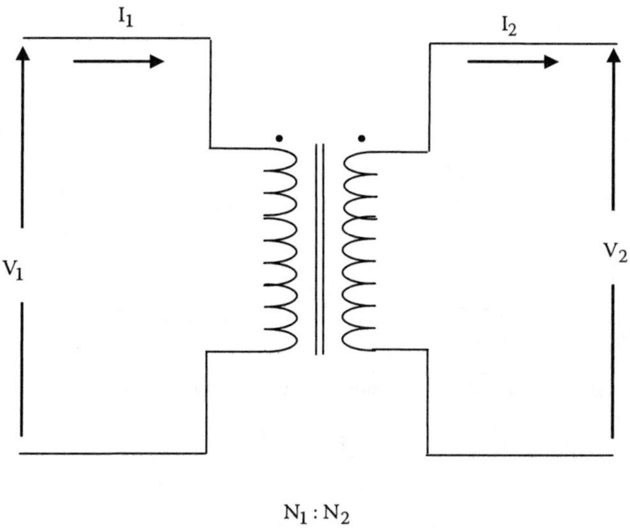

FIGURE 2.21
Ideal transformer.

V or 208 Y/120 V single phase. Distribution transformers can be single-type pole mounted, typically ranging from 15 to 100 kVA. They can withstand 200% overload for hours and last up to 50 years. Protection using fuses and lightning arrestors at the primary side provides for safety and economic security.

2.9.2.1 Principles and Operating Fundamentals

The single-phase ideal transformer is denoted in Figure 2.21 as a pair of insulated windings on a laminated soft iron core. The primary and secondary voltages and currents are shown, and the turn ratio of the coils is N_1:N_2. From Faraday's principles,

$$V_1 = L_1 \frac{di_1}{dt} = N_1 \frac{d\phi}{dt} \quad \text{and} \quad V_2 = L_2 \frac{di_2}{dt} = N_2 \frac{d\phi}{dt} \tag{2.88}$$

$$\frac{V_1}{V_2} = \frac{N_1 \frac{d\phi}{dt}}{N_2 \frac{d\phi}{dt}} = \frac{N_1}{N_2} = \frac{1}{a} \tag{2.89}$$

where a is the turn ratio of the transformer secondary winding relative to its primary windings. Because the transformer is assumed ideal and therefore lossless, then

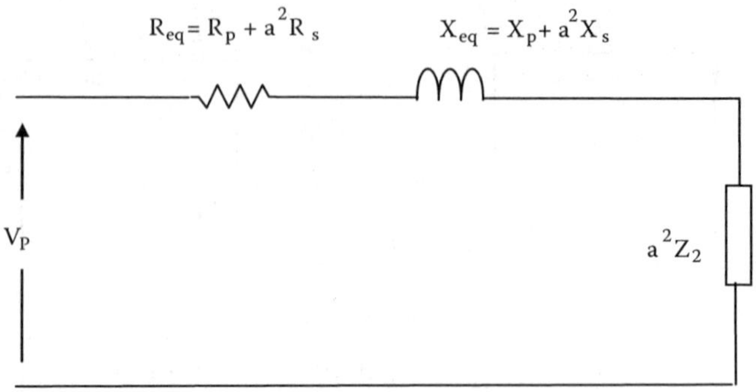

FIGURE 2.22
Equivalent diagram of an ideal transformer (parameters referred to primary side).

$$V_2 I_2 = V_1 I_1 \tag{2.90}$$

$$\frac{I_2}{I_1} = \frac{V_1}{V_2} = \frac{N_2}{N_1} = a \tag{2.91}$$

Neglecting the magnetization branch and referring all quantities to the primary side, the equivalent diagram of an ideal transformer is shown in Figure 2.22. Generally, we model transformers in terms of admittance given as

$$\begin{bmatrix} I_k \\ I'_k \end{bmatrix} = \begin{bmatrix} Y_k^{pr} & Y_k^{ps} \\ Y_k^{sp} & Y_k^{ss} \end{bmatrix} \begin{bmatrix} V_{k-1} \\ V_k \end{bmatrix} \tag{2.92}$$

Such that

$$V_{k-1} = \left(Y_k^{sp}\right)^{-1} \left[I'_k - Y_k^{ss}\right] \tag{2.93}$$

$$I_k = Y_k^{pp} V_{k-1} + Y_k^{ps} V_k \tag{2.94}$$

2.9.3 Autotransformer Model

An autotransformer is a voltage-regulating transformer that has a single side on a soft iron core and is commonly used in transmission systems. The primary and secondary sides are not isolated. Autotransformers are not used at distribution substations because the open exposure of secondary and primary contact to the customer could affect personal safety. Figure 2.23 shows the wiring configuration of an autotransformer.

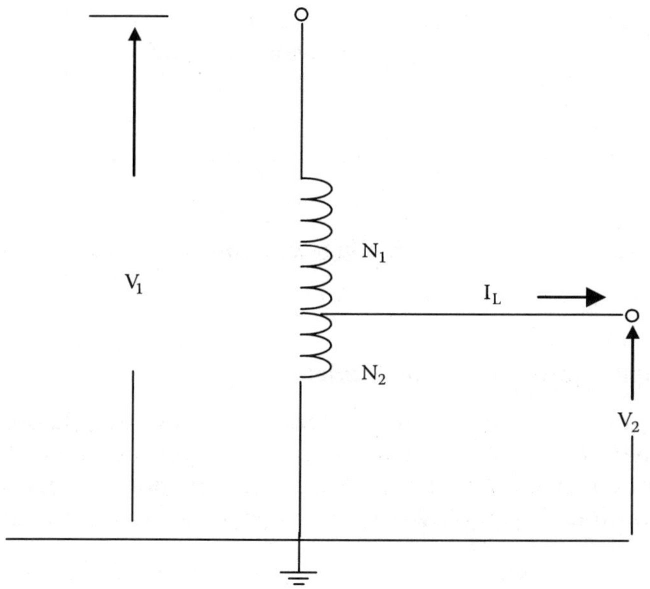

FIGURE 2.23
Auto transformer.

For an ideal transformer, $V_2 I_2 = V_1 I_1$, and the turns ratio can be shown to be

$$a = \frac{N_1 + N_2}{N_1} \tag{2.95}$$

such that

$$V_2 = \frac{V_1}{a} \quad \text{and} \quad I_2 = I_1 a \tag{2.96}$$

In general, we can rewrite

$$\frac{V_2}{V_1} = \frac{V_1 + \left(\frac{N_2}{N_1}\right) V_1}{V_1} = 1 + \frac{N_2}{N_1} \tag{2.97}$$

2.9.4 Cogenerator Model

Cogenerators are modeled as injected current, given as

$$I_{lk} = \left(\frac{S_{lk}}{V_k}\right)^* \text{ for a constant load} \tag{2.98}$$

or

$$I_{Gk} = \left(\frac{S_{Gk}}{V_k}\right)^* \text{ (voltage-controlled source)} \tag{2.99}$$

2.9.5 Synchronous Generator Model

If a generator is operated using automatic voltage regulators (AVRs) to regulate the terminal voltage at the specified voltage, the model is expressed by the photovoltaic-specified model using the fictitious model. In this case, a voltage-controlled or photovoltaic (PV) generator can be treated as follows:

$$\text{PQ} \Rightarrow \text{node, with } Q_{\text{spec}} = \left|V_i^t\right| \left[\frac{V_{\text{Re}(i)} - V_i^t}{X_{\text{fictitious}(i)}}\right] \tag{2.100}$$

where

V_i^t = calculated voltage at iteration t
$V_{\text{fictitious}_{it}}$ = calculated voltage at fictitious node at t iterations
$X_{\text{fictitious}(i)}$ = fictitious branch impedance of node i
Q_{spec} = fictitious reactive power at specified voltage

2.9.6 Inverter-Connected Generator in Photovoltaic Systems

In a distribution network, PV power sources are connected using inverters. Therefore, generators can be modeled as inverters with limited output values. They can be modeled as *PI*-specified buses, as *PQ*-specified buses, as *PV*-specified buses, or as synchronous generators, but the injection current is limited in values where active power output of generation and injection current are specified.

$$Q_{\text{spec}} = \pm\sqrt{|I|^2\left(e^2 + f^2\right) - P_{\text{spec}}^2} \tag{2.101}$$

where $V = e + if$ is used, and P_{spec} and Q_{spec} are given.

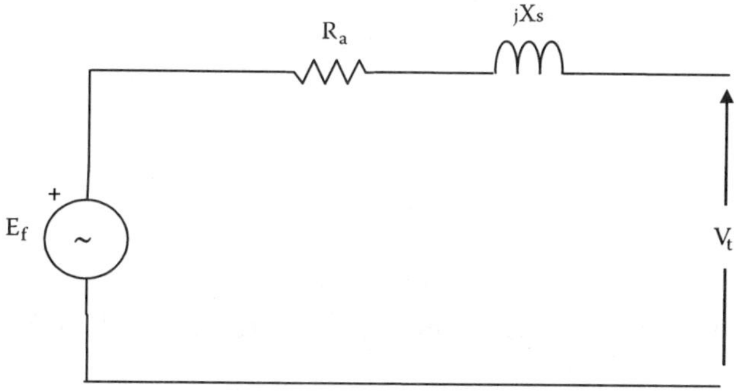

FIGURE 2.24
Equivalent model of a synchronous generator.

2.9.7 Synchronous Generator Model

The steady-state characteristics of synchronous generators can be expressed using the equivalent circuit illustrated in Figure 2.24. The reactive power of a synchronous generator is typically given as a function of voltage for a specified active power output.

2.10 Components Modeling

2.10.1 Line Model in Distribution Systems

The short-line model is given in Figure 2.25. Generally, the length of the distribution section is short, typically less than 800 km, and its voltage level is less than the transmission voltage. We cannot use a network representation model, since the short admittances are neglected.

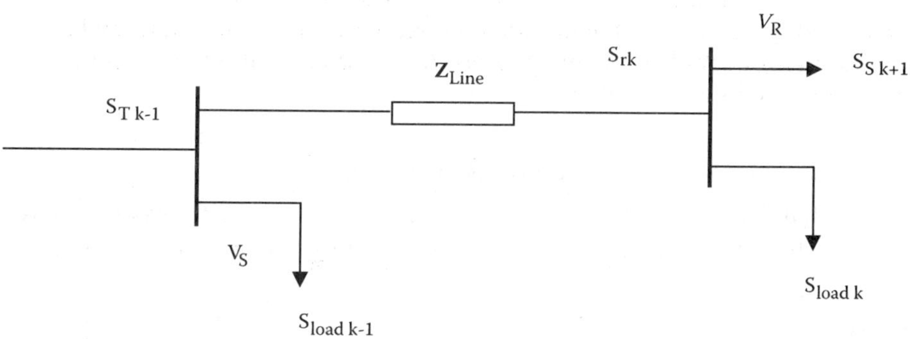

FIGURE 2.25
Short-line model.

$$S_{sK} = S_{rk} + S_{\text{loss } k} \tag{2.102}$$

$$S_{tk} = S_{s\ k+1} + S_{\text{load } k} \tag{2.103}$$

$$S_{\text{loss}} = \frac{|S_{rk}|^2}{|V_R|^2} Z_k \tag{2.104}$$

The quantities $Z = r(l) + jX(l)$ and Y/Z are represented as line charging. Hence, the line model is

$$\begin{bmatrix} V_{k-1} \\ \\ I_k \end{bmatrix} = \begin{bmatrix} V_k + Z_k(\frac{Y_k}{Z} V_k) - I'_k \\ \\ \frac{Y_k}{Z}(V_k + V_{k-1}) - I'_k \end{bmatrix} \tag{2.105}$$

2.10.2 Shunt Capacitor Model

Shunt capacitors are modeled as constant admittance. The injected current for the capacitor is modeled as $I_{ck} = Y_{ck}V_k$ at bus k.

2.10.3 Switch Model

Sectionalizing switches are modeled as branches with zero impedance. They are modeled as $V_{k-1} = V_k$ and $I_k = -I_k$.

2.10.4 Load Models

The distribution loads are ZIP modeled as $P = P_0V^{k1}$ or $Q = Q_0V^{k2}$, where P_0 and Q_0 are the specified active and reactive powers at nominal voltage, and V is the actual voltage magnitude in p.u.

2.10.4.1 Constant Power Loads ($k_1 = k_2 = 0$)

Given that $S_{Lk} = P_{Lk} - jQ_{Lk}$, where P_{Lk} and Q_{Lk} are constant values of active and reactive power at the bus k, the current injection to the load is given as $I_{Lk} = \left(\frac{S_{Lk}}{V_k}\right)^*$.

2.10.4.2 Constant Current Loads ($k_1 = k_2 = 1$)

Modeled as $S_{Lk} = V_k I^*_{Lk} = |V_k|(a_{Lk} + jb_{Lk})$, a_{Lk} and b_{Lk} are constant values of active and reactive currents.

2.10.4.3 Constant Impedance Loads ($k_1 = k_2 = 2$)

We use the following model

$$S_{Lk} = V_k\left(\frac{V_k}{Z_{Lk}}\right)^* = \frac{[V_k]^2}{Z_{Lk}} = \frac{[V_k]^2}{r^2_{Lk} + X^2_{Lk}}\left[r_{Lk} + jX_L\right] \tag{2.106}$$

where r_{Lk} and X_{Lk} are constant values of active and reactive load impedance, respectively. In distribution loads, $I_{Lk} = \dfrac{V_k}{Z_{Lk}}$ is used.

2.10.4.4 Composite/Nonlinear Loads

If $k_1 \neq k_2$, then the real and reactive power of the load is modeled as

$$P = P_0(a_0 + a_1V + a_2V^2 + a_3V^{1.38}) \tag{2.107}$$

$$Q = Q_0(b_0 + b_1V + b_2V^2 + b_3V^{3.22}) \tag{2.108}$$

$$a_0 + a_1 + a_2 + a_3 = 1 \tag{2.109}$$

$$b_0 + b_1 + b_2 + b_3 = 1 \tag{2.110}$$

2.10.5 SVC Device Model

Several flexible AC transmission (FACT) devices such as static VAr compensators (SVC), thyristor-controlled series compensators (TCSC), static synchronous compensators (STATCOM), static synchronous source series compensators (SSSC), and the unified power flow controllers (UPFC) are present in modern power systems.

In particular, within distribution networks, SVCs are connected at the substation level to provide appropriate voltage control, thereby serving the load at the customer end. SVCs installed in power systems are used to improve system performance in several ways, such as regulating system voltages, improving transient stability, increasing transmission capacity, reducing temporary overvoltages, increasing the damping of power oscillations, and damping the subsynchronous resonances and torsional oscillations of rotating machines. A typical SVC model is illustrated in Figure 2.26.

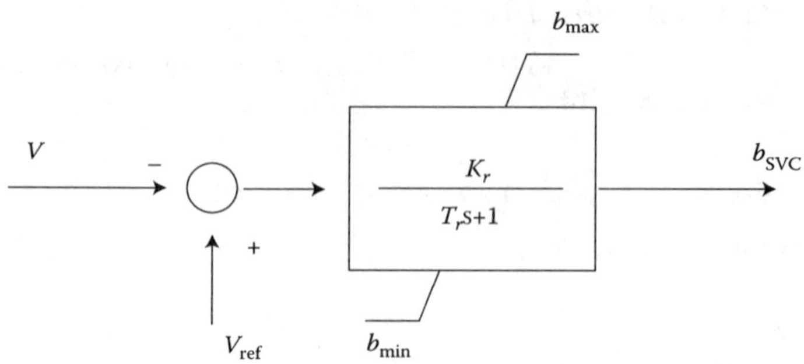

FIGURE 2.26
SVC model.

2.11 Distribution System Line Model

The representative component model is denoted in the following single-line diagram depicted in Figure 2.27. As stated previously, V is the nodal voltage, I is the branch current, θ^V and θ^l are the nodal voltage angle and load current angle, respectively, and I^l is the load current. From the figure at each node, applying current analysis,

$$I_m = I_{m+1} + I^l_{j+1} \tag{2.111}$$

$$|I_m| \angle\theta^l_m = |I_{m+1}| \angle\theta^l_{m+1} + |I^l_{j+1}| \angle\theta^l_{j+1} \tag{2.112}$$

where I^l_{j+1} is computed from $\dfrac{P^l_{j+1} - jQ^l_{j+1}}{V^*_{j+1}}$.

Using this method, we repeat the process to compute and finalize until the substation is reached. The nodal voltages are calculated by updating from the leaf node (end node). The voltage for node $j + 1$ is given in phasor form.

$$V_{j+1} = V_j - I_m Z_m \text{ (drop voltage)} \tag{2.113}$$

where

$$Z_m = R_m + jX_m \tag{2.114}$$

Thus

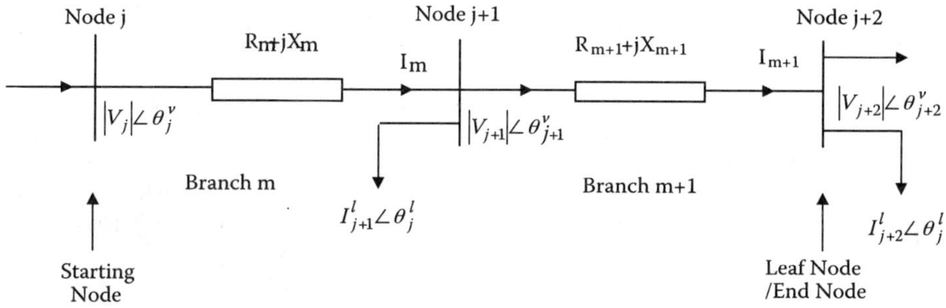

FIGURE 2.27
Representative component model: single-line diagram.

$$V_{j+1}\angle\theta_{j+1}^V = |V_j|\theta_j^V - |I_m||Z_m|\angle\phi_m \tag{2.115}$$

with

$$\phi_m = \theta_m^i - \tan^{-1}\frac{X_m}{R_m} \tag{2.116}$$

Simplifying, we get

$$V_{j+1} = \sqrt{|V_j|^2 + |I_m|^2|Z_m|^2 - 2|V_j||I_m|Z_m\cos\left(\theta_j^v - \phi_m\right)} \tag{2.117}$$

$$\theta_{j+1}^V = \tan^{-1}\left(\frac{|V_j|\sin\theta_j^V - |I_m||Z_m|\sin\phi_m}{|V_j|\cos\theta_j^V - |I_m||Z_m|\cos\phi_m}\right) \tag{2.118}$$

2.12 Distribution Power Flow Analysis

Power flow analysis is an important basic tool for the analysis of power systems. Planning and operation in distribution automation function, optimization, and repeated power flow analysis is needed, and these are represented in several software applications that facilitate efficient and fast determination of voltages, current losses, and analyses of system reconfiguration and performance.

Over the last three decades (since the 1960s), several advances have been made in load-flow computation for transmission systems. Its application to distribution has had limited success due to the radial structure of the distribution system topology, the low X/R ratio, and the dynamic condition of load in a distribution system. It involves using the classical power flow

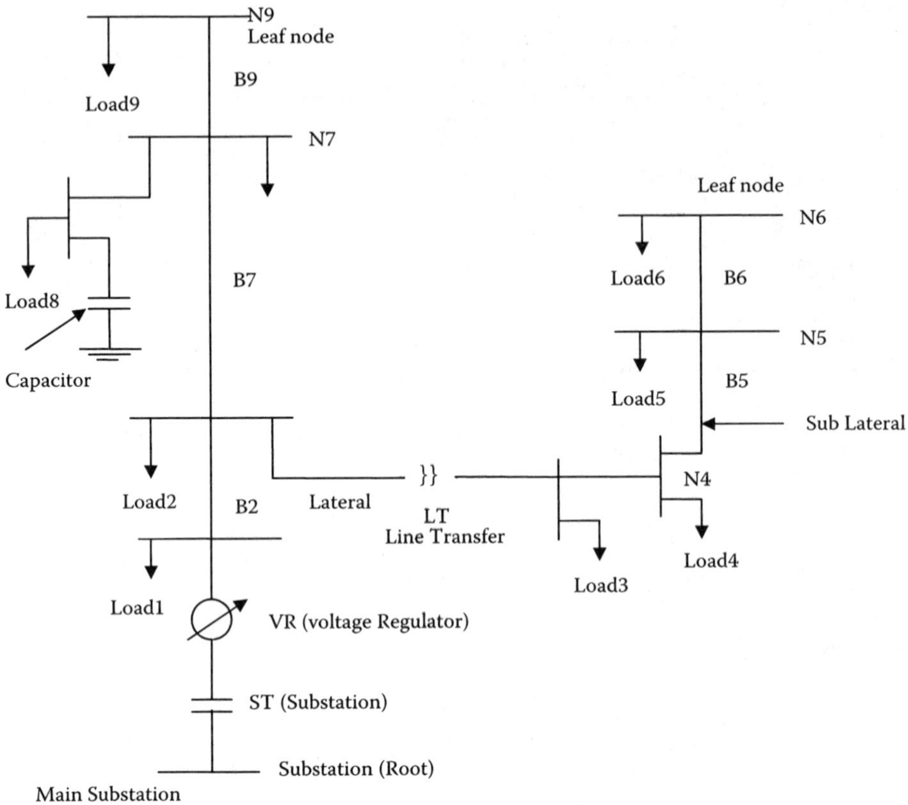

FIGURE 2.28
Typical radial-distribution network model.

analysis convergence problem while determining voltages and flows in the network.

Figure 2.28 is a radial distribution substation with all the components clearly shown, e.g., capacitor, transformer, voltage regulation, substation, and loads at different nodes. The system has the following features:

1. The main substation is designated as the root main feeder and is designated as the branch connecting the substation to the outside world.
2. The lateral branch represents the branch emanating from the main feeder.
3. The sublateral connects the lateral to many other nodes.
4. The leaf nodes represent the top of the highway from the substation to the far end of the service station.
5. The branches are electrical wires connecting nodes to nodes.
6. A node is a connection for tapping power off.

2.13 Distribution System Topology for Development of Load Flow

With the advent of distributed generation within a distribution system, the need for distribution power flow is imperative. Researchers have evaluated the well-known fast-decouple approach in load flow and developed an extension scheme to make it useful for distribution systems.

The following section reviews the traditional power flow technique used today for calculating distribution power flow. The following assumptions characterize distribution power systems, thereby enabling one to appreciate the differences in distribution power flow:

1. Distribution systems are radial or weakly meshed network structures.
2. They have high X/R ratios in the line impedances.
3. They consist of many single-phase loads that are handled by the distribution power flow program.
4. They may have distributed generation (DG) or other renewable generation sources and cogeneration power supplies installed in relative proximity to some load centers.
5. Distribution systems have many short line segments, most of which have low impedance values.

For the purpose of power flow study, we model the network of buses connected by lines or switches connected to a voltage-specific source bus. Each bus can have a corresponding load composite form (consisting of inductors, shunt capacitors, or a combination). Load and generator are connected to the buses.

2.14 Review of Classical Power Flow Methods

The classical methods of power flow used in the industry include:

1. Gauss-Seidal method
2. Newton-Raphson method
3. Fast-decouple methods

We summarize each of them in the following subsections for easy reference.

2.14.1 Gauss-Seidal Method

This method uses the nodal equations of Kirchoff's current law, given as $I_{\text{injection}}$ current at the node

$$I_{\text{inj}(j)} = \sum_{i=1}^{n} I_{ji} \tag{2.119}$$

where $I_{\text{inj}(j)}$ is the injection current at bus j, and I_{ji} = current flow from the jth bus to the ith bus. Rewriting, we obtain $I_{\text{inj}(j)} = Y_{\text{bus}} V_{\text{bus}}$, where the Y_{bus} admittance matrix is given as a V_{bus} vector of bus voltages. If we sum the total power at a bus, the generation and load is denoted as complex power. We have a nonlinear power flow equation given as

$$S_{\text{inj}-k} = P_g + jQ_g - (P_{\text{LD}} + jQ_{\text{LD}}) \tag{2.120}$$

$$= V_k \left(\sum_{j=1}^{n} Y_{kj} V_j \right)^* \tag{2.121}$$

This equation is solved by an iterative method for V_j if P and Q are specified. Additionally, it can be solved from

$$= \frac{1}{Y_{ii}} \left(\frac{P_{\text{L}}^{\text{sch}} - jQ_{\text{L}}^{\text{sch}}}{V_{\text{L}}^{*(k)}} - \sum_{j=1}^{n} Y_{ij} V_j^{(k)} \right) \tag{2.122}$$

where Y_{ij} are the elements of the bus admittance matrix, and P_i^{sch} and Q_i^{sch} are scheduled P and Q at each bus.

After a node voltage is updated within an iteration, the new value is made available for the remaining equations within that iteration and also for the subsequent iteration. Given that the initial starting values for voltages are close to the unknown, the iterative process converges linearly.

Notably, the performance of the classical method is worse in a radial distribution system because of the lack of branch connections between a large set of surrounding buses. It should be noted that the injection-voltage correction propagates out to surrounding buses on the layer of neighboring buses for every iteration.

2.14.2 Newton-Raphson Method

The most commonly used classical solution method of power flow is the Newton-Raphson method. It assumes an initial starting voltage that is used in computing the mismatch power ΔS, where $\Delta S = S_{ij-i}^{\text{sch}} - (V_i^{|k|})^*(\Sigma Y_{ij} V_j^k)$. To determine a convergence criterion given by $\Delta S \le \varepsilon$, where epsilon is a specific tolerance or accuracy index, a sensitivity matrix is derived from the inverse Jacobian matrix of the injected-power equations

$$P_i = |V_i| \Sigma |Y_{ij}||V_j| \cos(\theta_i - \theta_j - \psi_{ij}) \tag{2.123}$$

$$Q_i = |V_i| \Sigma |Y_{ij}||V_j| \sin(\theta_i - \theta_j - \psi_{ij}) \tag{2.124}$$

where θ_i is the angle between V_i and V_j, and ψ_{ij} is the admittance angle.

These expressions are followed by computation of the Jacobian matrix formulation, given as

$$J = \begin{bmatrix} \dfrac{\partial P}{\partial V} & \dfrac{\partial P}{\partial \theta} \\ \\ \dfrac{\partial Q}{\partial V} & \dfrac{\partial Q}{\partial \theta} \end{bmatrix} \tag{2.125}$$

which leads to solving for ΔV correction error in voltages

$$\Delta V^{(k+1)} = J(V^k)^{-1} \, \Delta S^{(k)} \tag{2.126}$$

The complex power ΔS can be expressed in polar or rectangular form

$$|\Delta V| = (\Delta e + \Delta f)$$

$$\Delta V = |\Delta V| \angle \theta_v$$

$$\Delta S = \Delta P + \Delta Q$$

Again, this method is excellent for large systems but does not take advantage of the radial structure of distribution, and hence it does not lend itself efficiently to the computational burden. Moreover, the method fails when the Jacobian matrix is singular or when the system becomes ill conditioned, as in the case where the distribution of X/R ratio is low.

2.14.3 Fast-Decouple Power Flow

The fast-decouple power flow simplifies the Jacobian matrix by using small-angle approximations to eliminate relatively small elements of the Jacobian.

For bus voltage angles $\delta_i - \delta_j \approx \delta_j = \delta_i$, and if θ_{ij} is the impedance angle, assuming X is much greater than R, then $\sin\theta_{ij} \to 0$ and $\cos\theta_{ij} \to 0$.

The Jacobian elements $\frac{\partial P_i}{\partial V_i}, \frac{\partial P}{\partial \theta}, \frac{\partial Q}{\partial V_i}$, and $\frac{\partial Q}{\partial \theta}$ are computed as follows based on these assumptions:

$$= |V_i||Y_{ij}| \cos(\theta_{ij} - \delta_i + \delta_j)$$

$$= |V_i||Y_{ij}| \cos 90° = 0 \ (\therefore\ \theta_{ij} = 90° \text{ and } \delta \text{ is small}) \tag{2.127}$$

$$= -|V_i||V_j||Y_{ij}| \cos(\theta_{ij} - \delta_i - \delta_j)$$

$$= -|V_i||V_j||Y_{ij}| \cos 90° = 0 \tag{2.128}$$

if

$$\frac{\partial P}{\partial \theta} \neq 0, \frac{\partial Q}{\partial V} \neq 0$$

For a flat start, all voltage magnitudes are set to 1.0 p.u. We can approximate $\frac{\partial P}{\partial \theta}$ and $\frac{\partial Q}{\partial V}$ as follows:

$$= |Y_{ij}||V| \sin 90° \Rightarrow \Delta P = |Y_{ij}||V|\Delta\delta \tag{2.129}$$

$$= |Y_{ij}||V| \cos\theta_{ij} \Rightarrow \Delta Q = |Y_{ij}||V|\Delta V \tag{2.130}$$

$$\Delta\delta = [Y]^{-1}\frac{\Delta P}{|V|} \text{ or } \Delta V = [Y_{ij}]^{-1}\frac{\Delta Q}{|V|} \tag{2.131}$$

If $Y_{ij} = G_{ij} + jB_{ij}$ and $G_{ij} << B_{ij}$, then $\therefore\ Y_{ij} \cong B_{ij}$, and

$$\Delta\delta = -[B']^{-1}\frac{\Delta P}{|V|} \text{ and } \Delta V = -[B']^{-1}\frac{\Delta Q}{|V|} \tag{2.132}$$

This fast-decouple method is one of the effective techniques used in power system analysis. However, it exhibits poor convergence for a high X/R-ratio system. The interaction of V and θ magnitudes with flows and (active and reactive) power cause poor convergence as well.

A variation of power flow is to solve current-injection equations instead of model power-injection equations. Again, the Newton-Raphson method is used for solving the system equations, except that the appropriate model of voltage-controlled or PV-type generation has to be specified.

2.15 Distribution Power Flow Methods

Due to the limitation of fast-decoupled power flow in solving ill-conditioned systems with high X/R ratio, the distribution power flow techniques require alternative methods. We summarize here the methods in current use:

1. Forward/backward methods to (a) solve for branch current or power flow by using the forward-sweeping method and (b) compute the nodal voltages by using the backward-sweep approach
2. Newton method, using power mismatches at the ends of feeders and laterals, to solve iteratively for the nodal voltage
3. Gauss method on the bus-impedance matrix equation to solve iteratively for the branch currents

2.15.1 Description of Distribution Power Flow Methodologies

2.15.1.1 Method 1: Forward/Backward Methods

The forward/backward-sweep method models the distribution system as a tree network, with the slack bus denoted as the root of the tree and branch networks as the layers that are far away from the root nodal. Weakly meshed networks are converted to radial networks by breaking the loops and computing injection currents. The backward sweep primarily sums either the line currents or power flows from the extreme feeder (leaf) to the root. The step-by-step application of the algorithm is as follows:

1. Select the slack bus and assume initial voltage and angle at the root, node, and other buses.
2. Compute nodal current injection at the kth iteration.

$$I_i^{(k)} = \left[\frac{S_i^{sch}}{V_i^{(k-1)}} \right]^* \tag{2.133}$$

3. Start from the root with known slack-bus voltages and move toward the feeder and lateral ends.
4. Compute the voltage at node j.

$$V_j^{k-1} = V_i^k - Z_{ij} I_{ij}^{(k)} \tag{2.134}$$

where Z_{ij} is the branch impedance between bus i and j, and V_i is the latest voltage value of bus j.

5. Compute the power mismatch and check the termination criteria using

$$\Delta S_i^{(k)} = S_i^{\text{sch}} - V_i^{(k)}\left(I_i^{(k)}\right)^* \leq \varepsilon \tag{2.135}$$

6. If step 5 is not reached, we repeat the previous steps until convergence is achieved.

Note that in step 2, from each known load power S, the lateral voltages are computed or assumed. This involves V_i^{k-1} as the $k-1$ past iteration of bus voltage and $I_i^{(k)}$ is the kth current iteration of injected current. We do this by starting from the last branch from the lateral feeder and moving backward through the tree node. This is done using the expression $I_i^{(k)} = \left[\dfrac{S_i^{\text{sch}}}{V_i^{(k)}}\right]$, as before, for all interconnected branches.

A feasible implementation flowchart of the forward/backward-sweep approach to solving the distribution power flow problem is shown in Figure 2.29.

2.15.1.2 Method 2: Power-Flow Method Based on Sensitivity Matrix for Mismatch Calculation

The distribution power flow is an improved forward/backward method utilizing a sensitivity-matrix scheme to compensate the mismatch between slack-bus power injection and the power flow at the feeder and lateral ends. This results in the so-called Newton-Raphson method for distribution power flow.

Now, consider a single feeder, as shown in Figure 2.30. The flow equations are found as follows:

$$P_{j1}^{(k)} = P_{ij}^{k} - r_{ij}\frac{\left(P_{ij}^{(k)2} + Q_{ij}^{(k)2}\right)}{V_i^{(k)2}} \tag{2.136}$$

$$Q_{j1}^{(k)} = Q_{ij} - X_{ij}\frac{\left(P_{ij}^{(k)2} + Q_{ij}^{(k)2}\right)}{V_i^{(k)2}} \tag{2.137}$$

Thus,

$$V_j^{(k)} = V_i^{(k)} - 2\left(r_{ij}P_{ij}^{(k)} + X_{ij}Q_{ij}^{(k)} + \frac{r_{ij}^2 + X_{ij}^2}{V_i^{(k+1)}}\left(P_{ij}^{(k)2} + Q_{ij}^{(k)2}\right)\right) \tag{2.138}$$

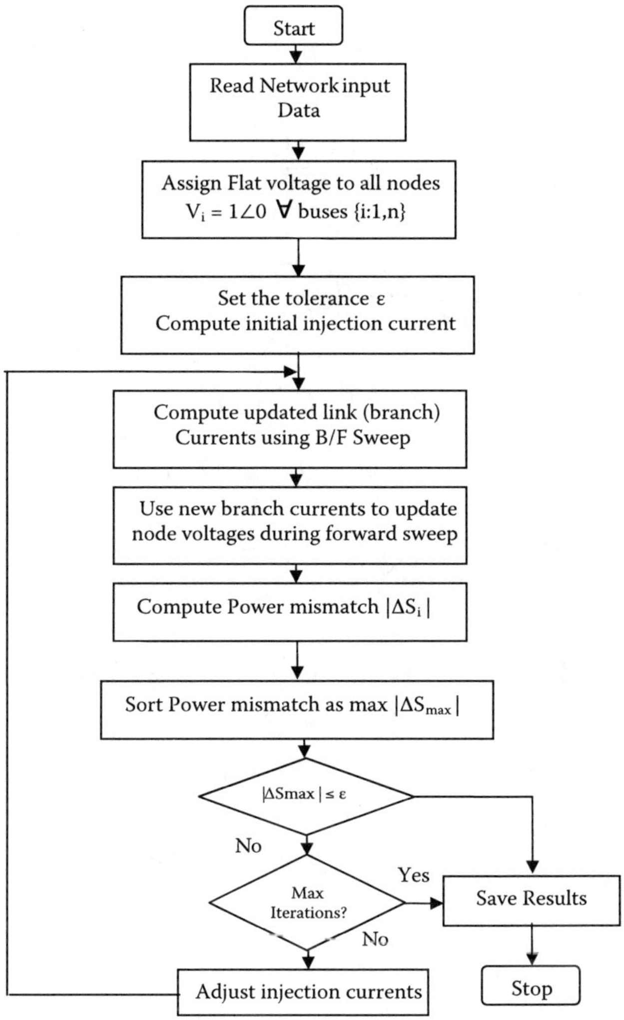

FIGURE 2.29
Flowchart for forward/backward-sweep algorithm.

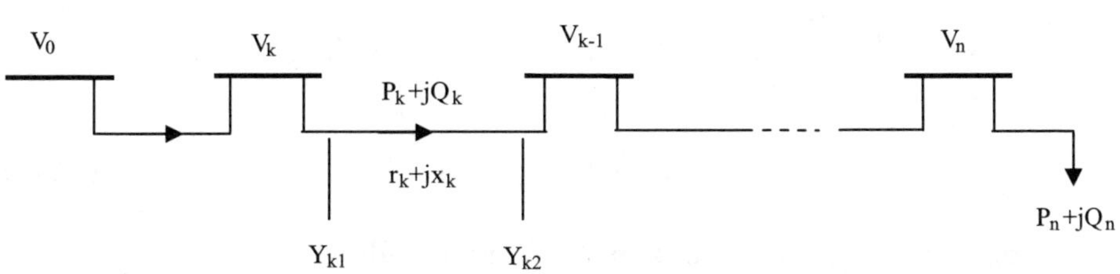

FIGURE 2.30
Single-feeder representation.

We compute

$$P_{k+1} = P_k - P_{\text{Loss }k} - P_{l\ k+1} \tag{2.139}$$

$$= P_k - r_k \frac{\left(P_k^2 + Q_k^2\right)}{\left|V_k\right|^2} - P_{l\ k+1}$$

$$= P_k - \frac{r_k}{\left|V_k\right|^2}\left\{P_k^2 + \left(Q_k + Y_k\left|V_k\right|^2\right)^2\right\} - P_{l\ k+1}$$

$$= f_p(P_k, Q_k, |V_k|^2) \tag{2.140}$$

$$Q_{k+1} = Q_k - Q_{\text{Loss }k} - Q_{l\ k+1} \tag{2.141}$$

$$= Q_k - \left\{X_k\left(P_k^2 + Q_k^2\right) - Y_{k1}\left|V_K\right|^2 - Y_{k2}\left|V_{k+2}\right|\right\} - Q_{l\ k+1}$$

$$= Q_k - \frac{X_k}{\left|V_k\right|^2}\left\{P_k^2 + \left(Q_k + Y_{k1}\left|V_k\right|^2\right)\right\} - Y_{k1}\left|V_k\right|^2 - Y_{k2} f_{v2}\left(P_k, Q_k + \left|V_k\right|^2\right) - Q_{l\ k+1}$$

$$= f_p(Q_k, P_k, |V_k|) \tag{2.142}$$

$$V_{k+1} = V_k - \left(V_k + jX_k\right)\frac{P_k + jQ_k}{V_k^*}$$

$$= f_v(P_k, Q_k, V_k) \tag{2.143}$$

$$\left|V_{k+1}\right| = \left|V_k\right|^2 + \frac{r_k + X_k}{\left|V_k\right|^2}\left(P_k^2 + Q_k^2\right) - 2r_kP_k + X_kQ_k \tag{2.144}$$

$$= \left|V_k\right|^2 + \frac{r_k^2 + X_k^2}{\left|V_k\right|^2}\left\{P_k^2 + \left(Q_k + Y_k\left|V_k\right|^2\right)^2 - 2r_kP_k + X_k\left(Q_k + Y_k\left|V_k\right|^2\right)\right\}$$

$$= f_{v2}(P_k, Q_k, |V_k|^2) \tag{2.145}$$

The procedures given for this method are as follows:

1. Assume the slack bus as the root node.

2. Assume P_0, Q_0 power injection at the slack-bus node equal to the sum of all the loads in the system.
3. Power flows in each branch are equal to the sum of downstream connected loads at the kth iteration starting from the root node with known voltage at the slack bus.
4. Obtain the latest V^k, P_{ij}^k, Q_{ij}^k (voltage and flows).
5. Compute power loss $= f\left(V^k, P_{ij}^*, Q_{ij}^*\right)$.
6. From the loss, compute receiving and power P_{ji}, Q_{ji}, and V_j.
7. The loads and shunt power are taken from the received power, and the remaining power is sent to the next feeder at lateral branches.
8. At network solutions $\Delta P_L, \Delta Q_L = 0$, when mismatch power is approximately zero, if power flow mismatch is less than the tolerance, $*\varepsilon$, then load flow has converged.
9. Update the slack-bus power from the sensitivity matrix given as follows:

$$\begin{bmatrix} \dfrac{\delta P}{\delta P_o} & \dfrac{\delta P}{\delta Q_o} \\ & \\ \dfrac{\delta Q}{\delta P_o} & \dfrac{\delta Q}{\delta Q_o} \end{bmatrix} \begin{bmatrix} \Delta P_o \\ \\ \Delta Q_O \end{bmatrix} = \begin{bmatrix} \Delta P_L \\ \\ \Delta Q_L \end{bmatrix} \tag{2.146}$$

This is a constant Jacobian.

The implementation algorithm for computing the power flow in distribution networks is shown in Figure 2.31.

2.15.1.3 Method 3: Bus-Impedance Network Method

This method uses the bus-impedance matrix and equivalent current injection to solve the network equation in the distribution system. It employs a simple superposition to find the bus voltage through the system. The voltage in each bus is computed after specifying the slack-bus voltage and then computing the incremental change ΔV due to current injection flowing into the network. The process involves the following steps:

1. Assume a no-load system.
2. Initialize the load bus voltage throughout the system using the value of the slack-bus voltage.
3. Modify nodal voltages due to current flows, which are a function of the loads connected.

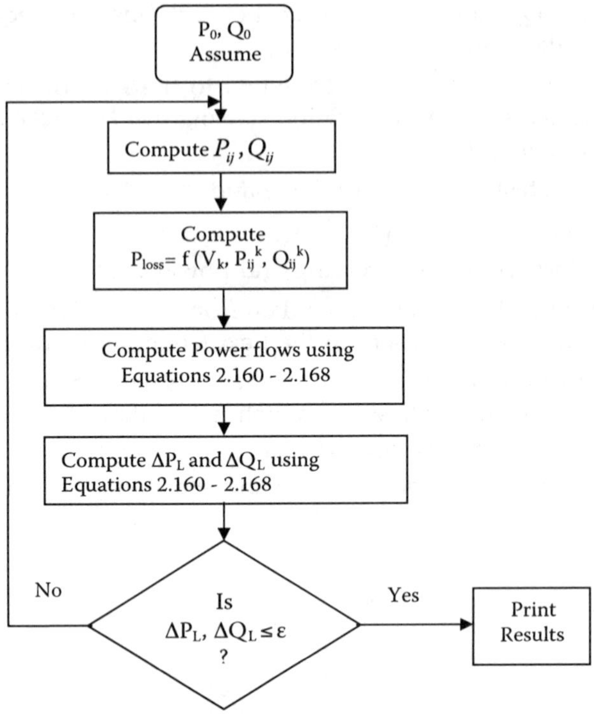

FIGURE 2.31
Flowchart for distribution-flow method.

4. The injection current is modified in the *k*th iteration as level changes.
5. Use $I_i^{(k)} = \left(\dfrac{S_i^{\text{sch}}}{V_i^{k-1}}\right)^*$ for the first equivalent current injection until we get $I_i^{(k)}$ at the last iteration $I_0^{(k)}$.
6. Compute the vector of voltage denoted as ΔV using $\Delta V_{\text{bus}}^{(k)} = Z_{\text{bus}} I_{\text{inj}}^{(k)}$, where Z_{bus} is an $n \times n$ bus-impedance matrix.
7. Determine the bus voltage updates throughout the network as $V_i^{k-1} = V_0 - \Delta V_i^{(k-1)}$, where V_0 is slack-bus voltage at the root node.
8. Check mismatch power at each load bus using specified and calculated values to obtain $\Delta S = S^{\text{spec}} - \sum V_i^{\text{calc}} I_{ij}^{\text{calc}}$ and stop if the value of $\Delta S \leq \varepsilon$.
9. Otherwise, go to step 3.

This method can easily be implemented using (a) sparsity techniques, (b) an implicit bus matrix, or (c) computational techniques. In summary, these load-flow techniques are well documented in the literature. Software packages are available to demonstrate their trade-offs and compatibility.

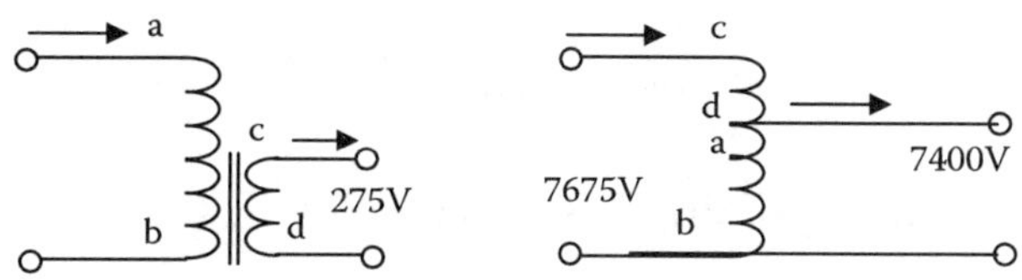

FIGURE 2.32
Schematic diagram of step-down and autotransformer connections.

2.16 Illustrative Examples

2.16.1 Distribution Transformer Considered for Use as a Step-Down Autotransformer

A 7400:275-V, 85-kVA distribution transformer is considered for use as a step-down autotransformer by connecting the 7400-V and 275-V coils in series (Figure 2.32):

a. What are the voltage, current, and kVA ratings when the transformer is connected as a two-winding transformer?
b. What are the voltage, current, and kVA ratings when the transformer is connected as the autotransformer?

Solution

a. The two-winding transformer ratings are presented in Table 2.1.

b. The autotransformer ratings are

$$V_{in} = 7400 + 275 = 7675 \text{ V} \quad (2.147)$$

$$V_{out} = 7400 \text{ V}$$

The current rating of the high-voltage coil is

TABLE 2.1
Two-Winder Transformer Ratings

High-Voltage Winding	Low-Voltage Winding
7400 V	275 V
10.22 A	309.09 A
85 kVA	85 kVA

$$I_{\text{H rated}} = \frac{85000}{7400} = 11.49 \text{ A} \tag{2.148}$$

The current rating of the low-voltage coil is

$$I_{\text{X rated}} = \frac{85000}{275} = 309.09 \text{ A} \tag{2.149}$$

The rated input current is limited to the rating of the 277-V coil, or 271 A. Therefore, the output current is

$$I_{\text{out}} = (7675/7400)275 = 285.22 \text{ A} \tag{2.150}$$

which means that the current in the common leg is

$$I_c = I_{\text{out}} - I_{\text{in}} = 285.22 - 275 = 10.22 \text{ A} \tag{2.151}$$

which is the rating of the 7400-V coil. The kVA rating of the transformer connected as an autotransformer is

$$\text{kVA}_{\text{in}} = 7675 \times 275 = 2110 \text{ kVA}$$

$$\text{kVA}_{\text{out}} = 7400 \times 285.22 = 2110 \text{ kVA}$$

This means that an 85-kVA, 7400:275-V, two-winding transformer could be used as an autotransformer to 2110 kVA at 7675:7400 V.

2.16.2 Transformer Short Circuit during an Open-Circuit Test

A single-phase two-winding transformer is rated 30 kVA, 480/120 V, 60 Hz. During a short-circuit test, where rated current at rated frequency is applied to the 480-V winding (denoted winding 1), with the 120-V winding (winding 2) shorted, the following readings are obtained: $V_1 = 45$ V, $P_1 = 350$ W. During an open-circuit test, where rated voltage is applied to winding 2, with winding 1 open, the following readings are obtained: $I_2 = I_2$ A, $P_2 = 220$ W:

a. From the short-circuit test, determine the equivalent series impedance

$$Z_{\text{eq1}} = R_{\text{eq1}} + jX_{\text{eq1}} \tag{2.152}$$

referred to winding 1. Neglect the shunt admittance.

b. From the open-circuit test, determine the shunt admittance

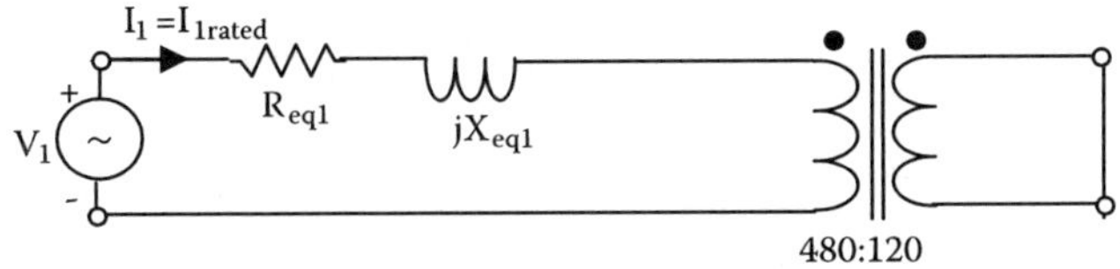

FIGURE 2.33(A)
Short-circuit test (neglecting shunt admittance).

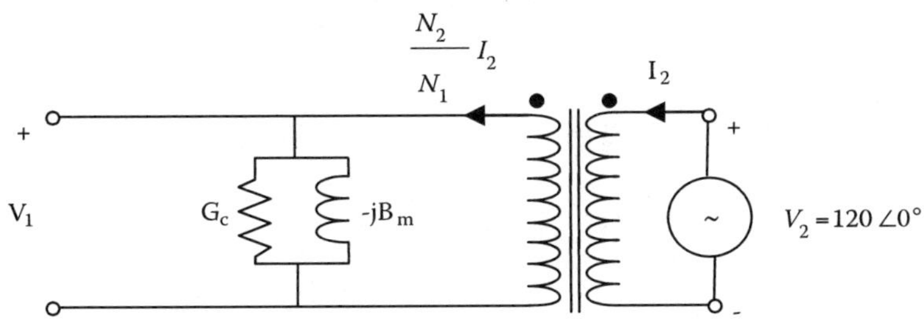

FIGURE 2.33(B)
Open-circuit test (neglecting series impedance).

$$Y_m = G_c - jB_m \tag{2.153}$$

referred to winding 1. Neglect the series impedance.

Solution

a. The equivalent circuit for the short-circuit test is shown in Figure 2.33(a), where the shunt admittance branch is neglected. Rated current for winding 1 is

$$I_{1\text{rated}} = \frac{S_{\text{rated}}}{V_{1\text{rated}}} = \frac{30 \times 10^3}{480} = 62.5 \text{ A} \tag{2.154}$$

R_{eq1}, Z_{eq1}, and X_{eq1} are then determined as follows:

$$R_{eq1} = \frac{P_1}{I_{1\text{rated}}^2} = \frac{350}{(62.5)^2} = 0.0896 \ \Omega \tag{2.155}$$

$$|Z_{eq1}| = \frac{V_1}{I_{1\text{rated}}} = \frac{45}{62.5} = 0.7200 \ \Omega \tag{2.156}$$

$$X_{\text{eq1}} = \sqrt{Z_{\text{eq1}}^2 - R_{\text{eq1}}^2} = 0.7256 \ \Omega \tag{2.157}$$

$$Z_{\text{eq1}} = R_{\text{eq1}} + jX_{\text{eq1}} = 0.0896 + j0.7200 = 0.7256\angle 82.91° \ \Omega \tag{2.158}$$

b. The equivalent circuit for the open-circuit test is shown in Figure 2.33(b), where the series impedance is neglected.

$$V_1 = E_1 = a_t E_2 = \frac{N_1}{N_2} V_{2\text{rated}} = \frac{480}{120}(120) = 480 \text{ V} \tag{2.159}$$

G_c, Y_{m}, and B_{m} are then determined as follows:

$$G_c = \frac{P_2}{V_1^2} = \frac{220}{(480)^2} = 0.00095486 \ S \tag{2.160}$$

$$|Y_{\text{m}}| = \frac{\left(\frac{N_1}{N_2}\right) I_2}{V_1} = \frac{\left(\frac{120}{480}\right)(12)}{480} = 0.00625 \ S \tag{2.161}$$

$$B_{\text{m}} = \sqrt{Y_{\text{m}}^2 - G_c^2} = \sqrt{(0.00625)^2 - (0.00095486)^2} = 0.00619 \ S \tag{2.162}$$

$$Y_{\text{m}} = G_c - jB_{\text{m}} = 0.00095486 - j0.00619 = 0.0063\angle -81.23° \ S \tag{2.163}$$

2.16.3 Unbalanced Set of Voltages

The zero and positive-sequence components of an unbalanced set of voltages are

$V_+ \quad = 4$

$V_0 \quad = 0.6 - j0.775$

The phase A voltage is

$V_A \quad = 5$

Obtain the negative-sequence component and the B- and C-phase voltages.

Solution

We have

$$V_A = V_+ + V_- + V_0 \tag{2.164}$$

$$5 = 4 + V_- + (0.6 - j0.775)$$

Thus

$$V_- = 0.4 + j0.775 = 0.872\angle 62.70°$$

In polar form, we have

$$V_0 = 0.6 - j0.775 = 0.98\angle -52.25°$$

Now, for phase B, we have

$$\begin{aligned} V_B &= \alpha^2 V_+ + \alpha V_- + V_0 \\ &= 4\angle 240° + 0.872\angle -177.30° + 0.98\angle -52.25 \\ &= 4.84\angle -117.95° \end{aligned} \tag{2.165}$$

For phase C, we have

$$\begin{aligned} V_c &= \alpha V_+ + \alpha^2 V_- + V_0 \\ &= 4\angle 120° + 0.872\angle -57.30° + 0.98\angle -52.25 \\ &= 2.16\angle 115.41 \end{aligned} \tag{2.166}$$

2.16.4 Newton-Raphson Method

Solve the following two equations in x_1 and x_2 using the Newton-Raphson method.

$$F_1(x) = x_1^2 + x_2^2 - 4x_1 = 0$$

$$F_2(x) = x_1^2 + x_2^2 - 8x_1 + 12 = 0$$

1. Find expressions for the elements of the Jacobian matrix and find the correction increments Δx_1, Δx_2.
2. Calculate the first five iterations to find estimates of the solution using the following initial guesses:
   a. $x_1 = 2$, $x_2 = 4$
   b. $x_1 = -5$, $x_2 = -5$
   c. $x_1 = -0.1$ $x_2 = 1$

Solution

1. The Jacobian elements are as follows

$$\frac{\partial F_1}{\partial x_1} = 2x_1 - 4; \qquad \frac{\partial F_1}{\partial x_2} = 2x_2$$

$$\frac{\partial F_2}{\partial x_1} = 2x_1 - 8; \qquad \frac{\partial F_2}{\partial x_2} = 2x_2$$

Now, using the formulation for the typical N-R problem, we have

$$\begin{bmatrix} \frac{\partial F_1}{\partial x_1} & \frac{\partial F_1}{\partial x_2} \\ \frac{\partial F_2}{\partial x_1} & \frac{\partial F_2}{\partial x_2} \end{bmatrix} \begin{bmatrix} \Delta x_1 \\ \Delta x_2 \end{bmatrix} = \begin{bmatrix} -F_1(x) \\ -F_2(x) \end{bmatrix} \tag{2.167}$$

$$\begin{bmatrix} 2x_1 - 4 & 2x_2 \\ 2x_1 - 8 & 2x_2 \end{bmatrix} \begin{bmatrix} \Delta x_1 \\ \Delta x_2 \end{bmatrix} = \begin{bmatrix} -(x_1^2 + x_2^2 - 4x_1) \\ -(x_1^2 + x_2^2 - 8x_1 + 12) \end{bmatrix}$$

2. The solution for Δx_1, Δx_2 is given by

$$\begin{bmatrix} \Delta x_1 \\ \Delta x_2 \end{bmatrix} = \begin{bmatrix} 2x_1 - 4 & 2x_2 \\ 2x_1 - 8 & 2x_2 \end{bmatrix}^{-1} \begin{bmatrix} -(x_1^2 + x_2^2 - 4x_1) \\ -(x_1^2 + x_2^2 - 8x_1 + 12) \end{bmatrix} \tag{2.168}$$

and power flow computations

$$\begin{bmatrix} \Delta x_1 \\ \Delta x_2 \end{bmatrix} = \begin{bmatrix} 3 - x_1 \\ -(x_1^2 + x_2^2 - 8x_1 + 12) \; 2x_2 \end{bmatrix} \tag{2.169}$$

As a result, the new estimates of the solution are given by

$$x_i^{n+1} = x_i^n + \Delta x_i^n$$

$$x_i^{n+1} = x_i^n + (3 - x_1^n) \tag{2.170}$$

$$x_i^{n+1} = 3$$

Note that x_1 is independent of the starting point. In addition, the iterative formula for x_2 is

$$x_2^{(n+1)} = x_2^{(n)} + \frac{x_1^{2(n)} - x_2^{2(n)} - 6x_1^{2(n)} + 12}{2x_2^n}$$

$$= \frac{x_1^{2n} + x_2^{2n} - 6x_1^n + 12}{2x_2^n}$$

with $x_i^n = 3$

$$x_2^{(n+1)} = \frac{x_2^{2n} + 3}{2x_2^n} \tag{2.171}$$

2.16.5 Polar Formulation of Load-Flow Equations

For the network shown in Figure 2.34, do the following:

1. Compute the Y-bus admittance matrix and state the initial bus voltages in polar form.
2. Write down the load-flow equations of the problem using the polar formulation.

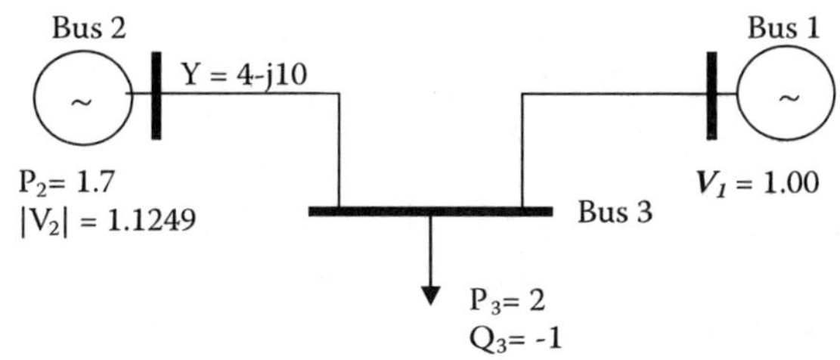

FIGURE 2.34
Single-line diagram for Example 2.16.5.

Solution

1. We have in polar form:
   The diagonal terms of Y_{bus} are

$$Y_{11} = 6.4031\angle -51.34°$$

$$Y_{22} = 10.77\angle -68.199°$$

$$Y_{33} = 17.00\angle -61.928°$$

   Off-diagonal terms of the Y_{bus} are

$$Y_{12} = 0$$

$$Y_{13} = 6.4031\angle 128.66°$$

$$Y_{23} = 10.77\angle 111.80°$$

   Furthermore, $Y_{ij} = Y_{ji}$, since the admittance matrix is inherently symmetrical.
   The bus voltages are represented as:

$$V_1 = 1\angle 0$$

$$V_2 = 1.1249\angle \theta_2$$

$$V_3 = |V_3|\angle \theta_3$$

2. We now can write the power-flow equation for the three-bus system shown in Figure 2.35. For bus 1, we have

$$P_1 - jQ_1 = V_1^*(Y_{11}V_1 + Y_{12}V_2 + Y_{13}V_3) \qquad (2.172)$$

$$\Rightarrow P_1 - jQ_1 = 6.4031\angle -51.34° + 6.4031|V_3|\angle 128.66 + \theta_3$$

   For bus 2, we have

$$P_2 - jQ_2 = V_2^*(Y_{12}V_1 + Y_{22}V_2 + Y_{23}V_3) \qquad (2.173)$$

$$\Rightarrow 1.7 - jQ_2 = (13.628\angle -68.199°) + (12.115V_3\angle 118) + \theta_3 - \theta_2$$

For bus 3, we have

$$P_3 - jQ_3 = V_3^*(Y_{13}V_1 + Y_{23}V_2 + Y_{33}V_3) \tag{2.174}$$

$$\Rightarrow -2 + j1 = V_3\angle -\theta_3$$

$$\left[6.4031\angle 128.66^\circ + (10.77\angle 111.8 * 1.1249\angle\theta_2) + 17\angle -61.928 * V_3\angle\theta_3\right]$$

which reduces to

$$-2 + j1 = (6.4031V_3\angle 128.66^\circ) + (12.11V_3\angle 111.8 - \theta_2 - \theta_2) + 17.00V_3\angle -61.928^\circ$$

We now separate the real and imaginary parts of the above equations such that for bus 1

$$P_1 = 4 + 6.4031V_3\cos(128.66^\circ + \theta_3)$$

$$-Q_1 = -5 + 6.4031V_3\sin(128.66^\circ + \theta_3)$$

for bus 2

$$1.7 = 5.6612 + 12.115V_3\cos(111.8 + \theta_3 - \theta_2)$$

$$-Q_2 = -12.653 + 12.115V_3\sin(111.8 + \theta_3 - \theta_2)$$

and for bus 3

$$-2 = 6.4031V_3\cos(128.66 - \theta_3) + 12.115V_3\cos(111.8 + \theta_2 - \theta_3) + 8V_3^2$$

$$1 = 6.4031V_3\sin(128.66 - \theta_3) + 12.115V_3\sin(111.8 + \theta_2 - \theta_3) - 15V_3^2$$

2.16.6 Gauss-Seidel Method

The circuit elements in the 138-kV circuit in Figure 2.35 are in per unit on a 100-MVA base with the nominal 138-kV voltage as base. The $(P + jQ)$ load is scheduled to be 170 MW and 50 MVAr.

1. Write the Y-matrix for the two-bus system.
2. Assume that bus 1 is a reference bus and set up the Gauss-Seidel correction equation for bus 2.

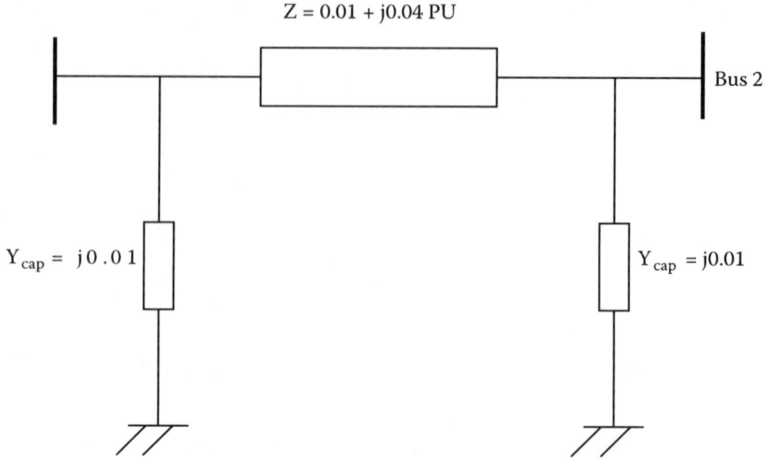

FIGURE 2.35
One-line diagram for Example 2.16.6.

Solution

1. The line impedance of this line is

$$z_{12} = 0.01 + j0.04 \text{ (p.u.)}$$

Then the line admittance can be expressed as

$$y_{12} = 1/z_{12} = \frac{1}{(0.01 + j0.04)} = 5.88 - j23.53 \text{ (p.u.)}$$

The bus admittance matrix components can be expressed as

$$Y_{11} = y_{12} + y_{10} = (5.88 - j23.53) + (j0.01) = 5.88 - j23.54$$

$$Y_{22} = y_{12} + y_{20} = (5.88 - j23.53) + (j0.01) = 5.88 - j23.54$$

$$Y_{12} = Y_{21} = -y_{12} = -(5.88 - j23.53)$$

Then the Y-matrix can be expressed as

$$Y_{bus} = \begin{bmatrix} 5.88 - j23.54 & -5.88 + j23.53 \\ -5.88 + j23.53 & 5.88 - j23.54 \end{bmatrix}$$

2. Taking bus 1 as reference,

$$V_1 = 1.0\angle 0.0 \text{ p.u.}$$

The Gauss-Seidel equation for bus 2 is

$$V_2^{(n+1)} = \frac{1}{Y_{22}}\left[\frac{P_2 - jQ_2}{V_2^{*(n)}} - Y_{21}V_1^{(n)}\right] \tag{2.175}$$

We have the load

$$P_2 - JQ_2 = (1.7 - j0.5) \text{ p.u.}$$

2.17 Summary

This chapter explained the major computational tools and concepts used in distribution system analysis, including different computations of voltage sag, power factor, power flows, and losses. The summary of procedures for different schemes of power flow, such as forward sweeping using Gauss-Seidel and Newton-Raphson methods of power flow solution, were discussed. Some illustrative examples using simple distribution systems were also provided for the reader.

Problem Set 2

2.1 Recall the various types of transformers discussed in the chapter.

a. Do a careful model and mathematical description of autotransformers, distribution transformers, and power transformers.

b. If 69 kV/12.5, 15 MVA, 3Ø Y:Y is operated at full load, calculate I_s and I_p assuming a well-grounded neutral.

c. State the purpose of transformer protection and identify major types of equipment used for protection.

2.2 Given a 227/480V 3Ø Y-connected system with the following loads:

M1 = induction motor 50 hp, η = 91, pf = 0.89 at full load
M2 = induction motor 25 hp, η = 0.9, pf = 0.9 at full load
R = heater load at pf = 1
M3 = synchronous motor driving a second 25-hp load

Calculate the leading kVArs the synchronous motor must provide to correct the power factor to unity. The synchronous motor will run at full load 25 hp, η = 0.94.

2.3 Distribution system engineers constantly monitor the "quality" of power.

a. Define the terms power factor correction, power quality, and voltage sag.

b. Construct a sample calculation procedure for computing power quality, voltage sag, and power factor correction for a given distribution system.

c. Construct your own simple example to illustrate your understanding of these terminologies. Please solve completely.

2.4 Consider the following circuit shown in Figure 2.36.

a. Solve for the DC currents in each loop.

b. Show that

$$V_1 = \sqrt{\left(100V_1 - 200 - 400\frac{V_1}{V_2}\right)}$$

$$V_2 = \sqrt{(V_1V_2 - 800)}$$

c. Solve for V_1 and V_2 using:

a. Gauss-Seidel method

b. Newton-Raphson method

if $V_1^0 = V_2^0 = 100$.

2.5 Using the load-flow algorithm, compute a distribution load flow for a typical three-bus system of your choice.

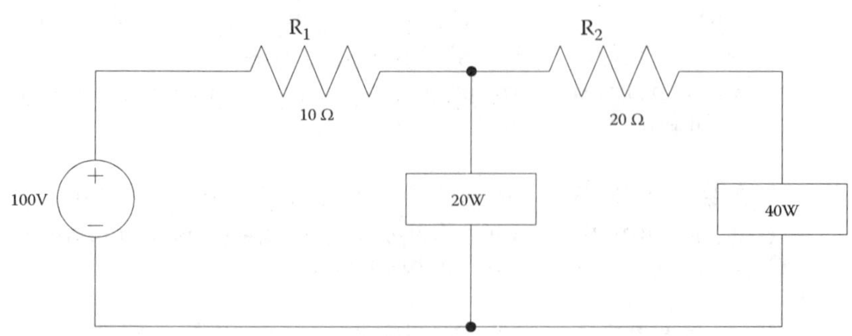

FIGURE 2.36
Diagram for Problem 2.4

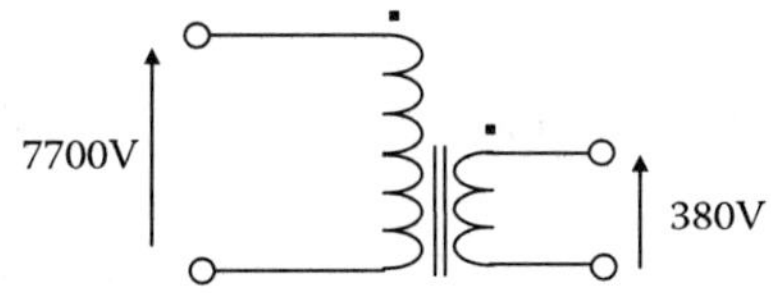

FIGURE 2.37(A)
Diagram for Problem 2.7.

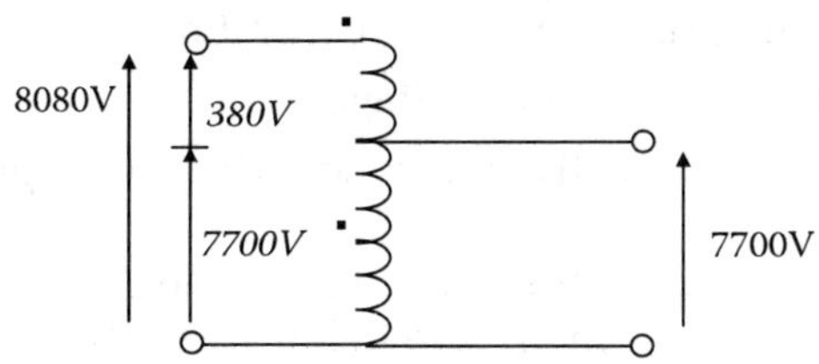

FIGURE 2.37(B)
Diagram for Problem 2.7.

2.6 Minimize $f = c_1x_1 + c_2x_2$

subject to
$$x_1 + x_2 \leq 1$$
$$x_1x_2 \geq 0$$

where $c_1 = 1$ and $c_2 = 2$.

2.7 Consider a 7,700:380V, 85 kVA distribution transformer that can be configured as a step down transformer or an autotransformer by appropriate connection of its 7.7kV and 380V windings.

a. What are the voltage, current, and kVA ratings when the transformer is connected as a two-winding transformer?

b. What are the voltage, current, and kVA ratings when the transformer is connected as the autotransformer?

(Hint: The rated input current is limited to the rating of the 380 V coil or 380 A).

2.8 Given the Y-connected, balanced 3-ϕ load, consisting of three impedance of 20 ∠ 30° ohms each as shown in the figure below is supplied with the balanced line-to-neutral voltages:

$$V_{an} = 220 \angle 0° \quad \text{V}$$

$$V_{bn} = 220 \angle 240° \quad \text{V}$$

$$V_{cn} = 220 \angle 120° \quad \text{V}$$

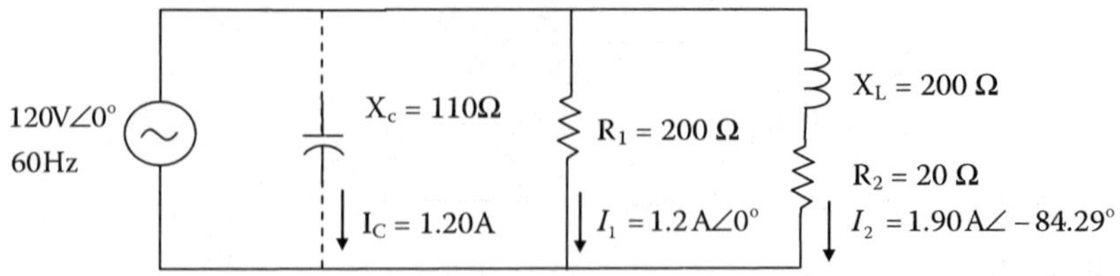

FIGURE 2.38
Diagram for Problem 2.9.

a. Calculate the phase currents in each line.
b. Calculate the line-to-line voltages.
c. Calculate the total active and reactive power supplied to the load.

2.9 Calculate the value of the capacitor needed to correct the power factor of the circuit in the figure below to unity.

2.10 Review two (2) methods of Distribution Power Flow.
a. Draw a detailed flowchart of the procedure for solving Distribution Power Flow problems using one of the methods reviewed.
b. Using a 3-bus network of choice, compute the load flow using 2 iterations.
c. Verify your results in (b) using MATLAB or any off-the-shelf distribution systems power flow program.

3

Distribution System Protection and Control

3.1 Introduction

Power system operation is ideally a balanced three-phase mode, with equal voltage and current and a phase-shift magnitude of 120° for each phase. For a balanced system connected in triangular or Y for a typical four-wire system, the fourth wire carries zero current when all of the phase impedances and voltages are equal.

For the unbalanced case, the situation is different. Short circuits occur, leading to single line-to-ground, line-to-line, double line-to-ground, and balanced three-phase faults. The path to ground for a fault is determined as Z_f, which is referred to as a bolted short circuit or a nonzero impedance.

The causes of faults are many. They include lightning, wind, objects falling on lines (trees, kites, airplanes, debris), vandalism, accidental break or short circuit, or inadvertent operation of a circuit breaker.

When a fault occurs, the three-phase network is decomposed into positive-sequence, negative-sequence, and zero-sequence networks. These networks are connected to represent the different types of fault classification for which the corresponding fault currents are decoupled into positive-, negative-, and zero-sequence currents that are computed for fault conditions. To simplify the power system model for deriving the equivalent sequence diagram, the following occurs:

1. We assume that, for a balanced system, the positive-, negative-, and zero-sequence networks are uncoupled before a fault occurs.
2. During fault, they are connected to represent a particular fault category.
3. Prefault load current is neglected so that prefault voltage for each fault location is equal to the internal voltage of all machines.
4. Shunt capacitance of line, shunt elements of line, and series resistance of line are neglected.

5. Synchronous machine armature resistance, saliency, and saturation effects are neglected.
6. We assume the internal voltage source is $1\angle 0$ p.u. (per unit) for the prefault voltage at its normal value prior to application of fault.

3.1.1 Introduction to Symmetrical Components

The general three-phase circuit for voltage and current is referred to in terms of its per-phase values. Under balanced conditions, $|V_a| = |V_b| = |V_c|$, and the voltage vectors are equally distributed by 120°. In the case of an unbalanced network, the magnitude of at least one phase voltage is different from another, and similarly for the phase currents. In this case, the theory of symmetrical components allows us to decouple the unbalanced voltage (or current) as follows:

$$V_a = V_{a0} + V_{a1} + V_{a2}$$

$$V_b = V_{b0} + V_{b1} + V_{b2}$$

$$V_c = V_{c0} + V_{c1} + V_{c2}$$

where

V_{a0}, V_{b0}, and V_{c0} are zero-sequence components of equal magnitude and of the same phase rotating in the same direction as V_a

V_{a1}, V_{b1}, and V_{c1} are negative-sequence voltage components rotating opposite to V_a

V_{a2}, V_{b2}, and V_{c2} are positive-sequence voltage components rotating in same direction as V_a

By using $\alpha = e^{j\frac{2\pi}{3}}$, we then define the phase quantities with respect to V_a as follows:

$$V_a = V_{a0} + V_{a1} + V_{a2}$$

$$V_b = V_{a0} + \alpha^2 V_{b1} + \alpha V_{b2}$$

$$V_c = V_{a0} + \alpha V_{c1} + \alpha^2 V_{c2}$$

For simplicity, we drop the "a" in the subscripts to obtain

$$\begin{bmatrix} V_a \\ V_b \\ V_c \end{bmatrix} = \begin{bmatrix} 1 & 1 & 1 \\ 1 & \alpha^2 & \alpha \\ 1 & \alpha & \alpha^2 \end{bmatrix} \begin{bmatrix} V_0 \\ V_1 \\ V_2 \end{bmatrix}$$

and

$$\begin{bmatrix} V_0 \\ V_1 \\ V_2 \end{bmatrix} = \begin{bmatrix} 1 & 1 & 1 \\ 1 & \alpha^2 & \alpha \\ 1 & \alpha & \alpha^2 \end{bmatrix}^{-1} = \frac{1}{3}\begin{bmatrix} 1 & 1 & 1 \\ 1 & \alpha & \alpha^2 \\ 1 & \alpha^2 & \alpha \end{bmatrix}\begin{bmatrix} V_a \\ V_b \\ V_c \end{bmatrix}$$

3.1.2 Sequence Networks Used in Fault Analysis

Consider the general three-phase system (Figure 3.1) that illustrates the fault representations in different categories as shown in Figure 3.2, Figure 3.3, and Figure 3.4, which depict a line-to-ground fault, a line-to-line fault, and a double line-to-ground fault, respectively. Each of these generalized networks can be developed into a positive, negative, and zero Thevenin equivalent from the terminals, as shown in Figure 3.5, Figure 3.6, and Figure 3.7, respectively.

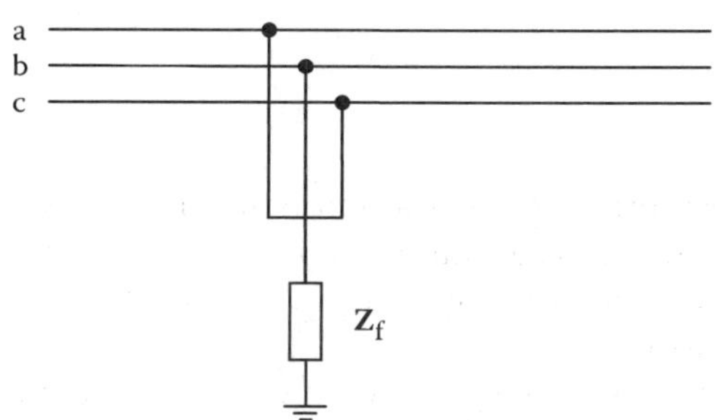

FIGURE 3.1
Three-phase fault.

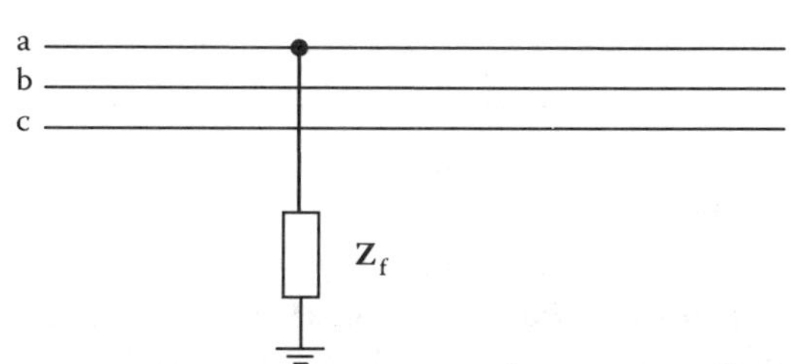

FIGURE 3.2
Single line-to-ground fault.

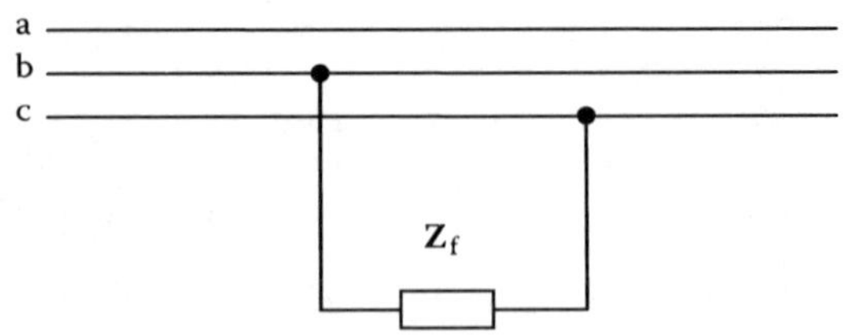

FIGURE 3.3
Line-to-line fault.

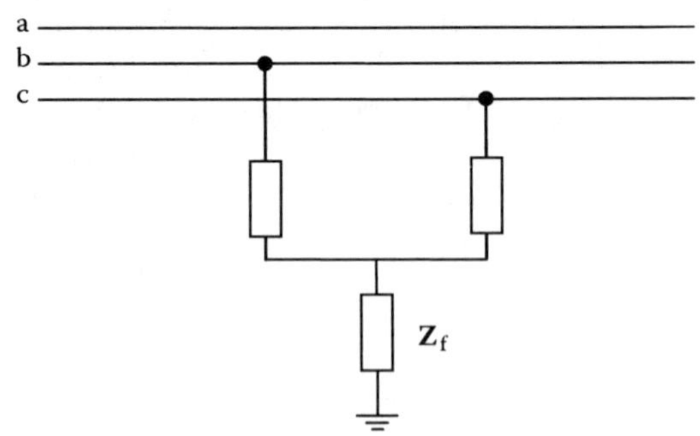

FIGURE 3.4
Double line-to-ground fault.

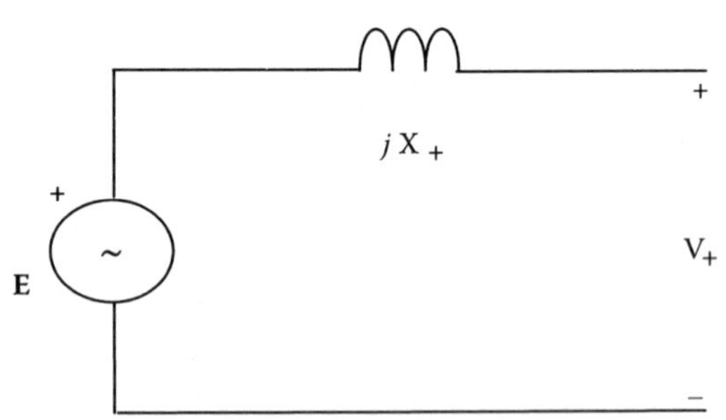

FIGURE 3.5
Positive-sequence network.

3.1.2.1 Computation of Phase and Total Power Using Sequence Networks

Let the total complex power to a three-phase load be defined as S_P, where

$$S_P = V_a I_a^* + V_b I_b^* + V_c I_c^* \tag{3.1}$$

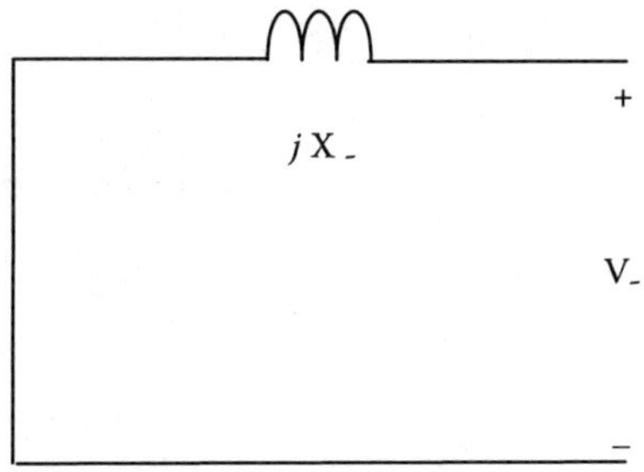

FIGURE 3.6
Negative-sequence network.

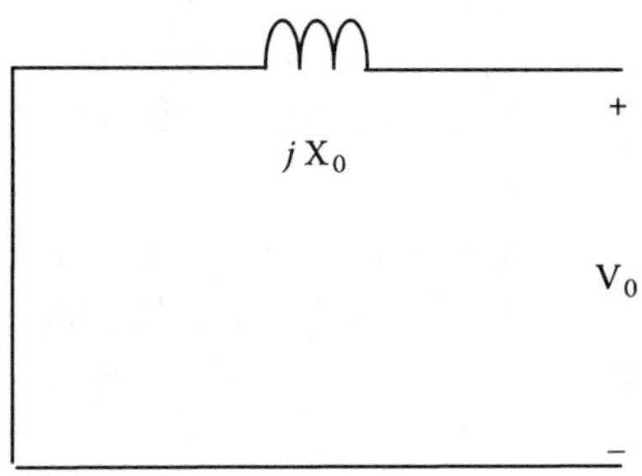

FIGURE 3.7
Zero-sequence network.

$$S_P = \begin{bmatrix} V_a & V_b & V_c \end{bmatrix} \begin{bmatrix} I_a^* \\ I_b^* \\ I_c^* \end{bmatrix} = V_P{}^T I_P{}^* \tag{3.2}$$

$$S_P = \begin{bmatrix} M & V_S \end{bmatrix}^T (MI_S)^* \tag{3.3}$$

$$= V_S{}^T (M^T M^*) I_S{}^* \tag{3.4}$$

$$= \begin{bmatrix} 1 & 1 & 1 \\ 1 & \alpha^2 & \alpha \\ 1 & \alpha & \alpha^2 \end{bmatrix} \begin{bmatrix} 1 & 1 & 1 \\ 1 & \alpha^2 & \alpha \\ 1 & \alpha & \alpha^2 \end{bmatrix}^* \tag{3.5}$$

$$= \begin{bmatrix} 1 & 1 & 1 \\ 1 & \alpha^2 & \alpha \\ 1 & \alpha & \alpha^2 \end{bmatrix} \begin{bmatrix} 1 & 1 & 1 \\ 1 & \alpha & \alpha^2 \\ 1 & \alpha^2 & \alpha \end{bmatrix} = \begin{bmatrix} 3 & 0 & 0 \\ 0 & 3 & 0 \\ 0 & 0 & 3 \end{bmatrix} = 3I \tag{3.6}$$

$$S_P = 3V_S{}^T I_S{}^* = 3\begin{bmatrix} V_0 & V_+ & V_- \end{bmatrix} \begin{bmatrix} I_0{}^* \\ I_+{}^* \\ I_-{}^* \end{bmatrix} \tag{3.7}$$

$$S_P = 3(V_0 I_0{}^* + V_+ I_+{}^* + V_- I_-{}^*) \tag{3.8}$$

The complex power S_P delivered to a three-phase network equals three times the total complex power S_S delivered to the sequence networks.

3.1.2.2 Development of Sequence Networks for Power Systems

Consider the two-machine system depicted in Figure 3.8, with pertinent data for each apparatus given in Table 3.1, which specifies the sequence components values in p.u. for the system.

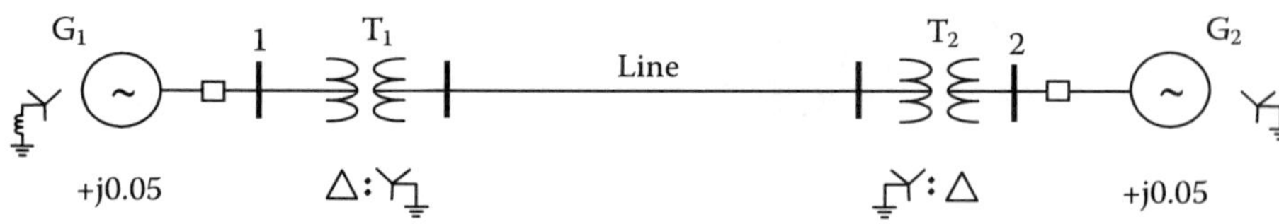

FIGURE 3.8
Single line diagram for sample power system.

TABLE 3.1

Sequence Components for the Power System in Figure 3.8

Network Component/Device	MVA Rating (MVAr)	Voltage Rating (kV)	Sequence Impedance (p.u.)		
			X_1	X_2	X_0
G1	100	13.8	0.150	0.170	0.050
G2	100	13.8	0.200	0.310	0.150
T1	100	13.8/138	0.100	0.100	0.100
T2	100	138/13.8	0.100	0.100	0.100
L1-2	100	138	0.315	0.105	0.105

Converting from actual to per unit quantities is done using $X_{p.u.} = X_{actual}/X_{base}$, where the base for the quantity, X_{base}, is referenced to the 100-MVA, 138-kV system bases.

Figure 3.8(a), Figure 3.8(b), Figure 3.8(c), and Figure 3.8(d) present an illustrative example, a positive-sequence network, a negative-sequence network, and a zero-sequence network, respectively. A summary of the equivalent sequence networks is given in Figure 3.8(e).

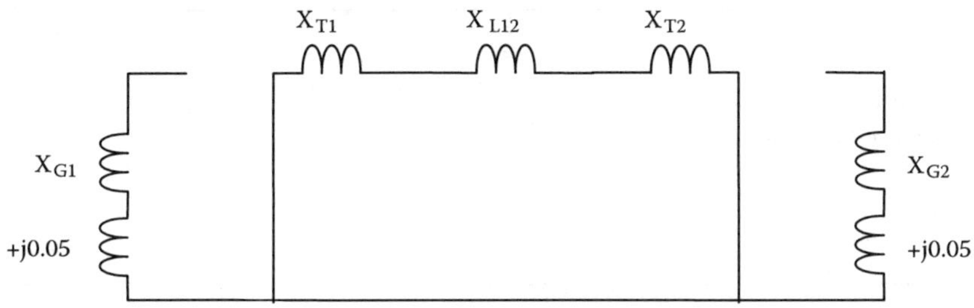

FIGURE 3.8(A)
Illustrative example.

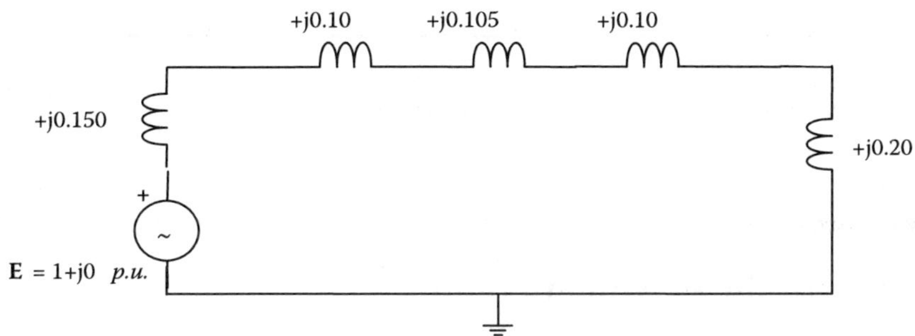

FIGURE 3.8(B)
Positive-sequence network.

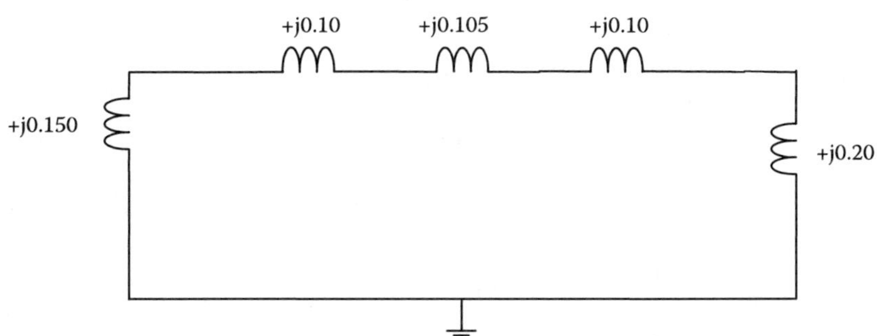

FIGURE 3.8(C)
Negative-sequence network.

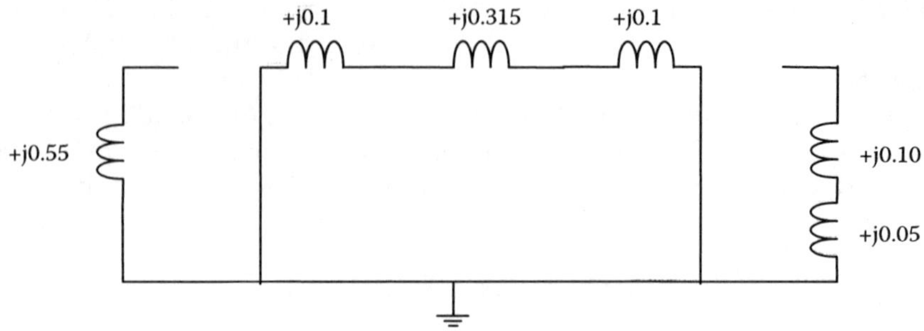

FIGURE 3.8(D)
Zero-sequence network.

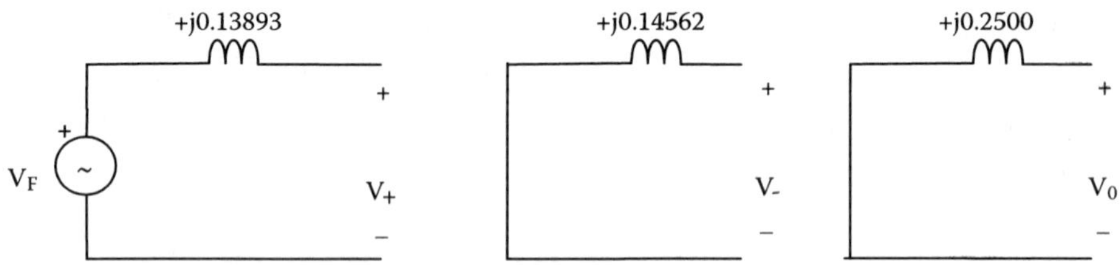

FIGURE 3.8(E)
Sequence networks for a fault in the network in Figure 3.8.

3.2 Single Line-to-Ground Fault

Consider a single line-to-ground fault on phase A to ground. We pose the following boundary conditions:

$$I_b = 0 \text{ and } I_c = 0$$

therefore

$$I_b = I_c \text{ and } I_a \neq 0$$

therefore

$$V_{ag} = I_a Z_f \tag{3.9}$$

for a bolted fault, $Z_f = 0$; otherwise it is nonzero. From the network interconnection of the sequence networks, we observed that

$$I_b = I_0 + \alpha^2 I_1 + \alpha I_2 \text{ and } I_c = I_b \tag{3.10}$$

which gives

$$(\alpha^2 - \alpha)I_1 = (\alpha^2 - \alpha)I_2 \Rightarrow I_1 = I_2 \tag{3.11}$$

and

$$I_c = I_0 + \alpha I_1 + \alpha^2 I_2 \tag{3.12}$$

$$I_1 = I_2$$

$$\therefore\ I_b = I_0 + \alpha^2 I_1 + \alpha I_2 \tag{3.13}$$

$$= I_0 + (\alpha^2 + \alpha)I_1$$

$$= I_0 - I_1$$

$$= 0$$

therefore, $I_0 = I_1$ or, simply put,

$$\begin{bmatrix} I_0 \\ I_1 \\ I_2 \end{bmatrix} = \frac{1}{3}\begin{bmatrix} 1 & 1 & 1 \\ 1 & \alpha & \alpha^2 \\ 1 & \alpha^2 & \alpha \end{bmatrix} * \begin{bmatrix} I_a \\ I_b \\ I_c \end{bmatrix} = \frac{1}{3}\begin{bmatrix} I_a \\ I_a \\ I_a \end{bmatrix} \tag{3.14}$$

because $I_b = I_c = 0$. This means that

$$I_0 = I_1 = I_2 \tag{3.15}$$

$$V_0 + V_1 + V_2 = 3(I_0 + I_1 + I_2)Z_f$$

$$= Z_f I_1 \tag{3.16}$$

If $I_0 = I_1 = I_2$, we have

$$I_0 = I_1 = I_2 = \frac{V_f}{Z_0 + Z_1 + Z_2 + 3Z_f} \tag{3.17}$$

Note, we compute $I_0 = I_1 = I_2$ from

$$\frac{1\angle 0}{(Z_0 + Z_1 + Z_2 + 3Z_f)}$$

where

$$3Z_f = 0 \tag{3.18}$$

$$= \frac{1\angle 0}{Z_0 + Z_1 + Z_2} \tag{3.19}$$

from which we get I_{abc}

$$\begin{bmatrix} I_a \\ I_b \\ I_c \end{bmatrix} = \begin{bmatrix} 1 & 1 & 1 \\ 1 & \alpha^2 & \alpha \\ 1 & \alpha & \alpha^2 \end{bmatrix} \begin{bmatrix} I_0 \\ I_1 \\ I_2 \end{bmatrix} \tag{3.20}$$

3.3 Double Line-to-Ground Fault on Phase B and C

Figure 3.9 depicts the equivalent double line-to-ground (DLG) fault. Using the boundary conditions for DLG

$$V_2 = V_1 \neq V_0 \tag{3.21}$$

$$I_a = 0,\ I_b \neq I_c \neq 0 \tag{3.22}$$

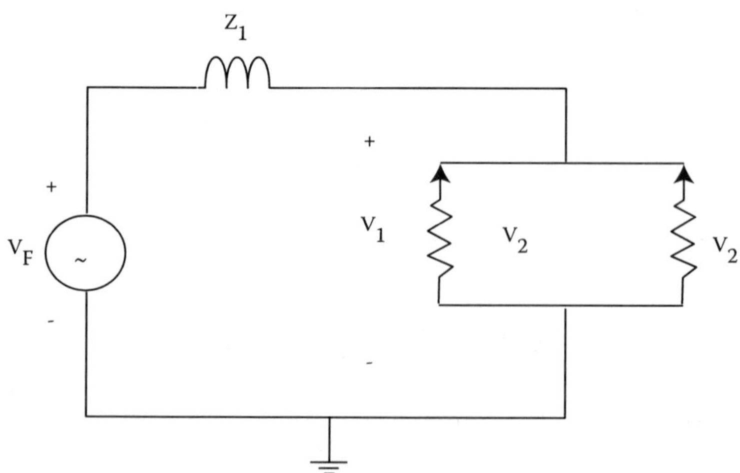

FIGURE 3.9
Double line-to-ground fault.

$$I_0 + I_1 + I_2 = I_a = 0 \tag{3.23}$$

but if

$$V_2 = V_1, I_a = 0$$

$$V_{bc} = V_c \tag{3.24}$$

$$V_b = (I_b + I_c)Z_f \tag{3.25}$$

then the equivalent circuit diagram for a DLG fault is as depicted in Figure 3.10:

$$V_0 + \alpha^2 V_1 + \alpha V_2 = V_0 + \alpha V_1 + \alpha^2 V_2 \tag{3.26}$$

$$\Rightarrow V_1 = V_2$$

$$V_0 + \alpha^2 V_1 + \alpha V_2 = (I_0 + \alpha^2 I_1 + \alpha I_2 + I_0 + \alpha I_1 + \alpha^2 I_2)Z_f \tag{3.27}$$

$$\Rightarrow V_0 - V_1 = (2I_0 + (\alpha^2 + \alpha)(I_1 + I_2))Z_f \tag{3.28}$$

$$[\therefore \alpha^2 + \alpha = -1] \tag{3.29}$$

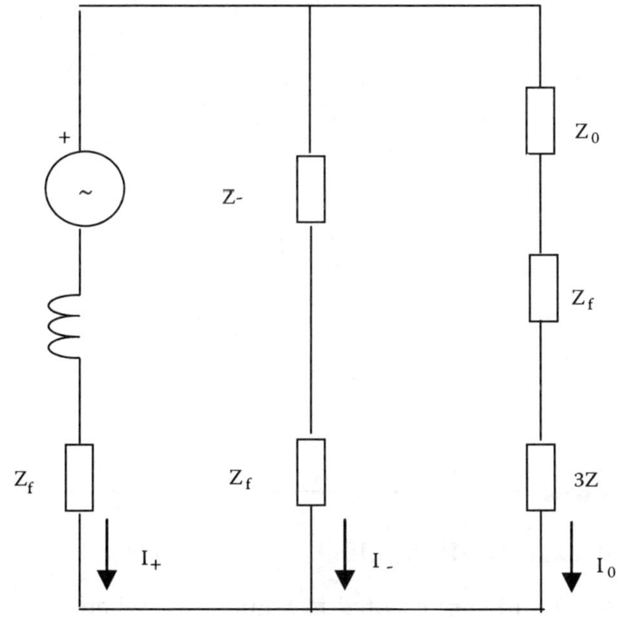

FIGURE 3.10
Equivalent circuit diagram: double line-to-ground fault.

$$I_0 = -(I_1 + I_2) \tag{3.30}$$

$$V_0 - V_2 = 3Z_f I_0 \tag{3.31}$$

From the equivalent circuit

$$I_+ = \frac{E_+}{Z_+ + Z_f + \left[\dfrac{(Z_- + Z_f)(Z_0 + Z_f + 3Z_g)}{Z_- + Z_0 + 2Z_f + 3Z_g}\right]} \tag{3.32}$$

where

$$Z_f = 0$$

$$= \frac{E}{Z_+ + \left[\dfrac{Z_2(Z_0 + 3Z_g)}{Z_- + Z_0 + 3Z_g}\right]} \tag{3.33}$$

$$I_- = -I_+\left[\frac{Z_0 + Z_f + 3Z_g}{Z_- + Z_0 + 2Z_f + 3Z_g}\right] \tag{3.34}$$

$$= (-I_+)\left[\frac{Z_0 + 3Z_g}{Z_- + Z_0 + 3Z_g}\right] \tag{3.35}$$

$$I_0 = -I_+\left[\frac{Z_-}{Z_0 + 3Z_f + Z_-}\right] \tag{3.36}$$

or

$$I_0 = -(I_+ + I_-) \tag{3.37}$$

3.4 Three-Phase Fault Analysis

The three-phase fault is represented in Figure 3.11, where

$$V_a = I_a Z_f \tag{3.38}$$

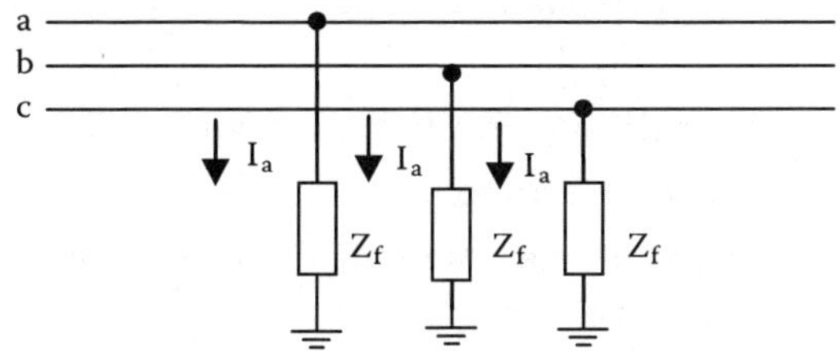

FIGURE 3.11
Three-phase fault.

$$V_b = I_b Z_f \tag{3.39}$$

$$V_c = I_c Z_f \tag{3.40}$$

Then V_{+-0} is as follows:

$$V_+ = \frac{1}{3}\left[V_a + \alpha V_b + \alpha^2 V_c\right] \tag{3.41}$$

$$V_+ = \frac{1}{3}\left[I_a + \alpha I_b + \alpha^2 I_c\right] Z_f \tag{3.42}$$

$$V_- = I_- Z_f \tag{3.43}$$

We continue computing using sequence components and impedances as follows:

$$V_+ = E - I_+ Z_+ \tag{3.44}$$

$$V_- = 0 - I_- Z_- \tag{3.45}$$

$$V_0 = 0 - I_0 Z_0 \tag{3.46}$$

$$V_+ = E - I_+ Z_+ = I_+ Z_f \tag{3.47}$$

$$I_+ = \frac{E}{Z_+ + Z_f} \tag{3.48}$$

$$I_- = 0$$

$$I_0 = 0$$

In summary, to detect a fault in a given part of a power system:

1. Define the equivalent Thevenin for each of the represented sequence networks.
2. Write boundary conditions in phase current/voltage.
3. Transfer to sequence quantities.
4. Manipulate to a form that facilitates synthesis for interpretation based on sequence networks.

3.5 Line-to-Ground and Line-to-Line Faults

3.5.1 Single Line-to-Ground Fault

Figure 3.12 depicts a single line-to-ground (SLG) fault, and Figure 3.13 depicts an equivalent circuit diagram for a single line-to-ground fault, where

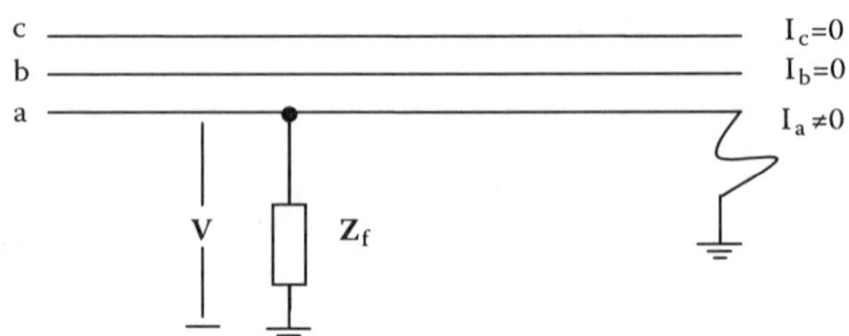

FIGURE 3.12
Single line-to-ground fault.

$$I_b = I_c = 0 \tag{3.49}$$

$$\text{SLG} = V_{ag} = Z_f I_a \tag{3.50}$$

$$\begin{bmatrix} I_0 \\ I_1 \\ I_2 \end{bmatrix} = \frac{1}{3}\begin{bmatrix} 1 & 1 & 1 \\ 1 & \alpha & \alpha^2 \\ 1 & \alpha^2 & \alpha \end{bmatrix} * \begin{bmatrix} I_a \\ I_b \\ I_c \end{bmatrix} = \frac{1}{3}\begin{bmatrix} I_a \\ I_a \\ I_a \end{bmatrix} \tag{3.51}$$

$$V_0 + V_+ + V_- = (I_0 + I_+ + I_-)Z_f \tag{3.52}$$

$$I_0 = I_+ = I_- \tag{3.53}$$

$$V_0 = V_+ = V_- = 3Z_f I_+ \tag{3.54}$$

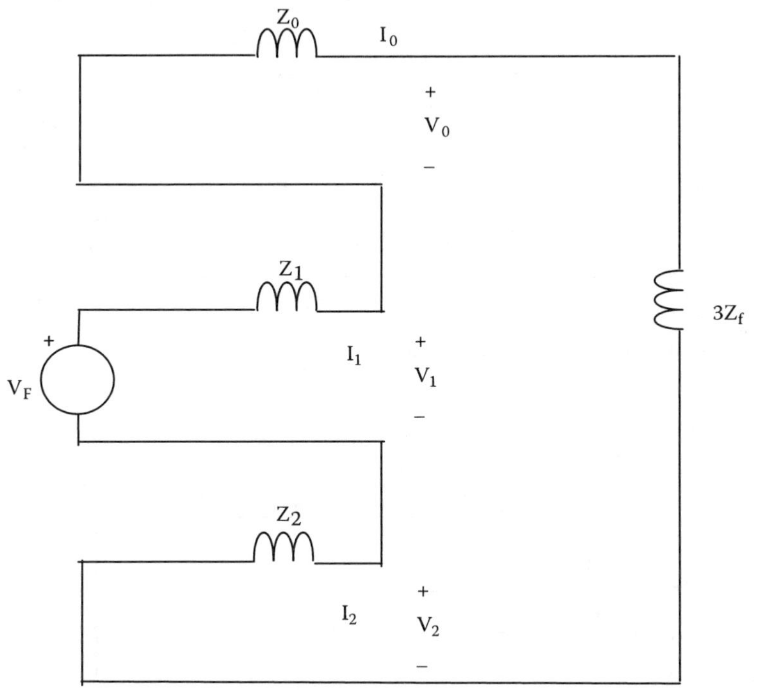

FIGURE 3.13
Equivalent circuit diagram: single line-to-ground fault.

$$\Rightarrow \frac{V_f}{Z_0 + Z_+ + Z_- + 3Z_f} = I_0 = I_+ = I_- \tag{3.55}$$

$$I_a = I_0 + I_+ + I_- = 3I_+ \tag{3.56}$$

$$I_b = I_0 + \alpha^2 I_+ + \alpha I_- = 0 \tag{3.57}$$

$$I_c = I_0 + \alpha I_+ + \alpha^2 I_- = 0 \tag{3.58}$$

3.5.2 Line-to-Line Fault

Figure 3.14 depicts an equivalent circuit diagram for a line-to-line fault. Figure 3.15 also shows an equivalent circuit diagram for a line-to-line fault.

The following shows fault correction in the phase domain:

$$I_a = 0$$

$$I_c = -I_b \tag{3.59}$$

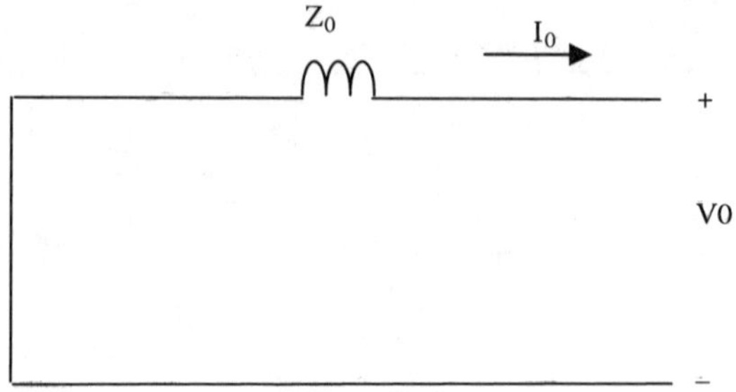

FIGURE 3.14
Equivalent circuit diagram: line-to-line fault.

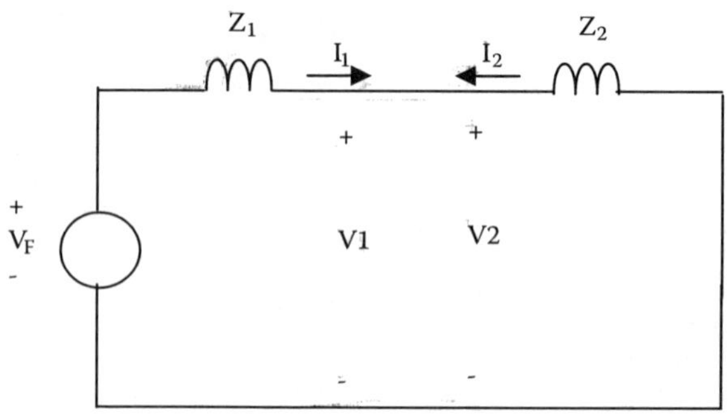

FIGURE 3.15
Equivalent circuit diagram: line-to-line fault.

$$V_{bg} - V_{cg} = I_b Z_f \tag{3.60}$$

$$I_0 = 0$$

$$I_- = -I_+ \tag{3.61}$$

$$V_+ - V_- = I_+ Z_f \tag{3.62}$$

$$\begin{bmatrix} I_0 \\ I_1 \\ I_2 \end{bmatrix} = \frac{1}{3}\begin{bmatrix} 1 & 1 & 1 \\ 1 & \alpha & \alpha^2 \\ 1 & \alpha^2 & \alpha \end{bmatrix}\begin{bmatrix} 0 \\ I_b \\ -I_b \end{bmatrix} = \frac{1}{3}\begin{bmatrix} 0 \\ (\alpha - \alpha^2)I_b \\ (\alpha^2 - \alpha)I_b \end{bmatrix} \tag{3.63}$$

$$V_+ - V_- = (V_0 + \alpha^2 V_+ + \alpha V_-) - (V_0 + \alpha V_+ + \alpha^2 V_-) \tag{3.64}$$

$$= (I_0 + \alpha^2 I_+ + \alpha I_-)Z_f \tag{3.65}$$

If

$$I_0 = 0 \text{ and } I_2 = -I_1$$

then

$$(\alpha^2 - \alpha)V_1 - (\alpha^2 - \alpha)V_2 = Z_f[(\alpha^2 - \alpha)I_+] \tag{3.66}$$

$$V_+ - V_- = I_+ Z_f \tag{3.67}$$

$$I_+ = -I_2 = \frac{V_F}{Z_+ + Z_- + Z_f}, \; I_0 = 0 \tag{3.68}$$

$$I_a = I_0 + I_+ + I_- = 0 \tag{3.69}$$

$$I_b = I_0 + \alpha^2 I_+ + \alpha I_- = (\alpha^2 - \alpha)I_+ \tag{3.70}$$

$$= -j\sqrt{3}I_+$$

$$= \frac{-j\sqrt{3}V_f}{Z_+ + Z_- + Z_f} \tag{3.71}$$

$$I_c = I_0 + \alpha I_+ + \alpha^2 I_- = (\alpha - \alpha^2)I_1 = -I_b \tag{3.72}$$

3.6 Protection Systems

As discussed in previous sections, faults resulting from a short circuit are caused by lightning or by switching surges that are accidental or the result of human error. The impact of the resulting currents and voltages can cause severe damage to insulation and cause conductor breakdown, leading to fire or explosion and the potential loss of life, property, or business. Consequently, any faults that are detected must be isolated quickly from the power system. Standard protection systems are designed to clear faults, to restore the system, or to isolate the faulty region to minimize the impact. In this section, we present the basic idea and construction of protection technology and the associated components (devices) used to achieve a safe and reliable protection scheme. The protection scheme includes fuses, reclosers, and relays coupled with circuit breakers that are used to protect primary distribution systems. In the following subsection, we summarize the definition of each of the protective devices commonly used.

3.6.1 Relay

A relay is defined as a device whose function is to detect defective lines or apparati or other power system conditions of an abnormal or dangerous nature and to initiate appropriate control action. In general, a relay is used to close a normally open circuit or open a normally closed circuit upon detection of an abnormality. Relays facilitate meeting the following objectives of protection design criteria:

1. Reliability: must detect and isolate the faults instantaneously and operate dependably
2. Selectivity: must discriminate between normal and abnormal system conditions
3. Speed: must operate speedily to minimize fault duration and equipment damage and to restore the system quickly
4. Economy: must provide maximum protection at minimum cost of equipment or operation
5. Simplicity: must be simple in design and usage of circuitry

Some compromises or imaginative engineering are required to satisfy these criteria.

A symbolic representation of a relay is illustrated in Figure 3.16. Figure 3.17 shows a schematic diagram of a relay circuit. Relays are used in low-voltage and high-voltage transmission. They show up in different forms as voltage-, current-, temperature-, or pressure-based relays. They are also used in domestic appliances such as air conditioners, washers, and dryers, in traffic control, in aerospace/telephone systems, etc.

Before discussing the construction of protection components, we briefly review the following commonly used devices that interact with relay operations and functions.

3.6.2 Instrument Transformers

The faulted current and voltage measured on an abnormal systems are usually very high and can damage the equipment being protected. To achieve

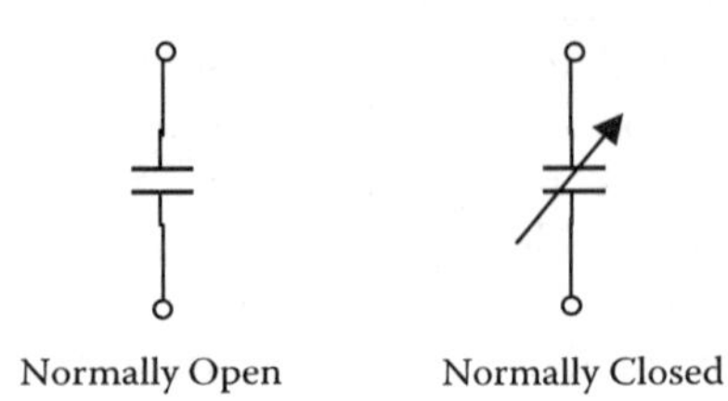

FIGURE 3.16
Symbolic representation of relay.

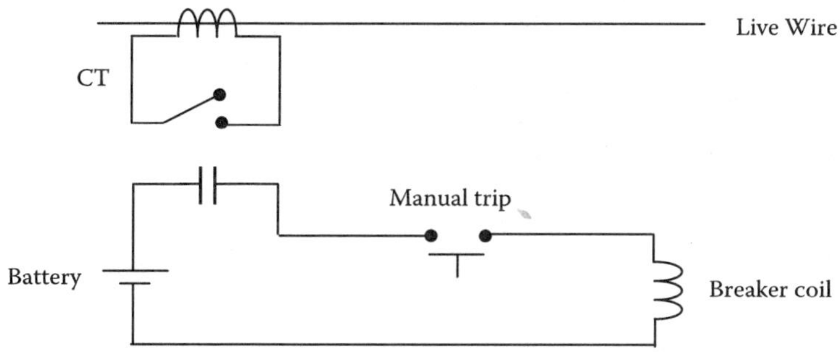

FIGURE 3.17
Schematic diagram of relay circuit.

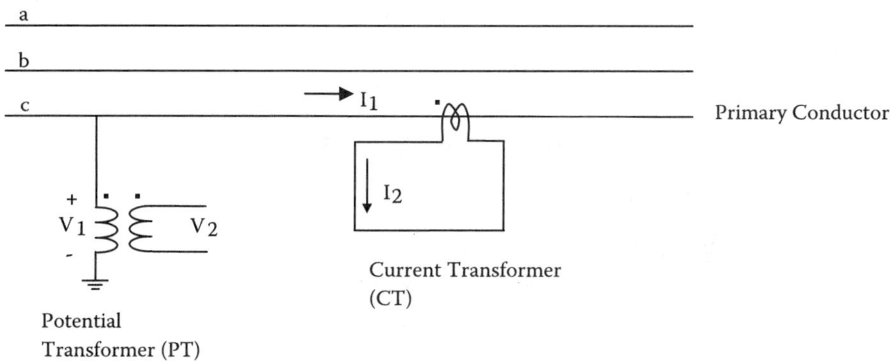

FIGURE 3.18
Instrument transformers.

safety, economy, and convenience of measurement, a step-down transformer is needed. The instrument transformers, namely potential and current transformers (PT and CT, respectively), are used for this purpose. The two basic types of instrument transformers are shown in Figure 3.18.

V_T is usually modeled as follows:

$$\frac{V_2}{V_1} = \frac{N_2}{N_1} \tag{3.73}$$

where V_2 is a scaled-down version of V_1 (primary conductor voltage)

$$N_1/N_2 = \left(\frac{1}{n}\right) \tag{3.74}$$

where n is specified.

Typical PT voltage ratios are 1, 2, 2.5, 4, 5, … , 20, … , 4500. For example,

$$\frac{N_1}{N_2} = \frac{240000}{120} = \frac{2000}{1}$$

For a current transformer (CT),

$$\frac{I_2}{I_1} = \frac{N_1}{N_2} \Rightarrow \frac{500}{5}$$

We also compute CT error from

$$\frac{I_2 - I_3}{I_2'} \tag{3.75}$$

where $I'_2 = I_2 + I_{\text{Excitation}}$

1. Typical ratios are current: 50:5, 450:5, 400:5, 100:5, 4000:5
2. Voltage ratio: $\begin{cases} 1:1 & 2:1 & 25:1 & 5:1 \\ 60:1 & 100:1 & 200:1 & 400:1 \\ 1000:1 & 2000:1 & 4500:1 & \end{cases}$

3.6.2.1 *Accounting for Saturation in CT*

CT equivalent has a mysterious curve performance that lends itself to a nonlinear characteristic. The actual model of the transformer reactance model is subject to an error, and the error is represented as

$$I_2 = I'_2 - I_e \tag{3.76}$$

Therefore, CT error is the deviation of I_2 from I'_2

$$\%\text{error} = \left(I_2' - \frac{I_e}{I_2'} \right) \times 100\% = \frac{I_e}{I_2} \times 100\% \tag{3.77}$$

3.6.3 Reclosers

Most faults (80 to 85%) on distribution/transmission lines are temporary, lasting only a few cycles. These special-purpose automatic circuit reclosers are used to protect distribution circuits from temporary faults. Reclosers are self-controlled devices that automatically interrupt overloads but not severe faults. They have built-in controls that allow temporary faults to clear and then restore service, quickly disconnecting a permanent fault. Reclosers can

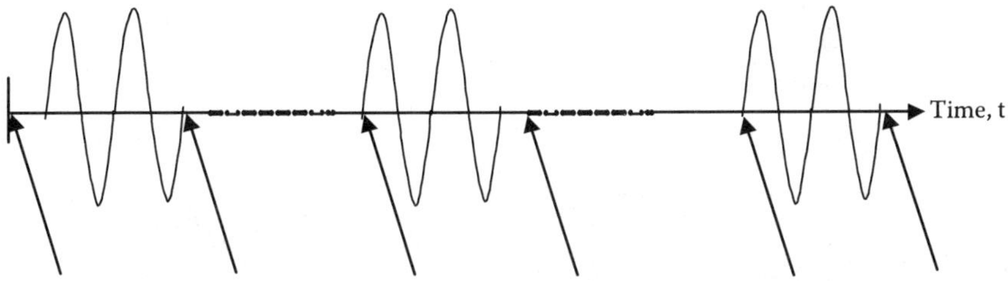

FIGURE 3.19
Process of reclosing and opening before lockout.

be programmed to sense an overcurrent and open the circuit, then reclose it after a preprogrammed time. They can open and reclose up to five times and, after a preset number of operations, remain open lockout.

Reclosers are used to protect feeders leaving the substation for important isolation of faulty areas and to minimize the area of service interruption. Reclosers are rated to carry a particular voltage, continuous current rating, and minimum fault current in the protected zone, and have the ability to coordinate with other protective devices such as relays, etc. For example, a load-side device should operate as a backup and clear a fault before the source side operates.

The process of reclosing and opening before lockout is shown in Figure 3.19. Two types of reclosers are currently in manufacture: the piston-controlled recloser used in hydraulic cylinders and the electronic circuitry (more expensive but feasible/accurate) used for heavy-duty three-phase reclosers. They are available for both single- and three-phase systems, and they use oil or vacuum for timing and interruption.

3.6.4 Fuses

Fuses are one-time devices for interrupting a fault current. They are nonreusable once the metallic element melts in response to an overload current and opens the circuit. Fuses are typically coordinated with reclosers and time-delay overcurrent relays. They are selected on the basis of maximum loads served from the taps and are rated in terms of voltage and interrupting current. Fuses actually clear before the reaction spurs. Coordination of fuses and reclosers is done using the fuse time-current curve and circuit fuse coordination curve, as seen in Figure 3.20.

Figure 3.21 shows a common protection scheme for a radial distribution circuit utilizing fuses, reclosers, and relays. For temporary faults, the recloser can be set for one or more instances of time-delayed trips and reclosers to clear the faults and restore service. If faults continue, the fuses operate for

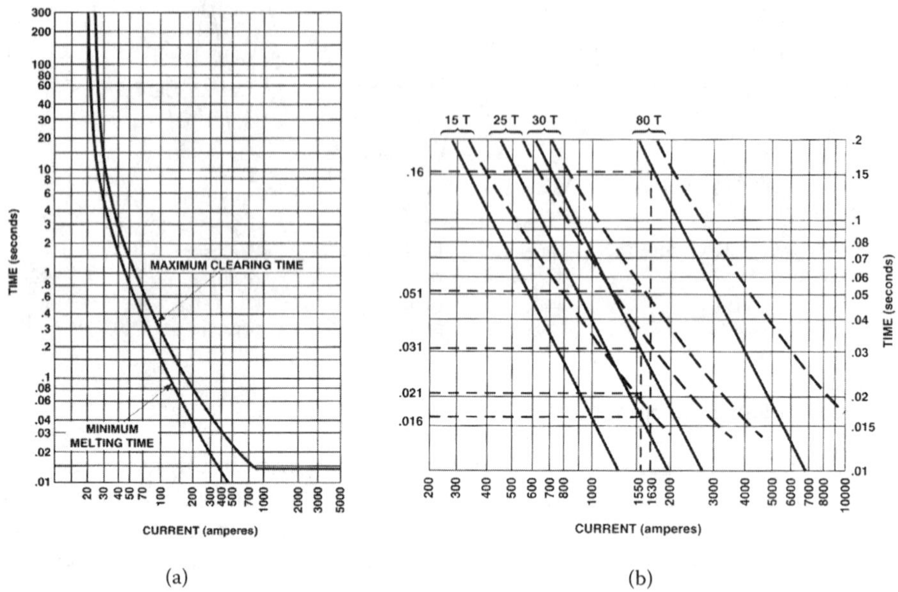

FIGURE 3.20
Fuse coordination (a) Fuse time-current curve, and (b) Time-current curves where the minimum melting curves are solid and maximum clearing curves are solid (courtesy of Copper Power Systems).

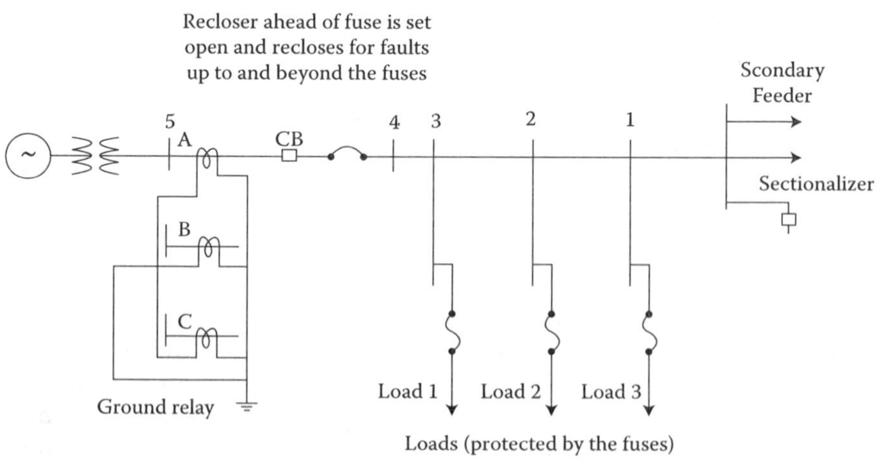

FIGURE 3.21
Protection scheme for radial distribution circuit utilizing fuses, reclosers, and relays.

downstream faults or the reclosers open after a time delay and lock out the faults between the restore point and the fuses. The relay opens the substation breaker.

3.6.5 Sectionalizer

A sectionalizer is a device that automatically isolates a faulted line segment from a distribution system. It senses any current above its actuating current, and the line is then deenergized by a recloser. It counts the number of overcurrents and, following deenergization sequences and after exceeding a preset number, it opens the circuit and locks it out.

A sectionalizer can reset itself to a zero count after the count exceeds some preset number. There are two types of sectionalizers: hydraulically operated and electronically operated, both of which communicate with the recloser. They are rated in terms of voltage (34.5 kV), continuous current (600 A), interruption current as high 1320 A, and Basic Insulation Level (BIL) as high as 150 kV.

The one-line diagram in Figure 3.21 shows the CB (Circuit Breaker), reclosers, and sectionalizers used to protect the lines.

3.7 Protective Relay Technology

A relay is an electromechanical or microprocessor-controlled electronic system that senses faulty or abnormal conditions within a distribution system (such as overcurrent, overvoltage, overfrequency, or undercurrent, undervoltage, or underfrequency); an excessive value generates a trip signal to a current breaker. Relays are classified into the following categories.

Monitoring relay: a device that monitors conditions within the power system and sends an alarm when conditions are unstable

Programming relay: a device that detects sequences of events; used to control and monitor synchronization

Regulatory relay: a device used to determine whether a parameter such as voltage, current, or impulse has exceeded its allowable limit; sends an alarm when parameter exceeds its limit

Auxiliary relay: a device that provides miscellaneous functions with other relaying systems, e.g., timers are examples of auxiliary relays

There are several ways to classify relays by their functions and applications. The two types of relay construction are electromagnetic and electronic (solid state). The electromagnetic types are based on the development of electromagnetic forces leading to torque, which causes movable members to operate physically to open or close contacts.

3.7.1 Digital Relaying

Solid-state method: provides a switching action (only) with no physical motion by changing the state of a serially connected solid-state component from a nonconducting to a conducting state (or vice versa).

Digital-computer relaying (1980s): This provides greater accuracy, improved sensitivity to faults, better selectivity, user friendliness, easier testing, and advanced relay-event monitoring/recording capabilities. Digital relaying can be updated in settings or shipping of signals from a remote software computer terminal or a control computer in real time by a relay engineer.

Microprocessor-controlled relay: The so-called microcomputer relay system has the ability to perform several relaying function with a single control relaying package in a very efficient/seasonal manner. The microcomputer relay calculates the information such as Z, I, and V from CT and PT and uses it compute VAr, Z, PQ (Power Quality), flow direction, trends over time, and running averages of quantities as needed.

The microcomputer-based relay (Figure 3.22) uses one of the computer techniques to compare a measured fault signal against a given specific signal-relaying threshold quantity to open or close the circuit breaker. The microcomputer can also regulate other parameters such as temperature, vibration, etc. The information, which is in analog, is fed to the microprocessor through an analog-to-digital converter and fed to the register in the microprocessor system. The output data, after comparing with the threshold, is in the form of alarm data to be sent to the control center (room) as a trip signal, in the case of overcircuit/overvoltage or overfrequency, to isolate the faulted region requesting a subsequent control action, such as tap changing of a transformer based on a new voltage or phase angle θ, values needed for safety.

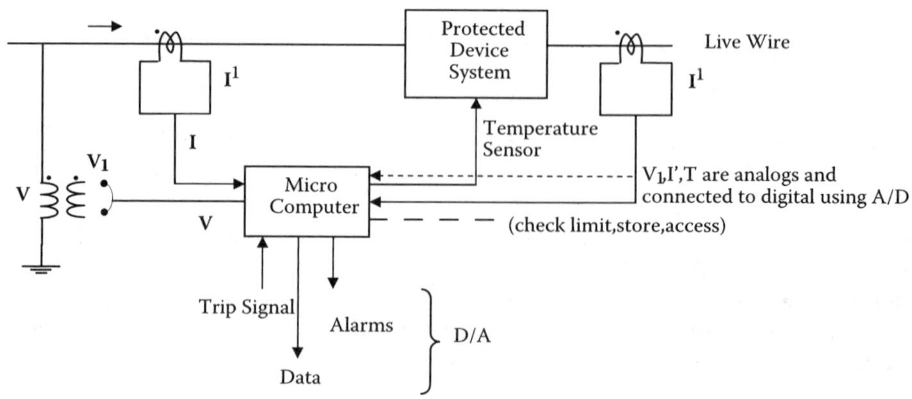

FIGURE 3.22
Schematic diagram of microcomputer-based relay.

3.7.2 Electromechanical Relay Technology

There are many varieties of electromechanical relays. The most widely used are the plunger-type (solenoid) relay, the instantaneous hinged armature (Clapper unit), and induction disc relays.

The plunger-type relay works under the principle of magnetic induction when the coil current is magnetized. It creates a Force αKI^2 and, using the air gap, Torque = $F \cdot r$ is created at pickup current I, and the trip signal operating the relay closes or opens the contacts. The plunger operates against the opposing force by the spring constant K_s creating $-F_s$, an opposing force.

The solenoid relay is used as a high-current instantaneous trip relay. The pull-in current can be varied by the magnetic core air-gap adjustment, which varies with the reluctance of the core. The closing action is from 5 to 50 msec, depending on overcurrent magnitude (slower with lower currents or relay size).

The instantaneous hinged armature (Clapper unit) with current through the coil magnetizes the core and the frame by magnetic induction principle. The magnetic attraction causes $F = KI^2$, which provides sufficient torque to adjust the pull-in current of the relay. The armature movement (Clapper) is used to drive the relay operation. The closing action is 2 to 4 msec operating time.

3.7.3 Induction Disc Relays

Induction disc relays (Figure 3.23) are used as inverse time-over-current relays where high overcurrent causes faster operation than lower current. It consists of an aluminum induction disc suspended between bearings with magnetic structure held in place on one side of the disc shaft. The movement of the disc is determined from

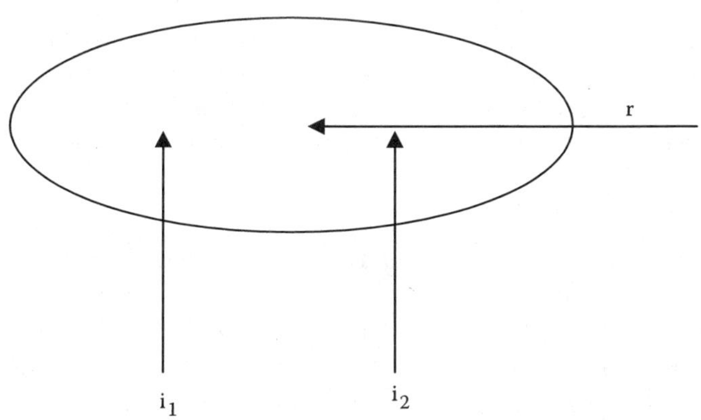

FIGURE 3.23
Induction disc relay.

$$F = KI_1I_2 \sin wt$$

where

$$i_1 = I_1 \sin wt$$
$$i_2 = I_2 \sin(wt - \theta) \quad (3.78)$$

at max

$$0 \leq \theta \leq 90 \quad (3.79)$$

The currents induced in the disc interact with the magnetic field to produce torque on the disc, which causes it to rotate (Torque = $F \cdot r$). It operates on an adjustable time delay, i.e., spin, making it possible to retrain by pulling the movable contact to stop or rotate, depending on the value of the current in the coil. The time delay can be introduced electrically or by mechanical or other methods. Commonly used adjustable time-delay methods adjust the time constant by using an RC network, bar magnet, bellows, etc.

Relays that are based on an induction disc operate like a wattmeter in principle. They have two adjustable settings defined as current tap settings (CTS), which are defined from the operating current to energize the relay to pick up, hence the name "pickup current." Time-dial settings (TDS) are used to adjust the amount of time (i.e., adjustable amount of time delay). These TDS settings are represented as characteristic curves with operating time in seconds versus relays shown in CO graphs with input as a multiple of the pickup current.

The CO-8 curves are usually based on operating time in seconds versus relay input current and multiples of pickup current. The curves decrease asymptotically with an increase in current (inverse). The inverse time characteristics can be adjusted up or down by the time-dial setting.

If the secondary current of I_0 of the CT exceeds the pickup current I_p, the relay contacts close instantly to energize the circuit breaker trip cord. If I' is less than the pickup current I_p, then the relay contacts remain open, blocking the trip coil.

3.7.3.1 Example 1, Coordinating Time-Delay Overcurrent Relays in a Radial System

Data for the 60-Hz radial system of Figure 3.24(a) are given in Table 3.2 (maximum loads), Table 3.3 (symmetrical fault currents), and Table 3.4 (breaker, CT, and relay data). Select current tap settings (CTSs) and time-dial settings (TDSs) to protect the system from faults. Assume three CO-8 relays for each breaker, one for each phase, with a 0.3-sec coordination time interval. The relays for each breaker are connected as shown in Figure 3.23(a), such that all three phases of the breaker open when a fault is detected on any one phase. Assume a 36.5-kV (line-line-line) voltage at all buses during normal operation. Future load growth *S* is also included in Table 3.2, such that maximum loads over the operating life of the radial systems are given in this table.

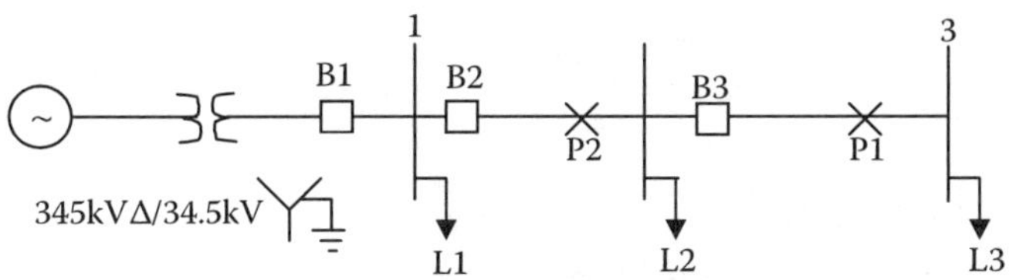

FIGURE 3.24(A)
60-Hz radial system for Example 1.

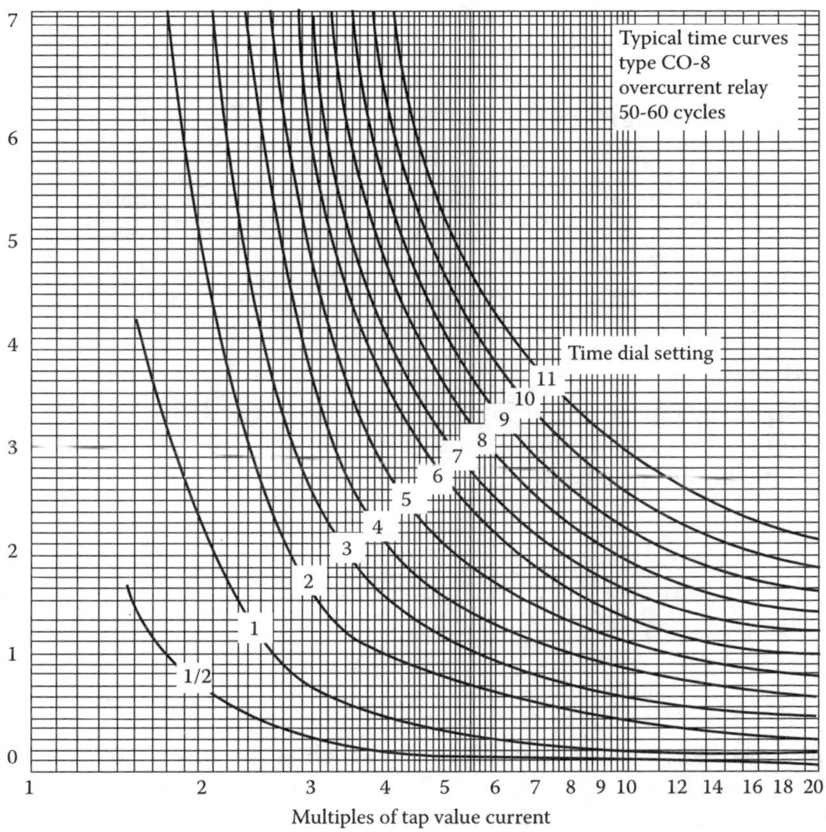

FIGURE 3.24(B)
CO-8 over-current relay time curves.

TABLE 3.2

Maximum Loads

Bus	S (MVA)	Lagging pf
1	11.0	0.95
2	4.0	0.95
3	5.0	0.95

TABLE 3.3

Symmetrical Fault Currents

Bus	Maximum Fault Currents (Bolted Three-Phase)	Minimum Fault Current (L-G or L-L)
1	3000	2200
2	2000	1500
3	1000	700

TABLE 3.4

Breaker, CT, and Relay Data

Breaker	Breaker Operating Time	CT Radio	Relay
B1	5 cycles	400:5	CO-8
B2	5 cycles	200:5	CO-8
B3	5 cycles	220:5	CO-8

Solution

First, select Tap Settings (TSs) such that the relays do not operate for maximum load currents. Starting at B3, the primary and secondary CT currents for maximum load L3 are

$$I_{L3} = \frac{S_{L3}}{V_3\sqrt{3}} = \frac{5\times 10^6}{(36.5\times 10^3)\sqrt{3}} = 79.089 \text{ A}$$

$$I'_{L3} = \frac{79.089}{(220/5)} = 1.80 \text{ A}$$

We select for the B3 relay a 3-A TS, which is the lowest TS above 1.80 A. Note that $|S_{L2} + S_{L3}| = |S_{L2}| + |S_{L3}|$ because the load power factors (pf) are identical, as seen in Figure 3.24(b).

3.7.3.2 Example 2, Radial System Protection

Consider the 60-Hz radial system shown in Figure 3.25. The following pertinent data for the three-phase short circuit are given:

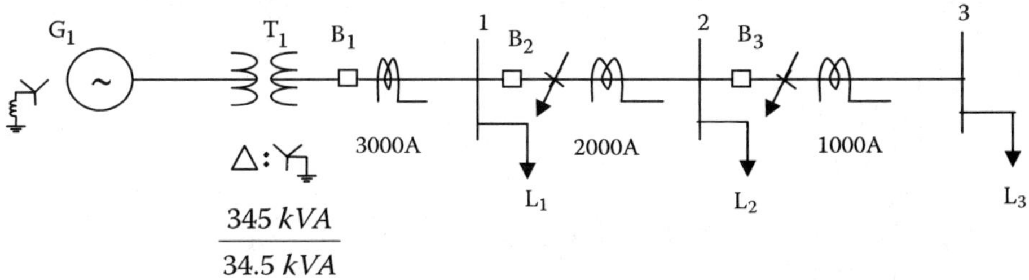

FIGURE 3.25
60-Hz radial system for Example 2.

I_{sc1} = 3000A, I_{sc2} = 2000A, I_{sc3} = 1000 A
B_1 = 400/5, C_2 = 200/5, C_3 = 200/5
S_{L1} = 11 MVA @ 0.95 pf lagging
S_{L2} = 4 MVA @ 0.95 pf lagging
S_{L3} = 6 MVA @ 0.95 pf lagging

Solution

Step 1: Let

$$I_{L3} = \frac{S_{L3}}{V_3\sqrt{3}} = \frac{6\times 10^6}{34.5\times 10^3\sqrt{3}} = 100.4 \text{ A}$$

$$I_{L2} = \frac{S_{L2}+S_{L3}}{V_2\sqrt{3}} = \frac{(4+6)\times 10^6}{34.5\times 10^3\sqrt{3}} = 167.3 \text{ A}$$

$$I_{L1} = \frac{S_{L1}+S_{L2}+S_{L3}}{V_1\sqrt{3}} = \frac{(11+4+6)\times 10^6}{34.5\times 10^3\sqrt{3}} = 354.4 \text{ A}$$

Now, using the CT ratios

$$I'_{L1} = \frac{351.4}{400/5} = 4.39$$

$$I'_{L2} = \frac{162.3}{200/5} = 4.18$$

$$I'_{L3} = \frac{100.4}{200/5} = 2.51$$

Step 2: Select from CO-8 the time setting (TS) for each relay

$B_1 = 5$ A (TS_1)
$B_2 = 5$ A (TS_2)
$B_3 = 5$ A (TS_3)

Step 3: Select the TDS to coordinate for maximum fault current at 3ϕ short circuit.

For Bus 2, we obtain 2000 A. Neglecting CT saturation error,

$$\frac{I'_{3\phi\,\text{fault}}}{TS_3} = \frac{2000/200/5}{3} = 16.7$$

To clear the fault as rapidly as we can, we must now select the smallest time-delay setting (TDS) for B_3 to operate. We choose TDS = $^1/_2$ = 0.5 from CO-8 with a corresponding relay operating time, $T_3 = 0.05$ sec.

Step 4: We add the breaker operating time (five cycles = 0.083 sec) = T_B.

Then the breaker (protection) open (close) of $T_s + T_B = 0.05 + 083 = 0.133$ sec.

Step 5: Next, we do fault-to-pickup current covered at B_2.

$$\frac{I'_{2\text{fault}}}{TS_2} = \frac{2000(200/5)}{5} = 10$$

Step 6: Add relay operation time $T_s = 0.05$ sec, breaker quantity = 0.083, and coordinate time = 0.3 sec

$$T_{\text{coordinate}} + T_{\text{breaker}} + T_{\text{relay op}} = [0.05 + 0.083 + 0.3] = 0.43 \text{ sec} = T_2$$

Step 7: Again using T_2 of 0.43 sec, approximating from 0.43 sec→10 A in Figure 3.24, gives TDS = 2.

Step 8: Next, select TDS at B_1. The largest current through B_2 is 3000 A.

For a 3ϕ fault at bus 1 (to the right of B_2), the fault-to-pickup current ratio at B_2 is

$$\frac{I'_{1\text{fault}}}{TS_1} = \frac{3000/400/5}{5} = 15.0$$

From TDS→T_2 at 15 A, then T_1 = 0.38 sec,

$$\frac{I'_{1\text{fault}}}{\text{TS}_1} = \frac{3000/400/5}{5} = 7.5$$

$$T_1 = T_{\text{coordinate}} + T_{\text{breaker}} + T_2 = [0.38 + 0.083 + 0.3] = 0.76 \text{ sec}$$

Select TDS = 3.

3.8 System Protection in General

Consider the power system single-line diagram in Figure 3.25. We consider the configuration of a power system consisting of generator, bus bar, transformer, transmission lines, monitors, and a newly emerging distribution generation system. The protection scheme is needed for handling of a fault anywhere within a zone of the system. We consider here the different zones and the overlapping zones of the protection scheme that are used to isolate the fault of a particular zone from the system. System protection zones are defined in Figure 3.25. Protection zones have the following features: zones are overlapped, and circuit breakers are located within the overlay regions (zero). For a fault anywhere within a zone, all circuit breakers in that zone open to isolate the fault.

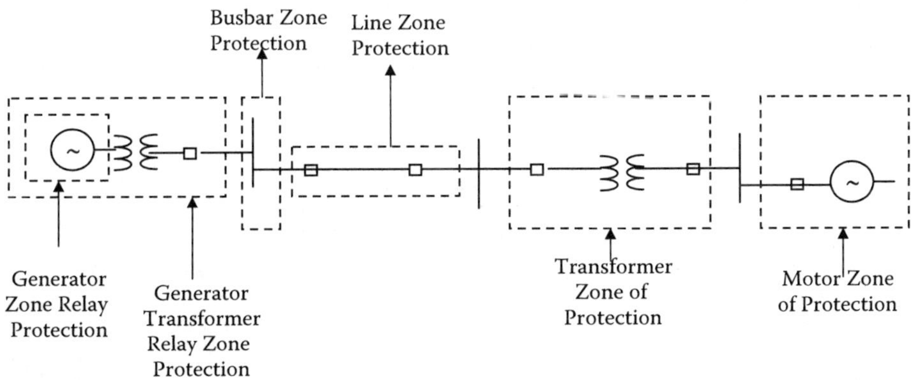

FIGURE 3.26
System protection zones.

3.9 System Protection for Different Power System Zone Components

3.9.1 Line Protection with Impedance Distance Relays

3.9.1.1 *Directional Overcurrent Relays*

Directional Overcurrent Relays are difficult to coordinate in transmission with multiple sources. To overcome these problems, relays that respond to voltage and current ratio are used. They operate on the basis of voltage-to-current ratio, called an impedance relay, a distance relay, or a ratio relay.

3.9.1.2 *Impedance Relay*

Consider a fault at a given point with two circuit breakers controlling that will detect substantially the same current to coordinate the two breaker devices. We used the impedance involved as follows, where $Z_{tap} = Z_d$ = tap setting is the total impedance of the line (Figure 3.26) to the protection distance, called the balance point.

If we put the CT and PT at *R*, we produce

$$Z = \frac{V}{I} = R + jX \tag{3.80}$$

Consider that if the relay responds to *Z*, we can use the fault to design a variety of relays for line protection. Assume a general universal relay equation

$$T = K_a A^2 + K_B B^2 + K_C AB\cos(\theta - \tau) + K_S \tag{3.81}$$

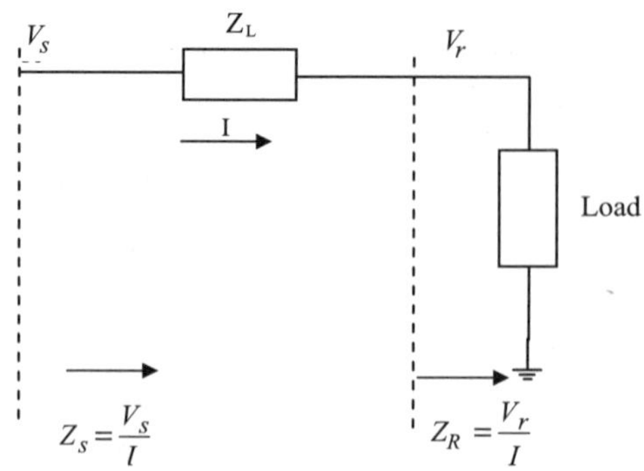

FIGURE 3.27
Schematic of short line.

where $K_C = K_S$, A = V, and B = I.

Therefore, for $K_aV^2 + K_bI^2 \geq 0$, or at the borderline where

$$K_aV^2 + K_bI^2 = 0 \tag{3.82}$$

$$\frac{V^2}{I^2} = \frac{-K_b}{K_a} \tag{3.83}$$

and

$$Z = \frac{V}{I} = \sqrt{\frac{V^2}{I^2}} = \sqrt{\frac{-K_b}{K_a}} = Z_r \tag{3.84}$$

the relay is on the verge of operation for different comparisons in terms of magnetism and phase. We have different types for ohmic and mhoic relays, respectively, for $K_a < 0$ and $K_b > 0$, as seen in Figure 3.28(a) ($Z_r = Z$) and Figure 3.28(b) ($Z < Z_r$).

3.9.2 Mho Relays

A mho relay is a modified impedance relay (amplitude comparison) obtained by offsetting the center of the impedance circle from the origin. This is done by modifying the impedance relay by appropriate setting of K_s, Z_1, Z_2, or ϕ. We obtain from

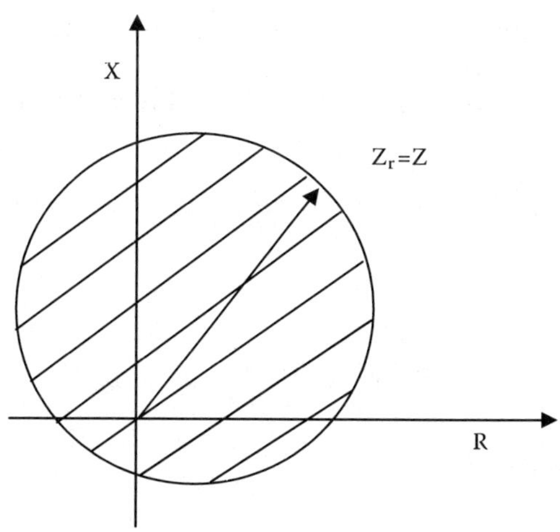

FIGURE 3.28(A)
$Z_r = Z$ for ohm relay.

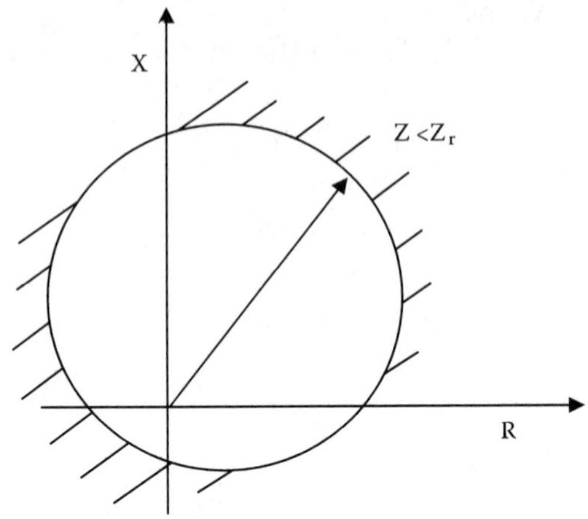

FIGURE 3.28(B)
$Z < Z_r$ for mho relay.

$$\left(R - \frac{|Z|}{K}\cos\phi\right)^2 + \left(X_L - \frac{|Z|}{K}\sin\phi\right)^2 \leq \frac{|Z|^2}{K^2} \tag{3.85}$$

an equation of a circle in the Z plane with off-center coordinates R_a, X_a and radius Z_r,

$$(R - R_a)^2 + (X_L - X_a)^2 = Z_r^2 \tag{3.86}$$

By matching the angle of the line-impedance locus, the resulting scheme is referred to as the mho relay (Figure 3.29).

Continuing, at amplitude $|V_2| \geq |V_1|$, at threshold $|V_2| = |V_1|$, and

$$|K_1||Z_L| + |Z_1\angle\phi_1 - \phi_L|^2 = |K_2||Z_L| + |Z_2\angle\phi_2 - \phi_L|^2 \tag{3.87}$$

using

$$|A + B\angle\beta|^2 = A^2 + B^2 + 2AB\cos\beta \tag{3.88}$$

the relay operates when

$$|K_1{}^2 - K_2{}^2||Z_L|^2 + 2|Z_L||K_1||Z_L|\cos(\phi_1 - \phi_L) = |K_2||Z_2|\cos(\phi_2 - \phi_L) \tag{3.89}$$

where K_1, K_2, Z_1, and Z_2 are adjustable.

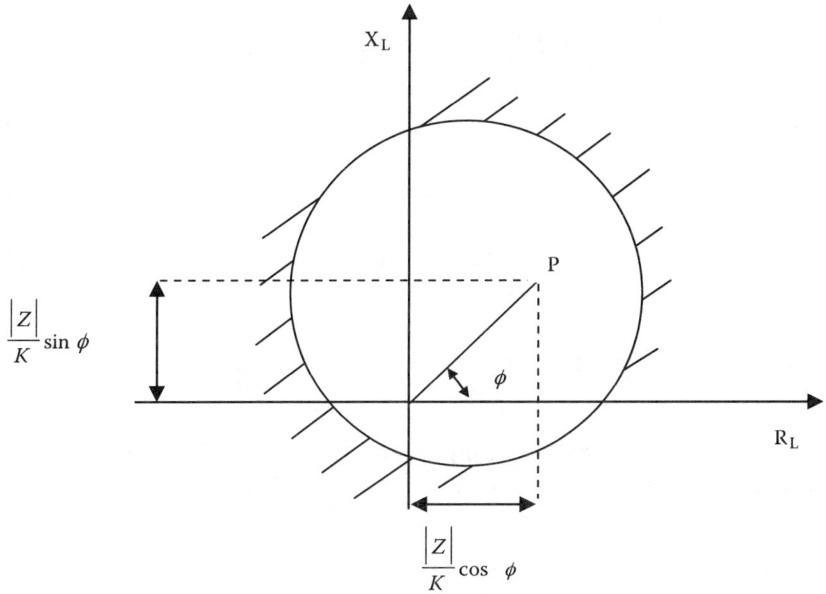

FIGURE 3.29
Mho relay characteristics.

3.9.3 Ohm Relays

See Figure 3.30, where $K_1 = K$, $K_2 = -K$, $Z_1 = 0$, $Z_L = 2$ and

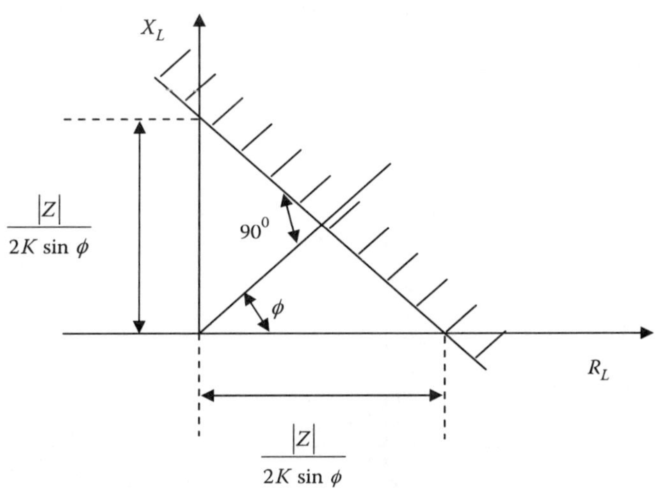

FIGURE 3.30
Ohm relay characteristics.

$$|Z_L|(\cos(\phi-\phi_L)+\sin(\phi-\phi_L)) \le \frac{|Z|}{2K} \quad (3.90)$$

$$Z_L = R_L + jX_L \quad (3.91)$$

$$R_L = Z_L \cos \theta_L \quad (3.92)$$

$$X_L = Z_L \sin \phi_L \quad (3.93)$$

$$R_L \cos\phi_L + X_L \sin\phi_L = \frac{|Z|}{2K\cos\phi} \quad (3.94)$$

The same is true for phase comparison, leading to ohm and mho relays, as shown respectively in Figure 3.31 and Figure 3.32.

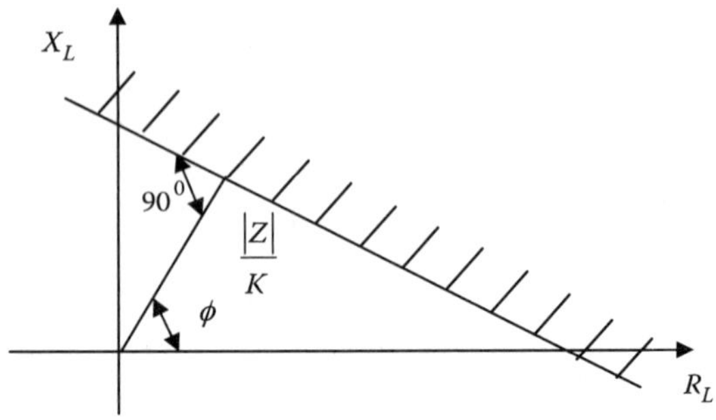

FIGURE 3.31
Ohm relay characteristic using phase comparison.

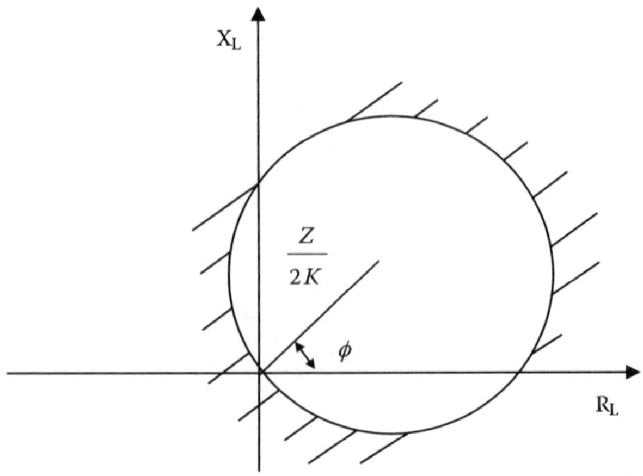

FIGURE 3.32
Mho relay characteristic using phase comparison.

$$|Z_L|\cos(\phi - \phi_L) \le \frac{|Z|}{K} \tag{3.95}$$

where

$$\left(R_L - \frac{|Z|}{2K}\cos\phi\right)^2 + \left(X_L - \frac{|Z|}{2K}\sin\phi\right)^2 \le \frac{|Z|}{4K^2}^2 \tag{3.96}$$

3.9.4 Generator, Buses, and Transformer

Differential relays are used to protect generators, buses, and transformers.

3.9.4.1 Generator Protection

Figure 3.33 illustrates generator protection with a differential relay. For no internal fault, conduction generator current windings, and $I'_1 = I'_2$ for current transformers, then $|I'_1 = I'_2| = 0$. However, if an internal fault such as phase-to-ground or phase-to-phase is shared within the generator, $I_1 - I_2$ and $I'_1 \neq I'_2$ occurs, leading to $\Delta I'$ and ΔI, which causes the relay to operate. Since the relay operates based on ΔI, we call this relay a differential relay.

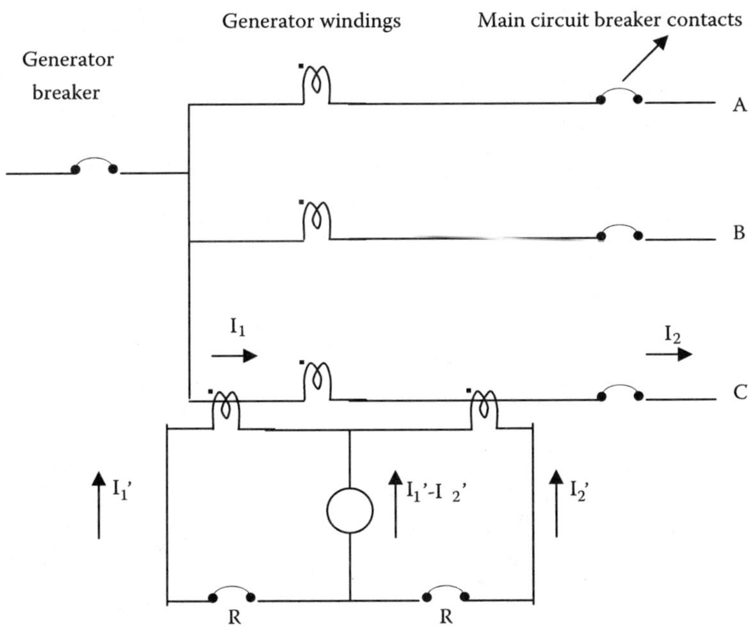

FIGURE 3.33
Generator protection with differential relay.

An electromechanical relay based on a balanced-beam principle is used to provide differential relaying for detecting faults. Generator fault types includes stator winding insulation failure, rotor damage, and bearing damage as well lack of synchronism and overheating. Many relay schemes based on the differential scheme are available for this purpose.

3.9.4.2 Bus Protection with Differential Relays

A differential relay scheme is also used to protect buses under fault. As in the generator case, a relay operation on one phase will cause all other three-phase circuit breakers to open or operate to ensure that three-phase bus is removed from service.

A single-line diagram of differential bus protection is given in Figure 3.34. For the case of no internal fault between current transformers, there is no bus fault, and $I_3 = -I_1 + I_2$.

For the case with bus $I'_1 + I'_2 - I'_3 = 0$, the relay does not trip.

For the case with bus $I'_1 + I'_2 - I'_2 \neq 0$, current flows in the operating coil to operate the relay with use of the restraining cord in the case of nonidentical current transformers (CTs).

Cases of saturation of CT can cause misoperation of the relay's operating coil. Various schemes to overcome this problem are described in the literature.

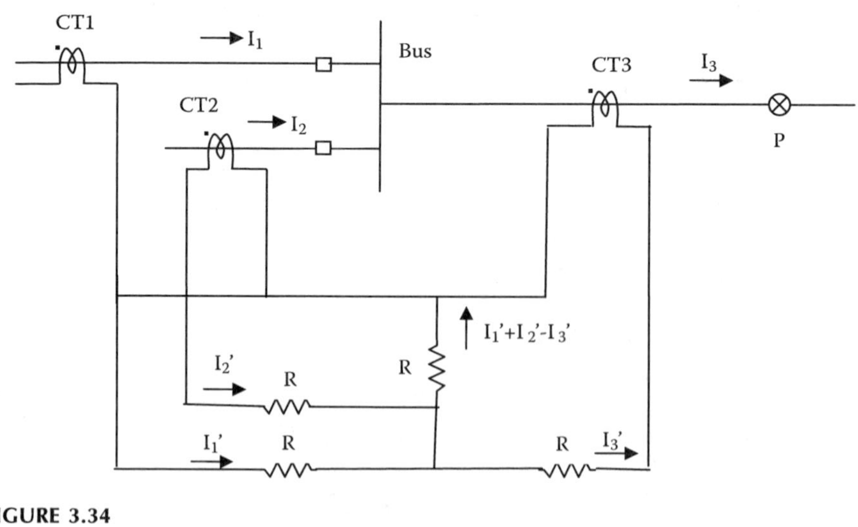

FIGURE 3.34
Single-line diagram of differential bus protection.

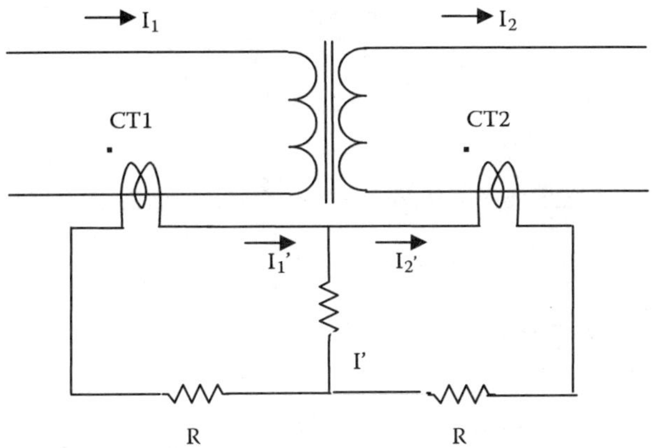

FIGURE 3.35
Transformer protection with differential relays.

3.9.4.3 *Transformer Protection with Differential Relays*

There are various forms of transformer protection (Figure 3.35), including cooling system, Buchholz (alarm) detection approach, and differential relays. The differential current from the CTs is used to balance or energize the cord to operate the relay during a transformer malfunction under fault.

For $I' + I'_1 - I'_2 \neq 0$, the relay will trip.
For $I' + I'_1 - I'_2 = 0$, the relay will not trip.

The values for I', I_1, and I_2, where $I' = \frac{I_1}{n_1} - \frac{I_2}{n_2}$, are obtained from the CTs.

3.10 Illustrative Examples

3.10.1 Example 1

Calculate S_P and S_S and verify that $S_P = 3S_S$

$$[V]_{abc} = \begin{bmatrix} 277\angle 0^\circ \\ 260\angle -120^\circ \\ 295\angle 115^\circ \end{bmatrix}$$

$$I_{abc} = \begin{bmatrix} 25.15\angle -46.76^\circ \\ 25.71\angle 196.34^\circ \\ 26.62\angle -73.77^\circ \end{bmatrix}$$

$$S_P = 277\angle 0^\circ \times (25.15\angle -46.76^\circ) + 260\angle -120^\circ \times (25.71\angle -196.34^\circ) + 295\angle 115^\circ \times (26.62\angle -73.77^\circ)$$

$$= 21{,}490\angle 43.78^\circ \text{ VA}$$

$$S_S = \begin{bmatrix} V_0 I_0^* & V_+ I_+^* & V_- I_-^* \end{bmatrix}$$

$$= 7163\angle 43.78^\circ \text{ VA}$$

Also,

$$3S_S = 3(7163\angle 43.78^\circ) = 21{,}490\angle 43.78^\circ$$

3.10.2 Example 2

Assume for the equivalent diagram in Figure 3.8(a) that

$$Z_+ = j0.13893,\ Z_- = j0.14652,\ Z_0 = j0.250$$

and

$$V = 1.5\angle 0$$

$$\therefore I_+ = \frac{V\angle 0}{Z_+ + Z_- + Z_0} = \frac{1.5}{\sum_{+-0} Z}$$

$$I_{012} = I_+ = I_- = I_0$$

With I_{012} known, we can use the following sequence networks to compute

$$V_+ = E - I_+ X_+ = 1.05\angle 0 - (j0.1383)(-j1.96427) = 0.7777\angle 0 \text{ (p.u.)}$$

$$V_- = 0 - I_- X_- = (-j0.14362)(-j1.96427) = -0.28654 \text{ (p.u.)}$$

$$V_0 = 0 - I_0 X_0 = (-j1.96427)(-1.96427) = -0.49107 \text{ (p.u.)}$$

$$V_{+-0} = \begin{pmatrix} 0.77710 \\ -0.28654 \\ -0.49107 \end{pmatrix}$$

$$\begin{bmatrix} V_a \\ V_b \\ V_c \end{bmatrix} = \begin{bmatrix} 1 & 1 & 1 \\ 1 & \alpha^2 & \alpha \\ 1 & \alpha & \alpha^2 \end{bmatrix} \begin{bmatrix} V_0 \\ V_1 \\ V_2 \end{bmatrix}$$

$$= \begin{bmatrix} 1 & 1 & 1 \\ 1 & \alpha^2 & \alpha \\ 1 & \alpha & \alpha^2 \end{bmatrix} \begin{bmatrix} -0.49107 \\ +0.77710 \\ -0.28604 \end{bmatrix}$$

$$= \begin{bmatrix} 0 \\ 1.179\angle 231.3^\circ \\ 1.179\angle 128.7^\circ \end{bmatrix}$$

3.10.3 Example 3

See Figure 3.36.

$$I_+ = \frac{1.05\angle 0^\circ}{\left[(j0.13813) + \dfrac{(0.14562)(0.25)}{(0.14562 + 0.25)} \right]}$$

$$= \frac{1.05\angle 0^\circ}{(j0.23095)}$$

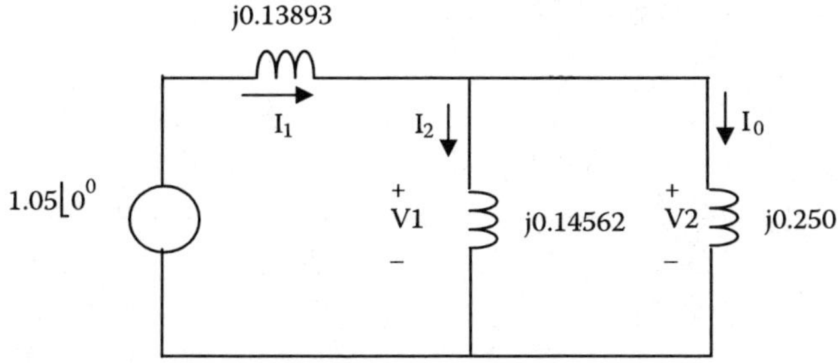

FIGURE 3.36
System for example 3.

$$= -j4.5464 \text{ (p.u.)}$$

$$I_- = j4.5464\left[\frac{0.25}{(0.25+0.14562)}\right] = j2.8730 \text{ (p.u.)}$$

$$I_0 = j4.5464\left[\frac{0.14562}{(0.25+0.14562)}\right] = j1.6734 \text{ (p.u.)}$$

Compute I_{abc}

$$\begin{bmatrix} I_a \\ I_b \\ I_c \end{bmatrix} = \begin{bmatrix} 1 & 1 & 1 \\ 1 & \alpha^2 & \alpha \\ 1 & \alpha & \alpha^2 \end{bmatrix} \begin{bmatrix} j1.6734 \\ -j4.5464 \\ j2.8730 \end{bmatrix} = \begin{bmatrix} 0 \\ 6.893\angle 158^\circ \\ 6.893\angle 21.34^\circ \end{bmatrix}$$

Therefore we have

$$V_+ = E - I_+X_+ = 105\angle 0 - (-j4.5464)(j0.138)$$

$$V_- = 0 - I_-X_- = (-j2.8730)(j0.14562)$$

$$V_0 = 0 - I_0X_0 = (-j1.6234)(j10.250)$$

and V_{abc} is given by

$$V_a = V_+ + V_- + V_0$$

$$V_b = \alpha^2V_+ + \alpha V_- + V_0$$

$$V_c = \alpha V_+ + \alpha^2V_- + V_0$$

3.10.4 Example 4, Three-Phase Fault

See Figure 3.37.

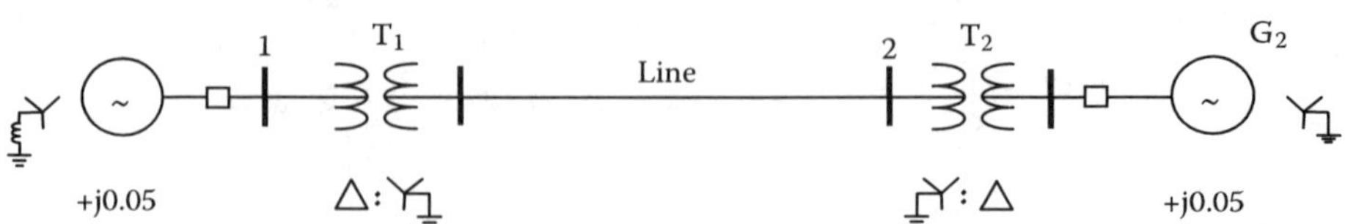

FIGURE 3.37
System for example 4.

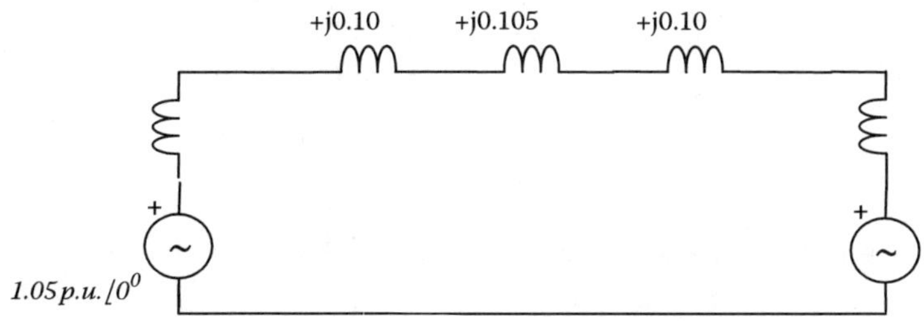

FIGURE 3.38(A)
Positive-sequence network.

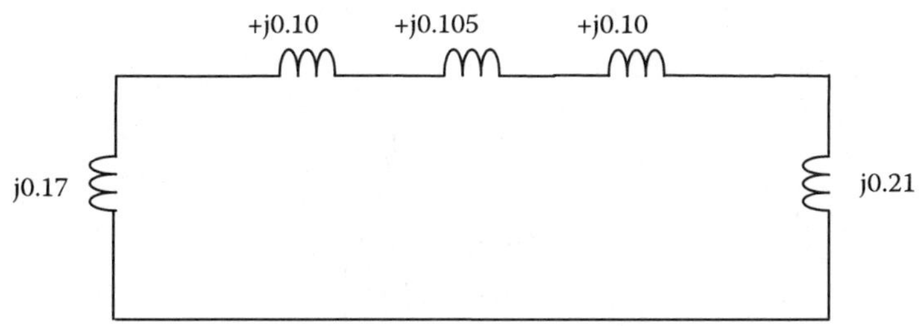

FIGURE 3.38(B)
Negative-sequence network.

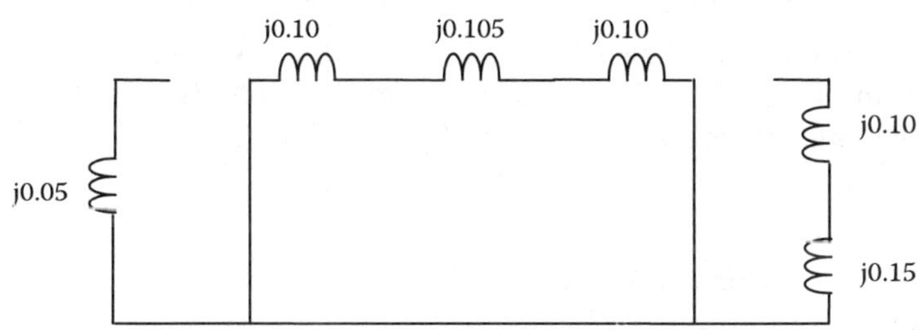

FIGURE 3.38(C)
Zero-sequence network.

For the positive-sequence prefault voltage (Figure 3.36[a]), $V_F = 1.05\angle 0°$. For the three-phase fault, we only work with the positive sequence, as follows:

$$I_0 = 0$$

$$I_- = 0$$

$$I_+ = \frac{V_F}{Z_+} = \frac{1.05\angle 0°}{j0.13893}$$

$$= -7.588 \text{ p.u}$$

Then I_{abc} are not zero, and we get them to follow

$$\begin{bmatrix} I_a \\ I_b \\ I_c \end{bmatrix} = \begin{bmatrix} 1 & 1 & 1 \\ 1 & \alpha^2 & \alpha \\ 1 & \alpha & \alpha^2 \end{bmatrix}$$

where

$$I_0 = I_- = 0$$

$$= \begin{bmatrix} 7.558\angle -90° \\ 7.558\angle 150° \\ 7.558\angle 30° \end{bmatrix}$$

3.10.5 Example 5, Single-Line-to-Ground (SLG) Fault

For the a faulted power system network, find the voltage and currents at the fault point for a Single Line-to-Ground fault where the sequence network connections and values are shown in Figure 3.39.

Solution

The sequence networks are connected in series for a single line-to-ground fault, as shown in Figure 3.39.

$$I_+ = I_- = I_0 = \frac{1}{j(0.2577 + 0.2085 + 0.14)}$$

$$= 1.65\angle -90° \text{ (p.u.)}$$

Therefore,

$$I_A = 3I_+ = 4.95\angle -90° \text{ (p.u.)}$$

$$I_B = I_C = 0$$

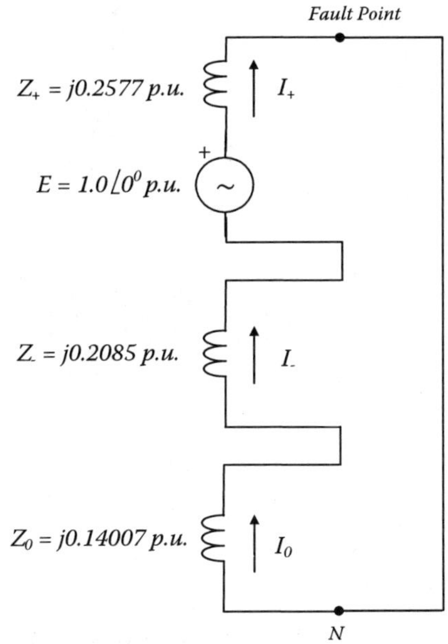

FIGURE 3.39
Sequence networks and their connections for exercise 5.

and the sequence voltages are as follows:

$$\begin{aligned} V_+ &= E_+ - I_+Z_+ \\ &= 1\angle 0 - (1.65\angle -90°)(0.269\angle 90) \\ &= 0.57\text{ (p.u.)} \\ V_- &= -I_-Z_- \\ &= -(1.65\angle -90°)(0.2085\angle 90°) \\ &= -0.34\text{ (p.u.)} \\ V_0 &= -I_0Z_0 \\ &= -(1.65\angle -90°)(0.14\angle 90°) \\ &= -0.23\text{ (p.u.)} \end{aligned}$$

The phase voltages are thus

$$V_A = V_+ + V_- + V_0 = 0$$

$$V_B = \alpha^2 V_+ + \alpha V_- + V_0$$

$$= (1\angle 240°)(0.57) + (1\angle 120°)(0.34) + (0.23)$$

$$= 0.86\angle -113.64 \text{ (p.u.)}$$

$$V_c = \alpha V_+ + \alpha^2 V_- + V_0$$

$$= (1\angle 120°)(0.57) + (1\angle 240°)(0.34) + (0.23)$$

$$= 0.86\angle 113.64° \text{ (p.u.)}$$

3.11 Summary

This chapter covered the basic functions of protection schemes and options for distribution systems. The discussion focused on different elements of relays and circuit breakers and their applications in the coordination and control of distribution system protection. The chapter also provided some illustrative examples showing the concepts of relay coordination under fault.

Problem Set 3

3.1 Figure 3.40 below depicts a voltage source connected to a parallel purely inductive load bank via a short cable. From the information provided on the figure, compute:

a. Sequence currents I_0, I_1 and I_2

b. Phase current vector of the source, I_{abc}

c. Sequence voltage vector across the load, V_{+0}

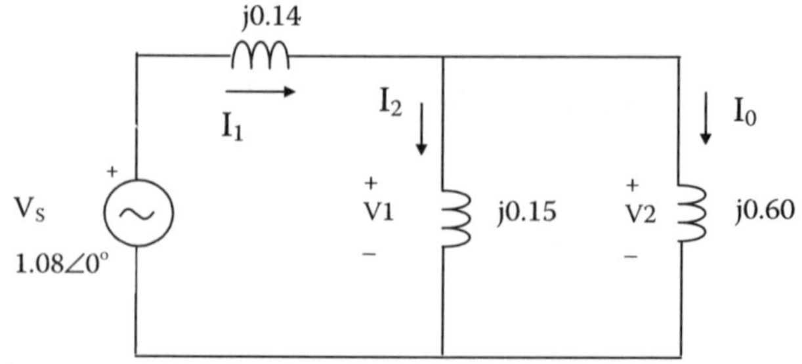

FIGURE 3.40
Network diagram for Problem 3.1.

3.2 Consider the following voltage and current vectors for a 3-phase power system.

$$[V]_{abc} = \begin{bmatrix} 250\angle 110^\circ \\ 270\angle -60^\circ \\ 285\angle -120^\circ \end{bmatrix} \text{Volts} \qquad I_{abc} = \begin{bmatrix} 23.87\angle -76.75^\circ \\ 26.10\angle 180^\circ \\ 25.15\angle -50^\circ \end{bmatrix} \text{Amps}$$

a. Compute the complex or active power, Sp for each phase
b. Compute the total active 3-phase power, S3-φ
c. Verify that $S_P = 3S_{3-\varphi}$

3.3 A distribution network has the following line diagram and corresponding positive, negative and zero sequence circuits as developed in the figures below. Selected sequence pre-fault voltage VF = $1.10\angle 0^\circ$ *p.u.* and compute the phase current vector, I_{abc} for the network. (*Hint*: Only one sequence network may be needed).

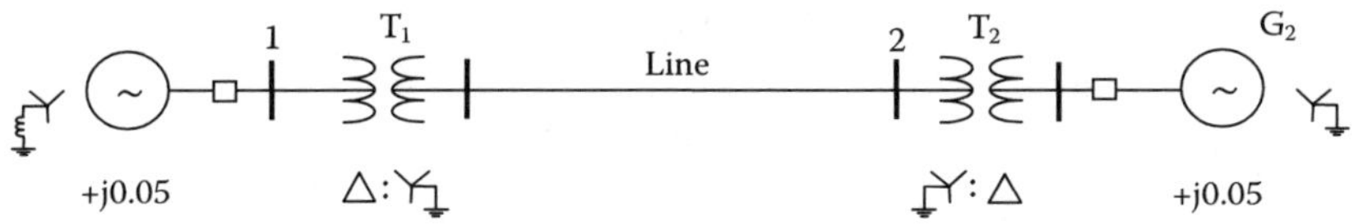

FIGURE 3.41
Single line diagram of the power system representation for Problem 3.3.

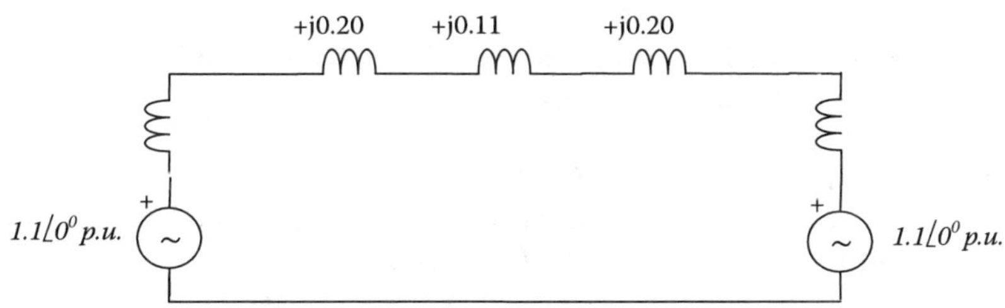

FIGURE 3.42(A)
Positive sequence Network.

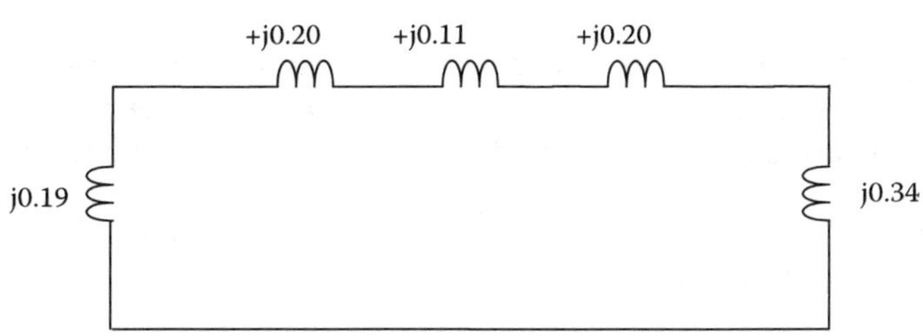

FIGURE 3.42(B)
Negative sequence Network.

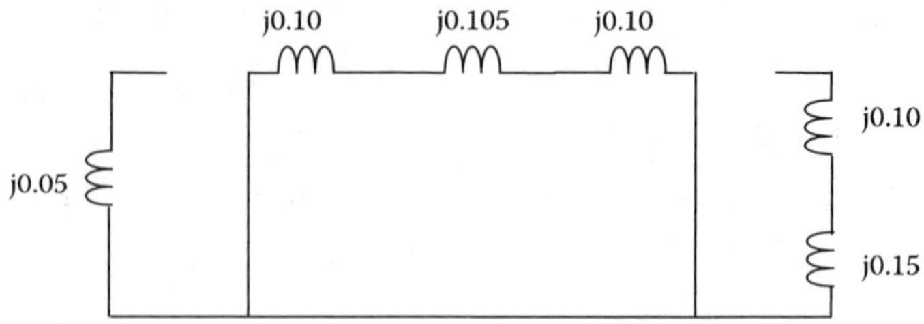

FIGURE 3.42(C)
Zero sequence Network.

3.4 Suppose that an unsymmetrical fault condition gave rise to the following data at the fault point:

$$E = 1.0 + j0.0 \ p.u.$$

$$I_+Z_+ = 0.10 \quad p.u.$$

$$I_-Z_- = 0.10 \quad p.u.$$

$$I_0Z_0 = 0.60 \quad p.u.$$

a. Find the phase voltages at the fault point.
b. Find the line-to-line voltages at the fault point.
c. Identify the type of fault.

3.5 What are the various type of relays? (List them and give a brief description of their operation).

3.6 A 40kVA single-phase transformer rated at 2400/240V is connected as an autotransformer. The rated voltage, $V_1^{rated} = 2{,}400$ kV is applied to the high voltage windings of the transformer.

a. Compute the voltage rating of the high voltage side when the transformer is connected as an autotransformer.
b. Compute the kVA rating of the autotransformer.

3.7 Corresponding to question 3.6 posed above, if the losses in the winding at rated load are given to be 1,007 W, determine the full-load efficiency as an autotransformer at 0.80 power factor, lagging.

3.8 What criteria are to be satisfied for breaker implementation and operation within a network?

4

Distribution System Reliability and Maintenance

4.1 Introduction

Distribution systems are critical subpower networks responsible for delivering power to the customer's doorstep. Outages and failures within a distribution system have an adverse effect on the customers. The ability of the power network to deliver uninterrupted power to its customers at a prescribed level of quality and security is defined as reliability. There is reliability related to the generation, transmission, and distribution subsystems. In this chapter, we provide a fundamental overview of reliability from the working definition that power system reliability addresses the ability of a system to provide an adequate supply of electric energy while satisfying load requirements as economically as possible with a reasonable level of continuity and quality.

There are multiple reasons for conducting reliability analysis, namely to:

- Characterize reliability measurements to a single customer or to the whole system satisfactorily
- Characterize reliability using reliability indices
- Determine cost of reliability or performance at a given cost
- Determine the worth of reliability and its impact on cost versus improvements to the system

Reliability analysis allows us to determine indices for supplying power to specific customers or to entire systems. These indices allow for comparison between customers in terms of system adequacy and improvements to be made. Therefore we embark on reliability analysis to:

1. Predict changes in reliability that will result from changes to the design, operation, and maintenance of a distribution system

2. Identify weaknesses in the design, operation, or maintenance; identify the dominant causes of unreliability; or ascertain whether major improvements such as tree trimming or replacement of broken insulators and missing grounds are needed or justified based on cost
3. Evaluate outage plans by using reliability indices to predict the reliability of system configurations with or without various maintenance activities or options and their associated costs
4. Recommend improvements to reflect utility and customer interruption costs based on cost-benefit analyses

Other pertinent reasons for reliability assessment are to:

- Provide a reliability history of individual circuits for discussion with prospective customers
- Satisfy regulatory reporting requirements
- Furnish management with performance data regarding the quality of customer service for each operating area

The need for these insights into system performance motivates the distribution analyst to use a variety of distribution reliability indices to arrive at conclusions or suggest recommendations in the light of sound engineering judgment. It should be noted that, in most cases, these analyses are based on forced outages, which can be temporary or transient, depending upon the duration of the outage and the steps that must be taken to clear it. Some fault events are random, and hence reliability analysis can also be based on the probability of known and unknown events.

4.2 Reliability Evaluation

Utilities typically keep records of various outages that have occurred within the system. Computational assessments of reliability involve the use of such outage information to determine reliability indices that can be used to assess past system performance. Such assessments are also commonly used to evaluate the reliability of the system on a yearly basis for regulatory authorities.

4.2.1 Inputs Required for Historical Assessment

Utilities use outage management system (OMS) software to compute reliability indices. Such systems typically record outages in terms of location, date/time, failed components, frequency of occurrence, number of customers affected, load interrupted, time to restore service, time to repair or replace failed components, cause of interruptions or anomalies, etc. Reliability indices are computed based on historical events classified in the following outage categories:

1. Maintenance: outages caused by vegetation, animals, or human error; such outages indicate the need for corrective action or maintenance improvements to the system.
2. Component failure: outages affected by the reliability of individual components such as fuses, auto-reclosers, switch gears, contactors, transformers, etc. Failure can be classified based on the components that fail and cause an outage.
3. Weather: outages caused by weather conditions prevalent in the system. The use of a reliability index can help in minimizing damage to the system.
4. Duration of outage: reliability indices based on outage duration are helpful in determining the impact on reliability of improvement with component switches by installing reclosers, automated switches, or changes in network reconfiguration or restoration that will improve the system's reliability at optimal cost.
5. Location of outage: reliability indices based on regional service areas and districts (e.g., urban, rural areas) are used to determine where future improvements should be applied.

4.3 Terminology/Definitions

A brief overview of IEEE definitions of relevant terms is presented to help the reader understand the nature of service interruption, its causes in the distribution system, and the nomenclature used to analyze the reliability of a distribution system.

Connected load: includes the connection of a transformer, a peak-load metered demand on the circuit, or a portion of the interrupted circuit

Interrupting device: a device that disconnects or restores service by automatic or manual control; such devices include transmission circuit breakers, feeder breakers, line reclosers and fuses, sectionalizers, and switches

Interruption: use of service to one or more customers connected to a distribution system that results in one or more component outages

Interruption duration: the time period from initiation of an interruption of service until it has been restored

Outage: the state of a component when it is not available to perform its intended function; an outage may or may not cause interruption of service, depending on the configuration of the system

Customer interruption: supply of power interrupted by component outages, system instability, thermal overloads, or undervoltages

Loss of service: a complete loss of voltage on at least one normally energized conductor to one or more customers

Momentary interruption: a single separation of an interrupting device that results in a zero voltage

Momentary interruption event: an interruption of duration limited to a period required to restore service by an interrupting-device switching operation; this operation must be completed within a specified time of 5 min or less; if a reclosing device operates multiple times within 5 min of the first operation, then the entire sequence is a momentary event

Planned outage: state of a component when it is not available to perform its intended function due to a planned event directly associated with that component

Sustained interruption: any unplanned interruption not classified as part of a momentary testing lasting more than 5 min

Unplanned interruption: interruption caused by an unplanned event/outage

Planned interruption: loss of power when a component is deliberately taken out of service or for construction/maintenance

4.4 Reliability Indices

Reliability indices are used to assess past performance or to predict future performance. Typically, they are measures of the average failure rate, the average outage duration, and the average annual outage time, which in turn determines the availability of power to customers. This section introduces a number of commonly used reliability indices for distribution systems. The customer-based indices are as follows:

1. SAIFI (system average interruption frequency index) is defined as:

$$\text{SAIFI} = \frac{\text{total number of customer interruptions}}{\text{total number of customers served}}$$

$$= \frac{\sum_{i \in R} \lambda_i N_i}{\sum_{i \in R} N_i} \tag{4.1}$$

where

λ_i = failure rate

N_i = number of customers at the load point interrupted by the *i*th interruption event

R = total number of customers and sustained outage rate at load point L_p.

SAIFI indicates how often the average customer experiences a sustained interruption over a period of time for the area. For a fixed number of customers, the only way to improve the SAIFI index is to reduce the number of sustained interruptions on the system. The SAIFI index has units in time^{-1}, and to calculate this index, it is important to have a clear and consistent definition of the outage type or length that characterizes the interruption. SAIFI cannot be used to compare the reliability of two utility companies unless they share some characterization of interruption.

2. CAIFI (customer average interruption frequency index) is used to calculate the failure rate of distribution systems or the interruption rate to which customers are subjected.

$$\text{CAIFI} = \frac{\text{total number of customer interruptions}}{\text{total number of customers affected}}$$

$$\text{CAIFI} = \frac{\sum_{i \in R} \lambda_i N_i}{\sum_{i \in R} M_i} \tag{4.2}$$

where M_i = number of customers affected at load point *i*.

The CAIFI index measures the interruption frequency for customers who are experiencing service interruptions, and this provides a means to recognize chronological trends of events in reliability of service to affected customer who have suffered interruptions (brownouts or blackouts).

3. SAIDI (system average interruption duration index) is used to determine the average duration of service interruptions for the system. It is defined as

$$\text{SAIDI} = \frac{\text{Sum of customer interruptions durations}}{\text{total number of customers}}$$

$$\text{SAIDI} = \frac{\sum_{i \in R} U_i N_i}{\sum_{i \in R} N_i} \quad (\text{in h}\cdot\text{yr}^{-1}) \tag{4.3}$$

where

U_i = annual unavailability or outage time at the load point i

N_i = number of customers affected by the ith event

The SAIDI distribution reliability index indicates the total duration for the outage customer during a period of time when there is a sustained service interruption leading to power loss. SAIDI has the dimension $\text{h}\cdot\text{yr}^{-1}$.

4. CAIDI (customer average interruption duration index) is defined as

$$\text{CAIDI} = \frac{\text{Sum of customer interruptions durations}}{\text{total number of customers interruptions}}$$

$$\text{CAIDI} = \frac{\sum_{i \in R} U_i N_i}{\sum_{i \in R} \lambda_i N_i} \quad (\text{in h}) \tag{4.4}$$

$$\text{CAIDI} = \frac{\text{SAIDI}}{\text{SAIFI}} \tag{4.5}$$

where

λ_i = frequency of the N possible losses of power events

N_i = number of customers affected by the ith event

U_i = annual unavailability or outage time at the load point i

The CAIDI reliability index represents the average time taken to restore service to the customer. This index can be improved by reducing the length of interruption (faster crew dispatch and response time, faster repair times, etc.).

5. ASAI (average service availability index) is a measure of the average annual outage time or the availability of the power supply.

$$\text{ASAI} = \frac{\text{Sum of hours of available service to customers}}{\text{customer hours service demanded}}$$

$$\text{ASAI} = \frac{\sum_{i \in R} 8760 N_i - \sum_{i \in R} U_i N_i}{\sum_{i \in R} 8760 N_i} \tag{4.6}$$

ASAI is the fraction of time that a customer has received power during the defined reporting time. These indices are based on average values of data collected on random events of interruption for all customers without regard to value or priority of customers. Higher ASAI reflects high levels of reliability.

6. EACI (expected annual cost of interruption) is an alternative index for measuring adequacy of customer service. The index, which is designed to provide an economic value to reliability or a cost of unreliability, is defined as

$$\text{EACI} = \sum_{\forall i} \lambda_i c_i \tag{4.7}$$

These distribution system reliability indices are based on average values of data collected on random events of interruption for all customers without regard to the value or priority of customers.

7. ASUI (average service unavailability index) is defined as

$$\text{ASUI} = \frac{\sum_{i \in R} U_i N_i}{\sum_{i \in R} 8760 N_i} \tag{4.8}$$

where

N_i = number of customers affected by the ith event

U_i = annual unavailability or outage time at the load point i (in h/yr)

8. ENS (energy not supplied), expressed in units of kW·h/yr, is defined as

$$\mathrm{ENS} = \sum_{i \in R} P_{ai} U_i \tag{4.9}$$

where

P_{ai} = average load (kW) connected to load point i

U_i = annual outage time (h/yr) at the load point i

9. AENS (average energy not supplied), expressed in units of kW·h/customer/yr, is defined as

$$\mathrm{AENS} = \frac{\mathrm{ENS}}{\sum_{i \in R} N_i} \tag{4.10}$$

10. ACCI (average customer curtailment index), expressed in units of kW·h/customer affected/yr, is defined as

$$\mathrm{ACCI} = \frac{\mathrm{ENS}}{\sum_{i=1}^{R} M_i} \tag{4.11}$$

where M_i is the number of customers affected at the load point. The customers affected should be counted only once, regardless of the number of interruptions per year.

4.5 Methods of Reliability Analysis

Distribution reliability analysis is based on two major methodologies, namely the analytical and simulation methods. The analytical method represents the power distribution system with a mathematical equation derived for solving or evaluating the reliability indices. It is transparent among results, model, and data simulation methods that simulate the random behavior of the system, addressing the actual failure repair and restoration of the system under disturbance. The simulation method provides records in the form of a distribution function indicating that the reliability is expected to vary. It can also handle complexity in data, procedures, and processes.

The proposed analytical methods to be discussed are used to evaluate transmission reliability as well as distribution system reliability. We are concerned more with determining the reliability based on systemwide measures of performance than on reliability based on specific supply loads.

In general, the variants of theoretical methods are equivalent. However, the selection of a particular method may be based on the practical application. This chapter provides an overview of the analytical methods in use. To start with, we consider distribution system component models to be normal operating states, faulted, opened/closed, or undergoing repair.

4.5.1 Analytical Methods

A distribution system can be modeled as a set of interconnected components. Various reliability indices can be defined to capture the effects of failure of each component on the system. The analytical methods used for reliability assessment provide the average performance of the system. The commonly used tools to evaluate distribution system reliability include state space diagrams, failure modes and effects analysis (FMEA), state enumeration, event trees, minimal cut sets, and fault trees. These are discussed in the following sections.

4.5.2 State Space Diagrams

A state space diagram represents a system by defining all possible states of interest that the system can adopt. The method uses transition between failed states of components to repaired normal states or partial states of restoration during an abnormal or adverse weather condition. Transitions between states are modeled as Markov processes. The state space diagram is referred to as the Markov diagram.

Two state transitions resulting from the failure and repair of typical power system components (buses, transformers, and circuit breakers) can be represented as follows:

$$\lambda = \text{failure rate}, \ \mu = \text{repair rate}$$

For the case when power is available to a customer at time $t = 0$, λ and μ do not vary with time, and the probability that power is available at time t is given as follows:

$$\text{Prob}\big(P(t)\big) = \frac{\mu}{\mu+\lambda} + \frac{\lambda^{e-(\lambda+\mu)t}}{\lambda+\mu} \tag{4.12}$$

As $t \to \infty$, the steady state of available power is

$$\begin{cases} \dfrac{\mu}{\mu+\lambda} & \text{, availability} \\ \dfrac{\lambda}{\mu+\lambda} & \text{, unavailability} \end{cases} \tag{4.13}$$

$$\cong \frac{\lambda}{\mu}\text{, unavailability for } \mu > \lambda$$

i.e., when the repair rate μ is greater than the failure rate λ. From the foregoing, the failure rate λ is based on the interruption frequency experienced by the customer.

4.5.2.1 Case A, Series Components

For the series-connected components,

$$\lambda = \sum_{i=1}^{3} \lambda_i T_i \text{ for } N = 3 \tag{4.14}$$

where $\Sigma\lambda_i T_i$ is the long-term unavailability, and T_i is the interruption or repair time.

4.5.2.2 Case B, Parallel Systems

In parallel systems, the frequency of a failure in a parallel network, where one interruption of power can cause loss failure in a second bus but not the second line, is deduced. The frequency of such a failure is computed as

$$\begin{aligned} 1 &= \lambda_2(\lambda_1 T_1) + \lambda_1(\lambda_2 T_2) \\ &= \lambda_1\lambda_2(T_1 + T_2) \end{aligned} \tag{4.15}$$

This is defined as the product of the failure rate of both components with summation of the repair times for both. Whereas the unavailability of both parallel paths when they both fail is defined as unavailability of power as determined by the product of the failure rate of components and the product of the times to repair both paths (given as $\lambda_1\lambda_2 T_1 T_2$), for distribution systems arranged in a parallel structure rather than services connected in a radial network.

4.5.2.3 Case C, Series and Parallel System

Most components in distribution can be connected in series or in parallel. Depending on the type of connection, reliability analysis will be different. For example, for a series network, a loss of one component such as a transformer or a circuit breaker can cause total loss of the feeder connecting them,

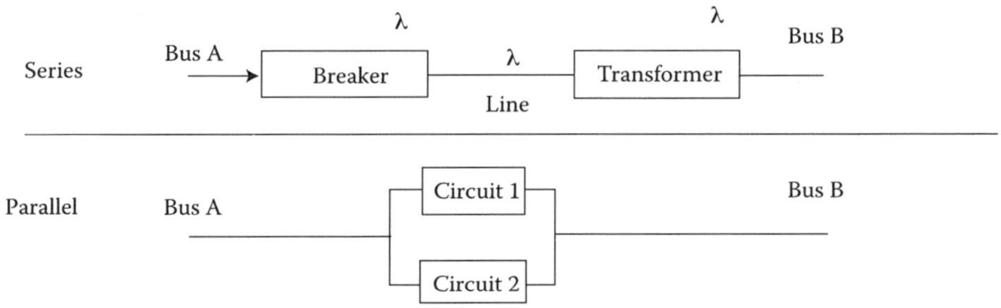

FIGURE 4.1
Distribution components in series and parallel.

whereas in a parallel system, the interruption of one of the components may not cause a total failure of the system, as shown in Figure 4.1

4.6 Failure Modes and Effects Analysis (FMEA) Method

This method is one of the simplest ways of estimating reliability. It is based on an inductive (what if?) analysis that can be used to identify the failure mode of components in a distribution system affected by changes in power or loss of power to a specified load caused by the states of breakers, circuit breakers, loads, and subsequent control actions to restore the system.

FMEA identifies single-component failure states that occur independently and are repaired before another occurs. It can be used as a repetitive survey of failure behavior or as a precursor to other analysis techniques, such as cut-set and fault-tree analyses. The failure states are recorded as the number of customers affected and duration of the event. To facilitate minimum cut set (first-order cut set, etc.) or a fault-tree approach for distribution reliability, a probability approach is used to weigh categories or put limits on them to be selected prior to determination of reliability. The principal advantage of FMEA is that it provides a detailed description of the failure behavior of the distribution system while evaluating the consequences of all failure modes of all components. The drawback of FMEA is that it is repetitive, and it is difficult to examine multiple failures in an efficient manner.

4.7 Event-Tree Analysis Method

The event-tree analysis method is used to provide a detailed examination of possible scenarios initiated by a fault event or a faulty component within a

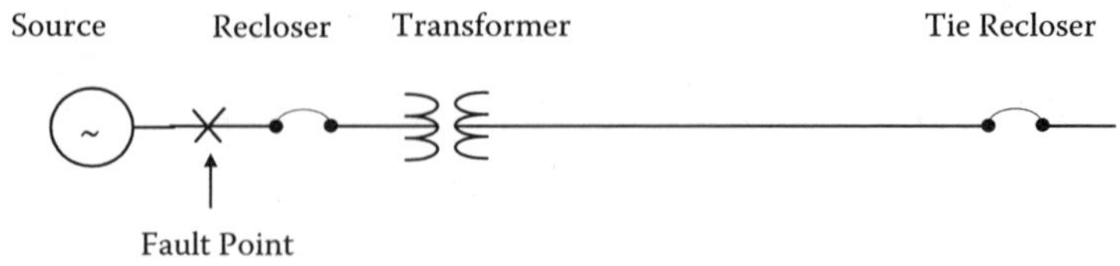

FIGURE 4.2
Distribution system with fault.

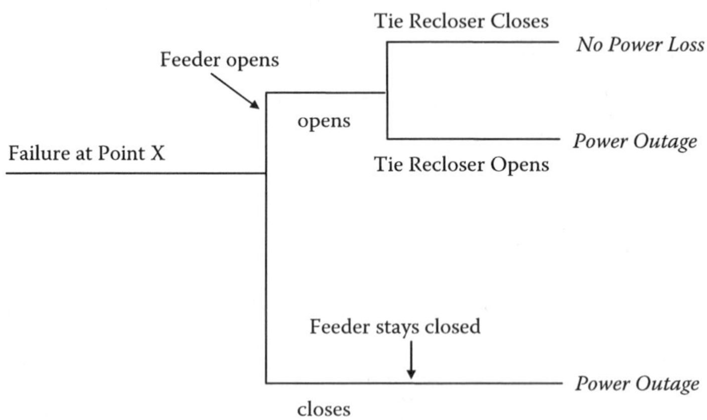

FIGURE 4.3
Event tree.

distribution system. The technique is well known in the nuclear industry for probabilistic work assessment. See Figure 4.2 for an example of a distribution system with a fault.

If a fault occurs at point X, the relocation between feeder and recloser at closing and opening conditions is analyzed using an event-tree diagram. The event-tree diagram is developed as shown in Figure 4.3, which shows the following sequence and operation of switches. For an open condition after a fault is identified in point X, the recloser opens, giving two cases: (a) tie closing and no power loss or (b) switch opens, indicating an outage, and customers will suffer interruption of power.

4.8 Fault-Tree Analysis Method

The combination of sequences illustrated using a graphical representation of the failure logic of a system is the so-called fault-tree analysis (FTA) method. It logically deduces the credible causes of events in determining a minimum cut set of events causing the problem. The FTA method uses each minimum cut set identifying the paths supplying power to a load point to determine the various outages that can cause interruption of service to load. Data are collected on load-point interruption frequency and duration, and reliability indices are computed for the systemwide vulnerability.

Four steps are used in fault-tree analysis for calculation of reliability indices:

1. Identify minimum cut sets.
2. Identify the paths by which power can be supplied to a load point, tracing to the substation to indicate the condition of switches/reclosers in the event of an overload or violation of voltage constraints. Load flow and transient stability studies may be required to decide which load should be dropped by evaluating the alternative paths that are viable.
3. Identify the minimum cut set that causes the outage. This involves determination of the interruption frequency reliability index, which is used to compute a measure of the unavailability of power at the load point of concern. Using normal paths through which power is supplied and alternative paths to the load through normally open connections that close, restoration/switching processes are completed without undue delay.
4. Calculate reliability indices or the unreliability cost of power supply to that point using the cut sets and failure and other data.

4.9 Unavailability of Power Calculations from the Cut Set

The calculation of power unavailability requires data about the duration of outages for each cause over time. The outage duration will be affected by repair time, switching times, and the possibility that switching is effective. The outage duration includes consideration of normal and alternative paths with opened/closed connections to be included in the calculations of unavailability for the earlier examples above. We have

$$\begin{aligned}\text{Unavailability} &= \lambda_b T_b + \lambda_c T_c + \lambda_d T_d + \lambda_e T_e + \lambda_f T_f + \lambda_g T_g \\ &\quad + \lambda_g T_a P_b + \lambda_a T_a P_e + \lambda_a T_a P_e + \lambda_a T_a \lambda_n T_n \end{aligned} \tag{4.16}$$

4.9.1 Fault Tree Based on Minimal Cut Set

4.9.1.1 Determine Power Interruption and Unavailability

Fault trees provide a graphical representation of the failure logic of distribution systems, accounting for all failure modes, maintenance activities, and weather conditions.

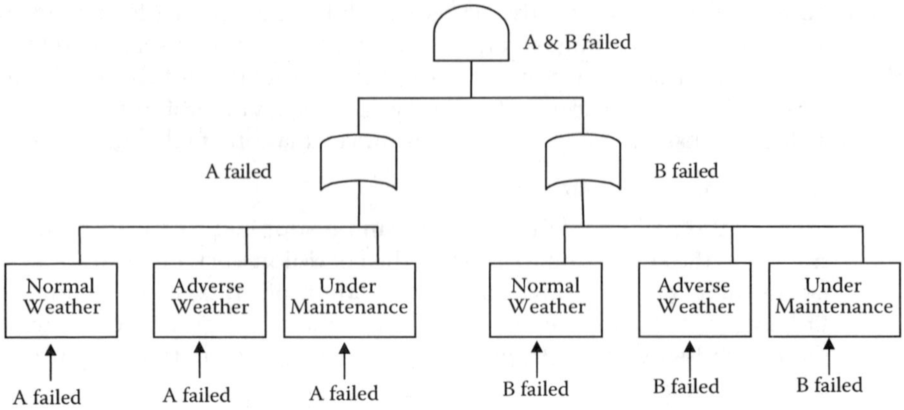

FIGURE 4.4
Fault tree.

Probabilities of events of interest (power interruption) are based on several analysis tools such as adequate probability density function and accumulation density function. Overall, the information from these analyses, as usual, provides benefit and cost accounting of reliability and improvement. Consider the fault tree depicted in Figure 4.4.

Components A and B are coordinated in Table 4.1 to yield a given minimal number of cut sets that result in cost failures.

Different contributions of cut sets in Table 4.1 will provide a decision based on engineering judgment:

1. Cut set at B will cause an overlapping maintenance of A and B.
2. Maintenance of components should start, if adverse weather cut at 6 and 8.

TABLE 4.1

Cause of Failures

Component A			Component B			Cut Set Number
Normal	**Adverse**	**Maintenance**	**Normal**	**Adverse**	**Maintenance**	**that Causes Failure**
X			X			1
X				X		2
X					X	3
	X		X			4
	X			X		5
	X				X	6
		X	X			7
		X		X		8
		X			X	9

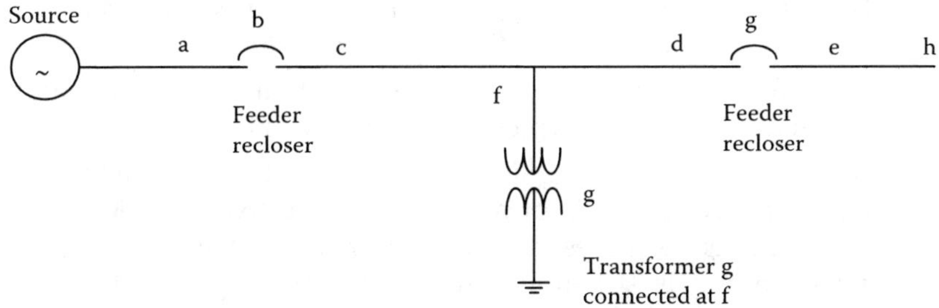

FIGURE 4.5
Sample system.

3. If an alternative path is targeted at cut sets 3 and 6, should maintenance be started or not?
4. Finally, if adverse weather persists, should cutting start or not?

4.9.1.2 Methodological Approach to Identifying Minimum Cut Set

Consider the system below, shown in Figure 4.5.

Minimum cut set:

a. If the supply path's transformer is out, then the power transformer will be out.
b. We look at power interruption to the customer by using all combinations or a minimal cut set included in the following paths in the logic write-up presented in Equation 4.17.

Loss of power to customers occurs in the following four minimum cut sets:

$$\begin{bmatrix} b+c+d+e+f+g \\ +ab_{\text{fails to open}} \\ +ae_{\text{fails to open}} \\ +ah_{\text{fails to open}}+a_{\text{fails to open}}h \end{bmatrix} \begin{matrix} (1) \\ (2) \\ (3) \\ (4) \end{matrix} \qquad (4.17)$$

The four minimum cut sets are represented herein:

a. Paths *a*, *b*, *c*, *d*, *e*, *f*, *g*, *h* represent faults in the same lines or pieces of equipment.
b. Paths *b*, *c*, *d*, *e*, *f*, *g* represent a single-event cut set.
c. Second-order cut set $ab_{\text{fails to open}}$ represents a fault failure at *a* while *g* recloser is open.

d. Second-order cut set $ge_{\text{fails to close}}$ again represents a fault failure at a followed by a failure of tie recloser e to close.
e. Second-order cut set ah_{failed} represents a fault failure of a subsequent to a failure in h, which precludes use of the alternative path.
f. Second-order cut set $a_{\text{failed}}h$ represents a fault failure at a followed by successive opening of recloser b and reclosing of tie recloser at e, leading to successful opening of recloser at e, leading to h fails while a is still under repair and b is open. Attention is required to track the path failures by paying special attention to the condition of the paths either during maintenance or prior to repairs.

Lastly, the identified minimum cut set is used to compute the frequency of power interruptions, which can be calculated by using the following frequency and conditional probability data in the equation. We write these as follows:

Frequency of loss of power is given for each path of failure as

$$\lambda_b + \lambda_c + \lambda_d + \lambda_e + \lambda_f + \lambda_g$$

$$+ \lambda_a P_b \text{ fails to open, } + \lambda_a P_c \text{ fails to close}$$

$$+ \lambda_a P_h, + \lambda_b P_h \tag{4.18}$$

where

P_h is unavailability of h, etc.
$P_h = \lambda_h T_h$ if maintenance is excluded

Other terms represent frequencies, unavailability, or conditional probabilities, as shown above. It should be noted that each cut-set analysis consists of one and only one event (first-order minimal cut sets). Common causes/failures that eliminate both normal and alternative paths may be initiated by an event that triggers service interruptions or an enabling event such as failed switches or other devices that can cause malfunctioning of the system. Second-order minimal cut sets are constructed from alternative paths that are important causes of power interruption.

4.9.2 Nonminimal Cut Set in Complete Unavailability

When a nonminimal cut set evolves in computing unavailability, we use a different equation for the example given in Figure 4.6. The minimal cut set equation is given for

$$\text{Unavailability} = a \text{ and } i \text{ switch} \tag{4.19}$$

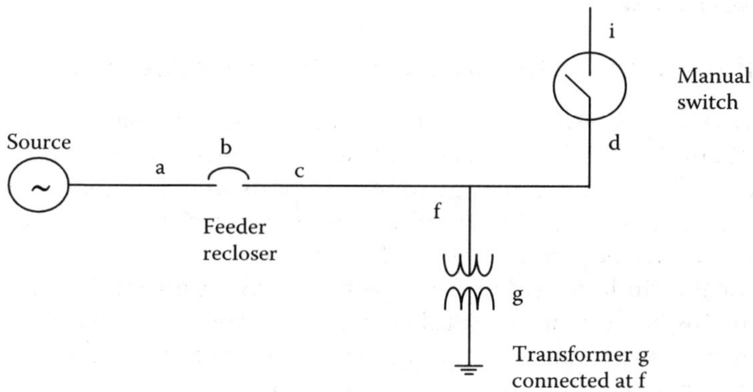

FIGURE 4.6
Sample system.

This equation can be quantified in this case by writing

$$\lambda_a T_{si} + \lambda_a p_i = \text{Min}\ (T_A - T_{si}, T_i) \tag{4.20}$$

where λ_a is the failure frequency of a switch and p_i is the probability that switch *i* fails to close. T_{si} is the time required to operate switch *i*, and T_a and T_i are the repair times for components *a* and *i*, respectively.

4.9.3 Summary of Findings Using Minimal Cut Sets to Identify Causes of Failures

- Minimal-cut-set analysis addresses only the causes of outages at specific loads or points. Based on the analyses for all loads, system-wide reliability analysis is done using the cut-set concept.
- Logic calculations based on a minimal cut set are done by hand for small systems and by computer methods for large systems.
- Cut-set analysis relies on predetermined patterns or procedures for restoration of power, regardless of how long an activity takes. The information (reliability distributions) is obtained using failure frequencies and restoration time.
- A review of cut sets, as in fault-tree application, is useful in identifying cut sets that violate system operating practices.
- Nonminimal cut sets may be of interest, but these can be difficult to create and handle.
- Incorporation of load models with seasonal power loads and a demand-side management program can make life a little more complex when using minimal-cut-set techniques or analyses.

4.10 Simulation Techniques for Reliability Analysis

In the previous section, analytical algorithms were proposed to evaluate system reliability. These methods are fast and effective in estimating distribution reliability under different time durations and frequency. However, they do not provide any information about the variability of the indices; rather they are based on statistical information.

Theoretical techniques suffer from the ability to capture and represent rare events. In this regard, it is useful to have historical data-based surveys/records over a time frame that could be seasonal or variable. The advantages of the simulation method include the following:

- Addresses all scenarios and occurrences, including "domino" effects
- Accommodates any failure and repair rate distribution, and is less restrictive, like the Markov process
- Estimates interruption frequencies for the entire system or for individual load points
- Accommodates complete maintenance strategies that are used to avoid outage combinations that can lead to overload or other problems
- Handles interruption costs
- Reflects the inherent uncertainty in failure and power restoration

We can use different distribution functions of reliability analysis to compare different designs and operating procedures. Finally, the simulation techniques can handle nonstandard and arbitrary probability distributions associated with component failure, repair, and power restoration. We can also view sequences of events, repairs, or switching strategies, depending on which action leads to a faster restoration plan.

An important follow-up of results by simulation methods is the use of load-flow models/analysis to determine whether they are viable or not, and to provide necessary rules for load shedding (as necessary). Additionally, chronological and aggregate load models can be employed in the simulation method to provide a probability that the load can be sustained with a given distribution configuration. Similarly, the cost of interruption with respect to an outage can be simulated.

4.11 Simulation Methods Utilized for Distribution Reliability Analysis

4.11.1 Monte Carlo Simulation Method

Monte Carlo methods simulate the failure, repair, and power restoration processes that characterize the operation of power distribution system. Monte Carlo methods generate probability distributions that can also be used in cut-set quantification. Two types of Monte Carlo simulation methods are discussed: sequential and nonsequential.

4.11.1.1 Sequential Monte Carlo Method

This method is based on the dynamic simulation of states that can evolve due to failed or repaired components in the system. Every component in the system is generally given in terms of failure and repair characteristics, which are stored as historical events and simulated using random number generators. The resulting repair and failure processes are used to model the actual failure phenomena. After simulation of these events for a long time, the response is used to determine reliability indices with statistical criteria such as standard duration, the maximum iterations are stopped, and the final results are computed.

4.11.1.1.1 Outline of Monte Carlo Methods Algorithm

The algorithm based on the sequential Monte Carlo method for reliability analysis is as follows:

1. Generate a random number for each element in the system and convert it to time to failure (TTF) corresponding to the probability distribution of the element parameter.
2. Determine the element with minimum TTF.
3. Generate a random number and convert this number into the repair time (RT) of the element with minimum TTF.
4. Generate another random number and convert the number into the switching time according to the probability distribution of the switching time if this action is possible.
5. Determine the load points that fail and record the outage duration for each failed load point.
6. Generate a new random number for the failed element and convert it into a new TTF and return to step 2 if the simulation time is less

than 1 year. If the simulation time (TTF + RT) of the failed components is greater than 1 year, go to step 9.
7. Calculate the number and duration of failures for each load point for each year.
8. Calculate the average values of the load point failure rate and failure duration of sample years.
9. Calculate the systems indices and record these indices for each year.
10. Calculate the average values of these system indices.
11. Return to step 2 if the simulation time is less than the specified simulation years; otherwise, output the results.

4.11.1.2 Nonsequential Monte Carlo Simulation

A sequential Monte Carlo method generates an artificial history of events to determine distribution system reliability based on the order of events occurring. The nonsequential Monte Carlo simulation (MCS) assumes that the contingencies occurring in a system are mutually exclusive and that system behavior does not depend on past events. In a nonsequential MCS, the list of possible contingencies and their times or duration over a specific time, for example, if the time to failure of each contingency is assumed to be exponentially distributed with parameter λ failure/yr, the number of times it fails within a specific interval of time (= 1 yr), we use a Poisson distribution and compute outage effects of each contingency by one of the analytical methods.

To determine reliability indices, the effect of each contingency is computed based on a weighted value. The procedure is repeated for many cycles to obtain reasonable reliability indices. The impact of the contingency on the customer and the duration of interruptions is evaluated using analytical techniques in a determination form.

4.11.1.3 General Statement: Monte Carlo Simulation

We must note the following features of setting up convergence and the stability of the results when simulating via MCS:

1. The initial system must be created to include:
    - Normal weather, components in their normal state, and weather patterns
    - The duration required for components to achieve their current state (fail or repair)
2. Repeat the process to ascertain when the next change in state occurs.
3. Simulate the behavior of the system over a period of time for an economic fashion (not all components will fail or repair or change state within the time period).

4. Keep a chronological record of changes and what does not change.
5. Since we are inspecting only transactions, we have to deal with a limited number of events and thus a limited number of times at which random numbers can be drawn to predict the future behavior.

Repeat the simulation until the estimated reliability index has convergence. In MCS, these indexes will fluctuate. To obtain stable results, fix the iteration set and the coefficient of variation to a set value and proceed with simulation until the coefficient is achieved.

4.12 Evaluation of Distribution Reliability Analysis Method

Regardless of the method used to determine reliability analysis, the predictions made and the conclusions should withstand the following expectations:

1. The results should not surprise anyone, and the conclusions drawn should withstand considerable scrutiny.
2. It should be possible to reconcile the predicted reliability with past behavior.
3. Where a cut set is identified, it should satisfy necessary and sufficient conditions to cause the predicted result.

4.13 Reliability Database Design

We discuss here the types of data needed for reliability analysis. In addition, to have access to qualify the data for reliability analysis, we also develop a database scheme for data storage consisting of failure rates, repair and restoration, and other components statistics. A brief discussion of available software and capabilities is presented.

4.13.1 DISREL

DISREL is designed to aid electric utility and industrial commercial distribution engineers with predictive reliability assessment of a distribution network. The DISREL software:

- Computes a set of reliability indices — including SAIFI, SAIDI, ASAI, load/energy curtailed, and the cost of outages — based on the component outage data and the cost of interruption to a customer

- Models time-sequenced switching actions taken by an operator/ repair-person following an outage event
- Provides typical outage data for major components; also provides the cost of interruption data for different types of customers.

4.13.1.1 General Information on DISREL

The majority of outages seen by customers are caused by failures in the distribution systems. It is therefore important to objectively assess the relative benefits of alternative distribution system schemes in computing customer-service reliability. The reliability targets and the means to achieve them should be based on the customers' needs and willingness to pay for a desired level of reliability so that the total cost (power supply cost plus customer outage costs) is minimized.

The objective is then to select an alternative that will satisfy a customer's desired level of reliability and is also within the budgeted funds. The alternatives available to a utility engineer and to a customer may include design modifications, reinforcements, allocation of spares, improvements in repair and maintenance policies, and alternative operating policies. The benefit per monetary unit expended and the merits of such alternatives can be compared by utilizing quantitative reliability techniques.

4.13.1.2 Main Features

- Computes both customer and system reliability indices
- Helps in monitoring/achieving performance-based ratemaking (PBR) plans and targets
- Provides a basis for risk/benefit analysis against investments
- Improves decision making for allocating limited capital
- Models user's specified switching strategy
- Provides typical outage data

4.13.1.3 Program Capabilities

The program, DISREL, provides the enhanced modeling capabilities needed to compute the reliability of a distribution system. DISREL calculates an array of indices, including SAIFI, SAIDI, and ASAI, load/energy not supplied, and expected outage cost, after recognizing the ability to transfer loads to alternative sources and after taking appropriate switching actions. The program's modeling capabilities provide a high degree of flexibility for analyzing distribution system configurations by taking the user's specified time-sequenced switching actions.

DISREL input consists of six data files:

1. Program control data

2. Component data
3. Network configuration data
4. Switching-action data (optional)
5. Reliability data (optional)
6. Load-curve data (optional)

The files can be created for a standard configuration and using a text editor, and they can easily be modified to reflect variations in different alternatives.

A user can create a master component data file that has information on typical system components. These components can be used to define other components in the system (inheritance property). The master components need not be connected in the configuration. The system topology is defined in a separate file. DISREL automatically traces the fault-interrupting devices (e.g., breakers and fuses) and the isolation points (normally open points). The input files use a free format, and a user can insert comment lines throughout a data file.

The DISREL software:

- Is capable of modeling time-sequenced switching actions for an outage event. The program follows the user's instructions to open/close a component at specified time intervals. This overrides the switching logic implemented in the program.
- Calculates frequency, duration, and load/energy interruption indices that reflect the manual/automatic switching time required to isolate a faulty component from the healthy system or to transfer loads to an alternative supply point.
- Computes outage-cost-related indices if the customer outage-cost data is provided for a customer. The indices include the impact of frequency, duration, and the amount of load interruption for a customer.
- Provides typical outage data for major components. The data is compiled based on the information published in various outage data reports.

4.13.1.4 Applications of DISREL

1. Reliability assessment of electric distribution supply system
2. Optimal allocation of funds in distribution facilities
3. Evaluation of customer service (load point) reliability
4. Quantification of risk/benefit against investments
5. Identification of weak links in the system

4.14 Maintenance and Reliability

The term "reliability" has been discussed earlier. Simply put, it relates to the ability of a system to perform its intended function for a given period of time under stated conditions. The transition diagram in Figure 4.7 illustrates the two states of normal and failure components. In this section, we introduce probabilistic parameters.

The transition to the normal state is called repair, whereas transition to the failed state is called failure. A repairable component remains in the failed state for a period and then undergoes a transition to the normal state when the repair is completed. We assume that the repairs restore the component to a good condition, as good as new, so we regard this process as repair or maintenance. The cycles of repair-to-failure and failure-to-repair processes explain the transition diagram in Figure 4.7. We briefly discuss each in the following subsections.

4.14.1 Repair-to-Failure Process

A life cycle is a typical repair-to-failure process; repair means birth and failure is equivalent to the death of a component.

Reliability $R(t)$ is the probability of survival to (inclusive or exclusive) age t and the number surviving at t divided by the total sample.

$$R(t) = P_r\{T \geq t\} = P_r\{T > t\} \tag{4.21}$$

Similarly, unreliability $F(t)$ is the probability of death to age t (inclusive or exclusive). It is obtained as

$$F(t) = P_r\{T \leq t\} = P_r\{T < t\} \quad \text{for } t \geq 0 \tag{4.22}$$

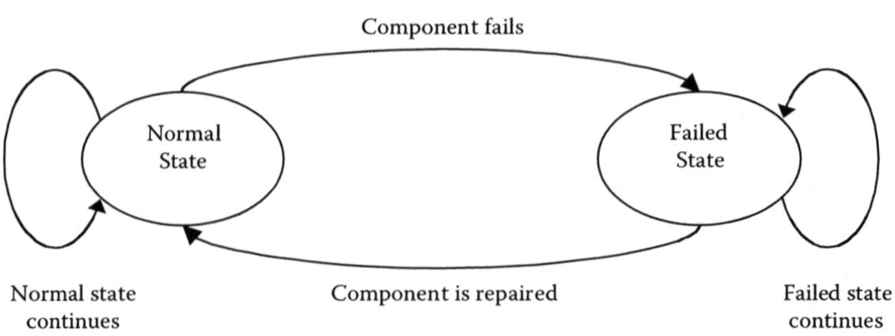

FIGURE 4.7
Transition diagram.

where $P_r\{T = t\} = 0$.

So we define reliability as

$$R(t) = P_r\{T \geq t\} \tag{4.23}$$

and unreliability as

$$F(t) = P_r\{T < t\} \tag{4.24}$$

and the reliability of a component as

$$= P(T \geq t) \tag{4.25}$$

where

$R(t)$ = reliability function; component will survive at time t

$F(t)$ = unreliability function

If

$$R(t) = 1 - F(t) = 1 - \int_0^t f(t)\,dt = \int_t^\infty f(t)\,dt \tag{4.26}$$

$$\therefore \int_{t_1}^{t_2} f(t)\,dt = \int_{-0}^{t_2} f(t)\,dt - \int_{-\infty}^{t_1} f(t)\,dt$$

$$= F(T_2) - F(T_1) \tag{4.27}$$

where $R(t)$ is the time to failure of the random variable T with probability density function $p(t)$.

In terms of reliability,

$$\int_{t_1}^{t_2} f(\tau) = \int_t^\infty f(\tau)\,dt - \int_{t_2}^\infty f(\tau)\,dt \tag{4.28}$$

$$= R(t_1) - R(t_2)$$

is called the rate of failure at (t_1, t_2), with the hazard rate of failure or failure rate during the interval being given as

$$h(t) = \frac{R(t_1) - R(t_2)}{(t_2 - t_1)R(t_1)} \tag{4.29}$$

with P\{a component of t with Δt if it has survived up to t\}

$$h(t) = \frac{f(t)}{R(t)} \tag{4.30}$$

or

$$R(t) = \frac{f(t)}{h(t)} = e^{-\lambda t} \tag{4.31}$$

A typical hazard graph is shown in Figure 4.8.

The general reliability function is given as

$$P(T \le t) = F(t) \tag{4.32}$$

where $t \ge 0$ and T is a random variable representing the failure time.

Here, the hazard function is called the bathtub curve, which illustrates the failure rate as a function of time. Period 1 represents the infant mortality period, which is the period of decreasing failure rate. This is called the back-in point, debugging period, early life period, or normal operating period

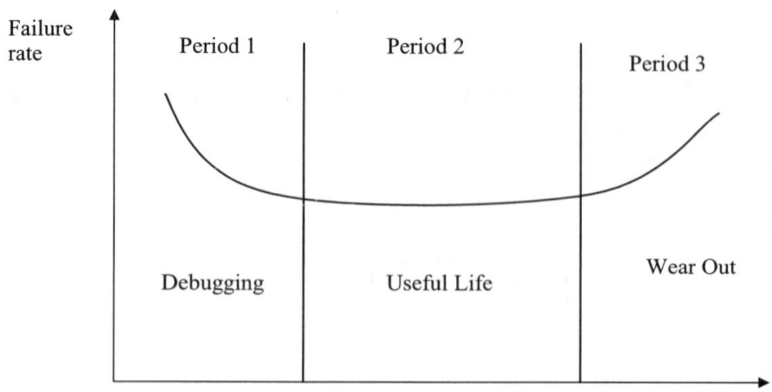

FIGURE 4.8
Typical hazard function graph.

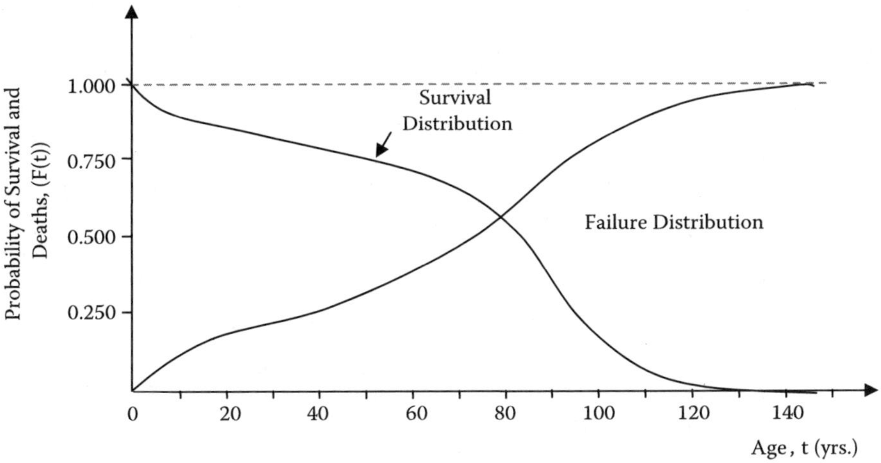

FIGURE 4.9
Probability of survival and deaths vs. age.

(failure may be caused by design). The second period is the useful life period or normal operating period. The failure rates of this period are constant, and failure rates are known as random failures/catastrophic failures. The third period is known as the wear-out period. The hazard rate increases as equipment deteriorates with age or as wear of components approaches the limit of useful life.

The probability of death at age t_1 and t_2 is the area under the curve in Figure 4.9.

$$F(t_2) - F(t_1) = \int_{\tau=t_1}^{\tau=t_2} f(\tau)d\tau \tag{4.33}$$

$$= \lim_{\Delta t \to 0} \left\{ \frac{F(t+\Delta t) - F(t)}{\Delta t} \right\} \tag{4.34}$$

Using approximate (data) points,

$$f(t) \cong \frac{n(t+\Delta t) - n(t)}{\Delta t N} \tag{4.35}$$

Consider individual survival at age t. The failure rate $r(t)$ is the probability of death per unit time at age t for an individual in the population.

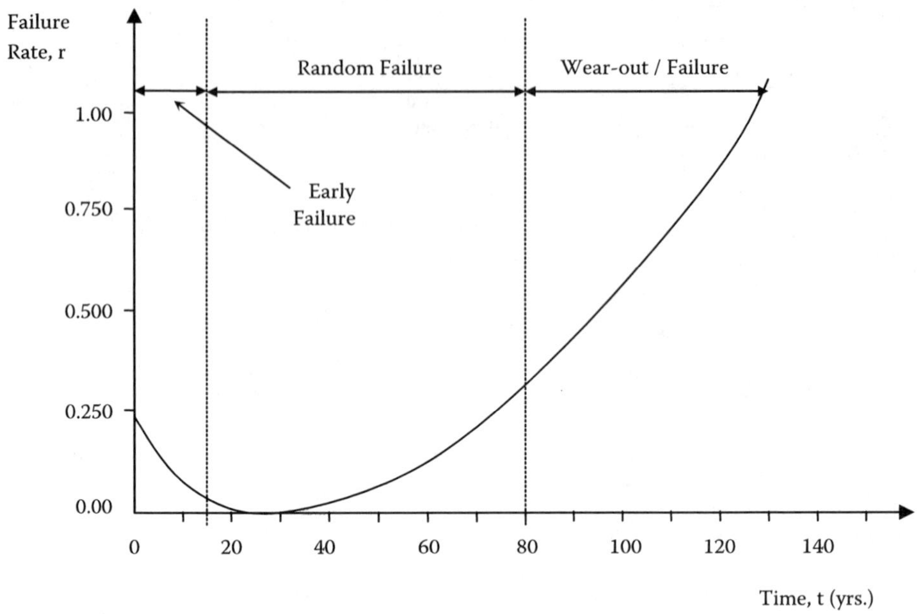

FIGURE 4.10
Bathtub curve.

$$\therefore r(t)\Delta t = \frac{f(t)}{R(t)}\Delta t \tag{4.36}$$

$$r(t) = \frac{f(t)}{R(t)} = \frac{f(t)}{L - F(t)} \tag{4.37}$$

The curve $r(t)$ is known as the bathtub curve, as shown in Figure 4.10.

4.14.2 Repair Failure: Repair Process

Definitions of key terms are summarized here. The key terms are in bold type.

$$\boldsymbol{R(t)} = \text{reliability at time } t$$

This is the probability that the component experiences no failure during the time interval $[0, t]$, given that the component was repaired at time zero. The curve $R(t)$ vs. t is a survival distribution given earlier. The following properties hold:

1. $\lim_{t \to 0} R(t) = 1$ shows that all components function near time zero.

2. $\lim_{t \to \infty} R(t) = 0$ shows that survival of components is very small.

$$\boldsymbol{F(t)} = \text{unreliability at time } t$$

This is the probability that the component experiences the first failure during the time interval [0, t] given that the component was repaired at time zero. $F(t)$ vs. t is called the failure distribution, and its monotonically increasing function of t leads to the following properties:

1. $\lim_{t \to 0} F(t) = 0$, few components fail just after birth (death)

2. $\lim_{t \to \infty} F(t) = 1$, asymptotic approval to complete failure

Because the component either remains normal or experiences its failure during the interval [0, t],

$$R(t) + F(t) = 1 \tag{4.38}$$

for $t_1 \leq t_2 \Rightarrow F(t_2) - F(t_1)$. From the probability that the component experiences its failure during the time interval [t_1, t_2], it follows that

$$f(t) = \text{failure density of } F(t)$$

and

$$f(t) = \frac{dF(t)}{dt} \tag{4.39}$$

yields

$$f(t)dt = F(t + dt) - F(t) \tag{4.40}$$

This is the probability that failure of the first component occurs during the small interval [t, $t + dt$], given that the component was repaired at time zero. Thus, the unreliability

$$F(t) = \int_{u=0}^{u=t} f(u)du \tag{4.41}$$

Similarly, $F(\infty) - F(t) = 1 - F(t)$ in the unreliability is the reliability

$$R(t) = \int_t^{\infty} f(u)du \tag{4.42}$$

$$r(t) = \text{failure rate}$$

This is the probability that the component experiences a failure per unit time at time t, given that the component was repaired at time zero and had survived to time t. The quantity $r(t)$ it is the probability that the component fails during $[t, t + \delta t]$, given that the component age is t. Note that failure rate is called hazard rate as well. A component with a constant failure rate $r(t)$ is considered as good as new if it is functioning.

TTF (time to failure): the span of time from repair to first failure. TTF is a random variable, since we cannot predict the exact time of its failure.

MTTF (mean time to failure): the expected value of TTF, which is obtained by

$$\text{MTTF} = \int_0^{\infty} tf(t)dt \tag{4.43}$$

where $f(t)dt$ is the probability that the TTF is approximately t, the average of all possible TTFs.

As $R(t)$ decreases to zero, i.e., as $R(\infty) = 0$, the MTTF of Equation 4.43 can be expressed as

$$\text{MTTF} = \int_0^{\infty} R(t)dt \tag{4.44}$$

Suppose a component has been normal to time u; then the mean regular time to failure is

$$\text{MRTTF} = \int_u^{\infty} \frac{(t-u)}{R(u)} f(t)dt$$

$$= \int_u^{\infty} \frac{(t-u)f(t)}{R(u)} dt \tag{4.45}$$

4.14.3 Failure-to-Repair Process

Consider a process with failure that ends at the completion of first repair. Assume that $t = 0$ is the time at which the component failed. The probability that the repair is completed before time t is based on the component having failed at $t = 0$. Then $G(t)$ = repair distribution at time t, and $G(t)$ has the same property as $F(t)$.

$$\lim_{t\to 0} G(t) = 0, \lim_{t\to\infty} G(t) = 1 \tag{4.46}$$

The repair density $g(t)$ of repair distribution $G(t)$ can be written as

$$g(t) = \frac{dG(t)}{dt} \tag{4.47}$$

or

$$g(t)dt = dG(t) = G(t + \Delta t) - G(t) \tag{4.48}$$

where we have

$$G(t) = \int_0^t g(u)du \tag{4.49}$$

$$G(t_2) - G(t_1) = \int_{t_1}^{t_2} g(u)du \tag{4.50}$$

where $G(t_2) - G(t_1)$ is the probability that the first repair is completed during $[t_1, t_2]$, given that the component failed at time zero.

Some other key definitions:

$m(t)$ = repair rate: probability that the component is repaired per unit time at time t, given that the component failed at time zero and has been failed to time t.

$m(t)dt \Rightarrow$: probability that component is repaired during $(t, t + d)$, given that the component's failure age is T.

Failure age t: means that the component failed at time zero and has been failed to time t; therefore, a component with constant repair rate has the same chance of being repaired whenever it is failed, and a nonrepairable component has a failure of zero. Therefore,

TTR = time to repair, the span of time from failure to repair completion; the time again is a random variable because the first repair occurred randomly

MTTR = mean time to repair; the expected value of the time to repair (TTR), defined as

$$\text{MTTR} = \int_0^{\infty} t\, g(t)dt \tag{4.51}$$

and if G(∞) = 1, then

$$\text{MTTR} = \int_0^{\infty} (1 - G(t))\, dt \tag{4.52}$$

4.14.4 Combined Reliability

$A(t)$ = combined process availability at time t

$A(t) \geq R(t)$ = availability larger than or equal to reliability $R(t)$

$A(t) = R(t)$ for nonrepairable component

$Q(t)$ = unavailability at time t

The probability that the system is unavailable is given as

$$Q = 1 - R(t) \tag{4.53}$$

$$= 1 - \int_0^{\infty} f(t)dt$$

$$= \int_0^{t} f(t)dt \tag{4.54}$$

where the component is in the failed state at time t, given that it was as good as new at time zero.

Unavailability is obtained from

$$A(t) + Q(t) = 1 \tag{4.55}$$

$$\therefore Q(t) \le F(t) \tag{4.56}$$

Therefore, for nonrepairable components

$$Q(t) = F(t) \tag{4.57}$$

Let $\lambda(t)$ represent the conditional failure rate intensity for number failure in $[0, t]$

$$\lambda(t) \ne r(t) \tag{4.58}$$

and for nonrepairable components

$$\lambda(t) = r(t) \tag{4.59}$$

Some other key definitions:

Expected life

$$E[T] = \int_0^\infty R(t)dt \tag{4.60}$$

$$= \int_0^\infty \left\{ \exp\left[1 - \int_0^t \lambda(t)\,dt \right] \right\} dt \tag{4.61}$$

$$= \int_0^\infty e^{-\lambda t} \mu = \frac{1}{\lambda} \tag{4.62}$$

Mean Time to Failure

$$\text{MTTF} = \bar{m} = \frac{1}{\lambda} \tag{4.63}$$

Mean Time between Failures

$$\text{MTBF} = \bar{T} = \bar{m} + \bar{r} \tag{4.64}$$

where $\overline{T}$ is mean cycle time, $\overline{m}$ is mean time to failure, and $\overline{r}$ is mean time to repair.

Mean Time to Repair

$$\text{MTTR} = \overline{r} = \frac{1}{\mu} \tag{4.65}$$

where μ is the mean repair rate; thus

$$\text{MTTF} = \overline{m} = \frac{\sum_{i=1}^{n} m_i}{n} \tag{4.66}$$

$$\text{MTTF} = \overline{r} = \frac{\sum_{i=1}^{n} r_i}{n} \tag{4.67}$$

$$\therefore \text{ MTBF} = \text{MTTF} + \text{MTTR} \tag{4.68}$$

Table 4.2, Table 4.3, and Table 4.4 summarize, respectively, equations for repair-to-failure process, failure-to-repair process, and combined processes.

4.15 Maintenance of Distribution Systems

Distributions systems are designed with the goal of providing quality, reliable, and efficient service at all times. To achieve this, maintenance plan actions must be taken to ensure that these criteria are met. The performance of the system must be maintained from life to death. Maintenance engineering analysis (MEA) is a systematic maintainability program developed to determine the effective useable condition of the equipment. There are two modes of maintenance: preventive and corrective.

4.15.1 Preventive Maintenance

Preventive maintenance is done on a scheduled basis for the purpose of retaining an item in a satisfactory condition. The process includes periodic test monitoring, servicing, and routine or scheduled inspection.

TABLE 4.2

Relations among Parameters for Repair-to-Failure Process

General Failure Rate, r(t)

$$R(t) + F(t) = 1$$

$$R(0) = 1, R(\infty) = 0$$

$$F(0) = 0, F(\infty) = 1$$

$$f(t) = \frac{dF(t)}{dt}$$

$$f(t)dt = F(t + dt) - F(t)$$

$$F(t) = \int_0^t f(u)du$$

$$R(t) = \int_t^\infty F(U)du$$

$$\text{MTTF} = \int_0^\infty tf(t)dt = \int_0^\infty R(t)dt$$

$$r(t) = \frac{f(t)}{1 - F(t)} = \frac{f(t)}{R(t)}$$

$$R(t) = \exp\left[-\int_0^t r(u)du\right]$$

$$F(t) = 1 - \exp\left[-\int_0^t r(u)du\right]$$

$$f(t) = r(t)\exp\left[-\int_0^t r(u)du\right]$$

Constant Failure Rate, r(t) = λ

$$R(t) = e^{-\lambda t}$$

$$F(t) = 1 - e^{-\lambda t}$$

$$f(t) = \lambda e^{-\lambda t}$$

$$\text{MTTF} = \frac{1}{\lambda}$$

4.15.2 Corrective Maintenance

Corrective maintenance is based on restoring equipment to an operable condition after failure or some other malfunction has occurred. Here, we briefly define a few of the terms used in maintenance work:

Routine: maintenance carried out in accordance with a predetermined policy or plan to prevent breakdown or reduce the likelihood of an item of the plant failing to meet an acceptable condition; also includes operational checks and diagnostic testing for acceptable positions

Prevention: planned maintenance carried out as a result of an inspection or report, but not the result of a breakdown

TABLE 4.3

Relations among Parameters for Failure-to-Repair Process

General Repair Rate, m(t)	
$G(0) = 0, G(\infty) = 1$	$\text{MTTR} = \int_0^\infty tg(t)dt = \int_0^\infty [1 - G(t)]dt$
$g(t) = \dfrac{dG(t)}{dt}$	
$g(t)dt = G(t + dt) - G(t)$	$m(t) = \dfrac{g(t)}{1 - G(t)}$
$G(t) = \int_0^t g(u)du$	$G(t) = 1 - \exp\left[-\int_0^t m(u)du\right]$
$G(t_2) - G(t_1) = \int_{t1}^{t2} g(u)du$	$g(t) = m(t)\exp\left[-\int_0^t m(u)du\right]$
Constant Repair Rate, m(t) = μ	
$G(t) = 1 - e^{-\mu t}$	$\text{MTTR} = \dfrac{1}{\mu}$
$g(t) = \mu e^{-\mu t}$	$\mu = 0$ (nonrepairable)

Breakdown: condition requiring repair or corrective maintenance to restore the system to an acceptable status

Postfault management: inspection and diagnostic tests to establish whether equipment is in acceptable condition and, if needed, corrective action to restore service

Overhaul: a minor overhauls is limited to lubrication and replacement of consumable points; a major overhaul involves major dismantling and replacement of items

Visual check: eyeball check to detect anything that might cause an item to fail due to an unacceptable position

Inspection check: careful scrutiny of an item carried out without dismantling and using all senses to detect the cause of an item's failure to operate

Monitor: inspection with partial dismantling of parts, measurement, and nondestructive tests for unsatisfactory performance of an item

Maintenance engineering analysis supports the design of equipment from both the planning and operation stages. It provides concepts for each

TABLE 4.4

Relations among Parameters for Combined Processes

Fundamental Relations	
Repairable	**Nonrepairable**
$A(t)+Q(t)=1$	$A(t)+Q(t)=1$
$A(t)>R(t)$	$A(t)=R(t)$
$Q(t)<F(t)$	$Q(t)=F(t)$
$w(t)=f(t)+\int_0^t f(t-u)v(u)du$	$w(t)=f(t)$
$v(t)=\int_0^t g(t-u)w(u)du$	$v(t)=0$
$W(t,t+dt)=w(t)dt$	$W(t,t+dt)=w(t)dt$
$V(t,t+dt)=v(t)dt$	$V(t,t+dt)=0$
$W(t_1,t_2)=\int_{t_1}^{t_2} w(u)du$	$W(t_1,t_2)=F(t_2)-F(t_1)$
$V(t_1,t_2)=\int_{t_1}^{t_2} v(u)du$	$V(t_1,t_2)=0$
$Q(t)=W(0,t)-V(0,t)$	$Q(t)=W(0,t)=V(0,t)$
$\lambda(t)=\dfrac{w(t)}{1-Q(t)}$	$\lambda(t)=\dfrac{w(t)}{1-Q(t)}$
$\mu(t)=v(t)/Q(t)$	$\mu(t)=0$

subsystem and component, facilitates determination of resource requirements, and results in system design requirements for reliability and maintainability using MEA, which must attend to the following:

- Personnel requirements for the project
- Skill level and training requirements
- Maintenance function and technical data requirements, including maintenance manuals, records, and other technical support tools

- Maintenance facility requirements, such as waste power, clean room, repair shops, spare parts, and others

To carry out the maintenance effectively, we have derived the following criteria:

1. Identification of equipment components that will affect the functionality of the plant using equipment-impact analysis on plant operations, safety, and the replaceability of interface equipment given the possible configurations for system operation
2. Evaluation of equipment in an effort to select equipment requiring maintenance based on historical analysis of future availability, present-condition failure analysis, and failure-use analysis, which concentrate on electrical-life-continuation analysis and any engineering modifications needed
3. Prediction of remaining equipment life using visual inspection, generic data, and specific data used for failure diagnosis

4.16 Reliability-Centered Maintenance

The largest business cost for the electric utility industry is the budget for the operation and maintenance of distribution and transmission systems. There is pressure to control costs and balance the trade-offs between the following:

1. Cost and impact of equipment failure and safety
2. Cost of achieving power quality for a given maintenance investment
3. Cost of extending equipment life and reliability

We discuss each of the trade-offs here:

1. *Safety*: Reliability is centered around a maintenance program that must ensure that applicable safety codes and regulatory policies on safety are adhered to. For example, an effective maintenance program must monitor vegetation, sagging of wires, and equipment age while providing early detection of faults.
2. *Reliability*: In a reliability maintenance program, reliability assessment is a priority. The procedure must be used to identify wear and tear, degradation, and incipient failure.

To achieve reliability, different analysis tools are employed to identify specific problem areas for follow-up and possible maintenance. Some of the preventive/corrective actions can include management of vegetation and soil characteristics and identification of critical components for repair or replacement.

Documentation of the results of a detailed inspection of the system state/outage history, located in a database, is recommended. It is used to generate a list of failure modes and benchmarks against future reference in a maintenance schedule.

A decision matrix scheme is developed to compare the system attributes and weightings for comparison. This must include such parameters such as age, length of segment, number of customers directly/indirectly served by the circuit, total instantaneous outages to date, total extended outages to date, line voltage, and the type of connection.

Each of these attributes is weighted according to relative importance as determined by engineering judgment. Heavy emphasis is placed on customers served and outage history. Secondary attributes are used to assess distribution lines, including structure standards, wire sagging, microenvironment characteristics, and others.

In summary, reliability implementation must be based on adequate discrimination of inspection, interpretation of data, and a decision matrix to help in ranking the priority of feeders or lines for assessment procedures.

4.17 Security and Reliability-Centered Maintenance

Distribution and transmission system security depends on the ability of the components of the distribution/transmission system to constrain or prevent a progressive and catastrophic collapse of the system after a failure of one or more of the weakest components during a fault.

Assessment of a distribution system reveals how an upgrade or addition of key components can present cascading or failure of the distribution system infrastructure. A cost-benefit analysis scheme that balances reliability and cost of maintenance is used to develop key long-range maintenance plans. The information from the security assessment of the distribution system helps in planning future upgrades, improvements, and investments.

Security assessment steps for the distribution system include the following:

1. Listing of structures and components in need of an upgrade, determination of the cost implications and public perceptions upon a loss of system availability, and determination of maintenance costs
2. Ranking the assessment results in terms of cost-benefit ratio to defer projects with lower probability of failure until funds permit further system improvements

4.18 Implementation Plan for Various Component-Maintenance Techniques

The proposed maintenance-management system will systematically replace existing methods or reliance on outside contractors. The existing methodology for implementing various maintenance standards includes:

1. Formulating and establishing criteria, guidelines, and methodologies to determine maintenance standards and frequencies
2. Producing detailed work specifications for all major plants
3. Developing and installing a practical and effective reporting system for the maintenance system that provides information on the operation and identifies problems
4. Determining existing and future requirements for spare parts to achieve an effective maintenance philosophy
5. Continuously monitoring the progress of the maintenance program and regularly updating procedures to ensure that specifications are met

Some recommended maintenance programs for different system components are summarized here and can be updated according to utility policies and practices.

4.18.1 Overhead Lines

Maintenance of overhead lines typically consists of inspection and testing, followed by implementation of recommended actions. Patrol inspections at ground level by foot or vehicle and a thorough monitoring scheme are recommended. Maintenance guidelines should be periodic and should be followed according to established procedures for quarterly, yearly, and seasonal periods.

4.18.2 Circuit Breakers

A recommended diagnostic test in all types of circuit breakers includes a timing test using a reliable apparatus. For a closed operation, an open operation, a closed/open operation, and the time from initiation to operation of the contacts recorded, all interruption should be timed at once to enable comparison tests and to detect deterioration in contacts or connections using voltage-drop resistance measurements.

Checks should be carried out on the circuit breaker operating mechanism to determine running hours of operating the motor, automatic start-up pressure, low pressure alarm, and lockout pressure. Additional diagnostic test are recommended, such as checks on gas leakage, rate of SF_6 gas-based circuit breaker, air insulation, temperature, dielectric strength of the insulating oil, etc. All should be checked according to company policy and practices.

4.18.3 Transformers

The principal objective of transformer maintenance is to maintain the insulation in good condition. Diagnostic testing is needed to obtain an indication of the equipment's condition as well as all of the items associated with the main function. The various parts of the transformer requiring maintenance include the main transformer, cooling equipment, tap-changer bushings, protective devices, control gear, reactors, earthing transformer, neutral earthing resistors, lightning arrestors, oil retaining compound, etc.:

- Maintenance scheduled on a monthly or yearly basis should be carried out at the specified time intervals.
- Diagnostic testing of transformer oil and a follow-up analysis should be performed to identify potential problems over time.
- Analysis of gases collected from a typical Buchholz relay will help in determining whether there is an internal fault and in diagnosing the location of the fault.
- Resistance values of the transformer windings, together with measurements of insulation resistance, will give an indication of their electrical condition. Ratio checks through the complete range of tap positions will prove consistency and the sequential stepping of the tap changer.
- Measurement of oil and winding temperatures provides information about transformer condition. Check the calibration of the oil-temperature instrumentation to verify correct operation of pumps, false alarms, and tripping of the tap changer. Check the effectiveness of limit switches and mechanical override defenses. Regular monitoring of the tap-changer operation for all automatic tap-changer installations should be carried out.

4.18.4 Substation Equipment

Regular and periodic diagnostic checks or power maintenance actions are recommended for current transformers (CT) and voltage transformers (PT). Here we carry out procedures to check oil levels, oil seals, dielectric "loss angle" value of insulators, bus bars, fittings, and connections. Here checks should be carried out to determine resistance values across the contacting surfaces and the temperature of joints under loading conditions. Other pieces of auxiliary equipment are also diagnosed, and the maintenance procedure is applied accordingly.

The costs associated with this maintenance are computed using cost-benefit analysis to justify the investment and value of reliability-centered maintenance. Case studies using cost-benefit analysis to support maintenance and system upgrades are discussed in the subsequent section.

4.19 Illustrative Examples

4.19.1 Example 1

Consider a series-connected system, consisting of up to n components connected in series, that yields the results shown in Figure 4.11. For a radial distribution system, the failure rate is given as

$$\lambda = \sum_{i=1}^{N} \lambda_i \tag{4.69}$$

where λ_i is the failure rate of the ith component, and λ is the failure rate of the entire system, assuming there are no multiple failures.

Similarly, the repair rate of the system is given as

$$\mu = \sum_{i=1}^{N} \mu_i \tag{4.70}$$

which is the steady-state probability of power being unavailable because of failure of any component, given from $\frac{\lambda}{\mu + \lambda_i}$. Now, let the unavailability of power be $\sum_{i=1}^{N} \frac{\lambda_i}{\mu_i}$ with a repair rate of $\mu = \frac{\lambda_i}{\sum_{i=1}^{N} \frac{\lambda_i}{\mu_i}}$.

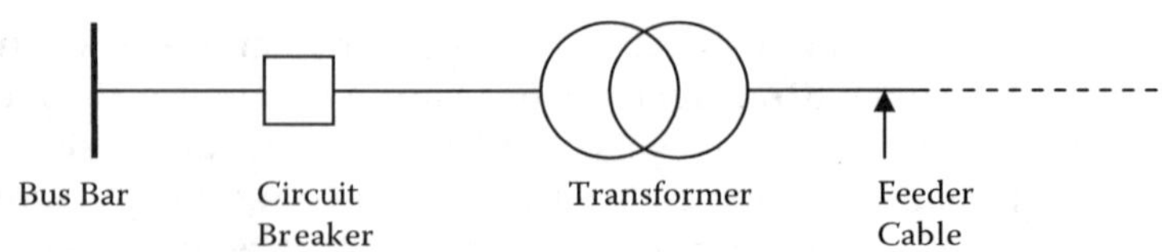

FIGURE 4.11
Series-connected system for Example 1.

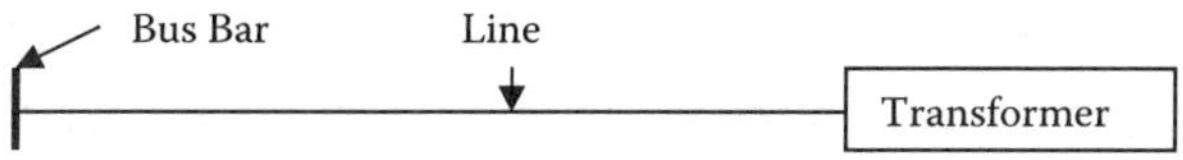

FIGURE 4.12
Sample system for Example 1.

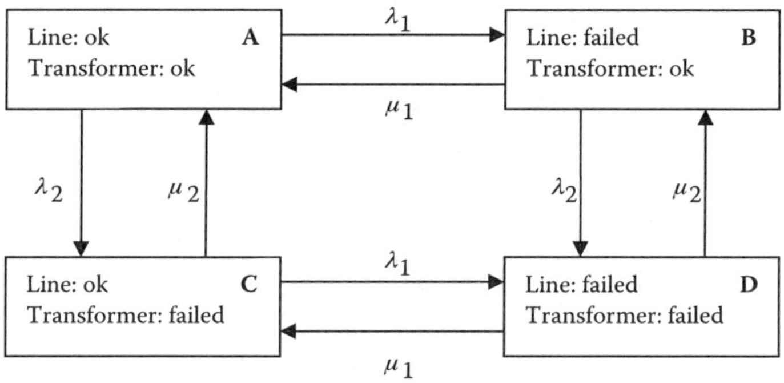

FIGURE 4.13
Markov diagram for Example 1.

Multiple-state transitions are possible, such as a four-state model of a line and transformer. The failure and repair of a line and transformer are shown as a possible combination of failed and separating states in Figure 4.12, which is also represented in the Markov diagram in Figure 4.13.

The resulting transitions from one state to another are put in matrix form using the well-known Markus process model to yield

$$= \begin{bmatrix} -(\lambda_1+\lambda_2) & \mu_1 & \mu_2 & 0 \\ \lambda_1 & -(\mu_1+\lambda_2) & o & \mu_2 \\ \lambda_2 & 0 & -(\lambda_1+\mu_2) & \mu_1 \\ 0 & \lambda_2 & \lambda_1 & -(\mu_1+\mu_2) \end{bmatrix} \cdot \begin{bmatrix} N_A \\ N_B \\ N_C \\ N_D \end{bmatrix} - \begin{bmatrix} \dot{N}_A \\ \dot{N}_B \\ \dot{N}_C \\ \dot{N}_D \end{bmatrix} \tag{4.71}$$

The matrix equation is written in compact form $\dot{N} = P_{1xm}N_{mx1}$, where N is the set of probabilities of each state, and P is the matrix of probabilities P_{ij} that define transition from state i to j and not the previous history, and constant failure and repair states are assumed. This is of course the limitation of the Markov process for reliability evaluation work, where restoration of functionality is more important than restoration of the prior state. Thus, from the foregoing, the Markov process is complex for solving huge practical problems. Assumptions are generally made by using appropriate equivalent components and solving states that are of no concern. However, the state space analytical method is considered to be accurate as an explicit model of states that can be used for failure and repair state transition in a distribution system. At best, state space is preferred for reliability analysis of small systems.

4.19.2 Example 2

If experimental distribution (constant risk) is assumed for the failure and repair of components and for success in switching actions, the time that lapses before a component changes state can be defined using the following simple equations using random numbers (taking a value between 0 to 1) for the computations:

$$\text{Time of failure} = -\frac{1}{\lambda}\ln\big(F(t)\big) \tag{4.72}$$

$$\text{Time of switch} = -\frac{1}{\mu_{s\omega}}\ln\big(Q(t)\big) \tag{4.73}$$

$$\text{Time of repair} = -\frac{1}{\mu_r}\ln\big(G(t)\big) \tag{4.74}$$

where $F(t)$, $Q(t)$, and $G(t)$ are random numbers between 0 and 1. Their equations can be derived. Assuming a constant rate λ, the cumulative probability density function describing the probability that the component has failed by time x is

$$u = 1 - e^{-\lambda x} \tag{4.75}$$

treating u as a random variable with uniform distribution [0, 1].

The time to failure is

$$\text{TF} = -\frac{1}{\lambda}\ln(1 - u) \tag{4.76}$$

If u is random, so is $(1 - u)$

$$\text{TF} = -\frac{1}{\lambda}\ln(u) \tag{4.77}$$

Random numbers are generated mathematically, physically, or as a pseudo-random number. The critical properties these numbers should possess is uniform distribution [0,1], independence, and a long repeat period. A useful feature of MCS is that, while the error band or confidence range decreases as the number of iterations or simulations increases, the number of iterations required for accuracy is independent of system size.

TABLE 4.5
TTR for Repair of Electric Motors

Repair No.	Time (h)	Repair No.	Time (h)
1	3.3	10	0.8
2	1.4	11	0.7
3	0.8	12	0.6
4	0.9	13	1.8
5	0.8	14	1.3
6	1.6	15	0.8
7	0.7	16	4.2
8	1.2	17	1.1
9	1.1		

4.19.3 Example 3

The repair times, TTR, for an electric motor used to run a cooling system has been logged in every time a repair is performed. Table 4.5 shows a sample of repairs performed and recorded by the maintenance personnel.

Using these data, we obtain the values of

Repair probability at time t, $G(t)$

Repair density of $G(t)$, $g(t)$

Repair rate $m(t)$

Mean time to repair, MTTR

Total system out-of-service cost for the entire period (assuming a system unavailability cost of $15 per hour)

Solution

$N = 17$ = total number of repairs

t(TTR)	Number of Completed Repairs, $m(t)$	$G(t)=\dfrac{m(t)}{N}$	$g(t)=\dfrac{G(t+\Delta)-G(t)}{\Delta}$	$m(t)=\dfrac{g(t)}{1-G(t)}$
0.0	0	0	0	0
0.5	0	0	0.9412	0.9412
1.0	8	0.4706	0.58822	1.110
1.5	13	0.7647	0.2354	1.0004
2.0	15	0.8824	0	0
2.5	15	0.8824	0	0
3.0	15	0.8824	0.1176	1
3.5	16	0.9412	0	0
4.0	16	0.9412	0.1176	2.0000
4.5	17	1	…	…

$$\text{MTTR} = (0.25\times 0 + 0.75\times 0.9412 + \ldots + 4.25\times 0.1176)\times 0.5$$

$$= 1.3676$$

The total out-of-service cost is

$$\text{Total cost} = \text{N} \times \text{MTTR} \times \text{hourly cost}$$

$$= 17 \times 1.3676 \times 15 = \$348.738$$

4.19.4 Example 4

In a system there is a critical unit that requires spares to maintain a specified unit reliability of 99% over a period of 250 h. The unit has an MTBF (mean time between failures) of 1250 h and exhibits a constant failure-rate characteristic. How many spares would be required to achieve this 99% reliability if the faulty part is easily accessible and can be replaced almost immediately by inserting an identical spare when the functioning unit fails?

Solution

The solution can be found by using the Poisson distribution and answering the equivalent question, "How many failures, equal to the number of spares, can be tolerated to attain a 99% reliability?"

$$F(k) = \sum_{j=0}^{k} \frac{e^{-\lambda t}(\lambda t)^j}{j!}$$

$$= e^{-\lambda t}\left[1 + \frac{\lambda t}{1!} + \frac{(\lambda t)^2}{2!} + \frac{(\lambda t)^3}{3!} + \ldots + \frac{(\lambda t)^k}{k!}\right]$$

where $\lambda t = \dfrac{1}{\text{MTBF}} \times t = \dfrac{1}{1250} \times 250 = 0.2$

The problem is now restated as: determine the value of k such that

$$0.99 = e^{-0.2}\left[1 + \frac{0.2}{1!} + \frac{(0.2)^2}{2!} + \frac{(0.2)^3}{3!} + \ldots + \frac{(0.2)^k}{k!}\right]$$

With one spare, the reliability is $F(1) = 0.98248$, and with two spares, the reliability is $F(2) = 0.99885$; hence $k = 2$ is the correct number.

4.20 Summary

Different reliability indices were computed in this chapter using frequencies of fault, duration, and time to capture different failure rates of each relevant index based on data collection. Furthermore, based on concepts of maintenance, the implications on reliability are summarized. Illustrative examples and software computational tools available for reliability analysis and design are defined.

Problem Set 4

Table 4.6 presents outage data for a two-feeder system. Feeder 1 has a total of 1000 customers along with a load of 2000 kVA, and Feeder 2 has 1900 customers with a load of 3800 kVA.

4.1 Using the data in Table 4.6, calculate the SAIFI index.

4.2 Using the data in Table 4.6, calculate the SAIDI index.

4.3 Using the data in Table 4.6, calculate the CAIFI index.

TABLE 4.6
Historical Outage Data for a Two-Feeder System

Date	Feeder	No. Customers Affected	Load (kVA)	Interruption Type
3/23/2006	F1	1000	2000	momentary
4/15/2006	F1	550	1100	momentary
5/5/2006	F1	400	800	sustained
6/12/2006	F2	400	800	sustained
7/6/2006	F2	1900	3800	momentary
8/20/2006	F1	450	900	sustained
8/31/2006	F2	900	1800	sustained
9/3/2006	F2	950	1900	sustained
10/2/2006	F2	1850	3700	sustained
10/31/2006	F2	900	2600	sustained
11/23/2006	F1	550	1100	sustained
12/13/2006	F2	1850	3700	momentary

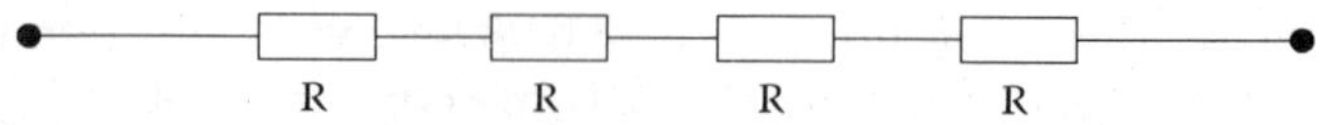

FIGURE 4.14(A)
Block diagram of subsystem 1 for Problem 4.6.

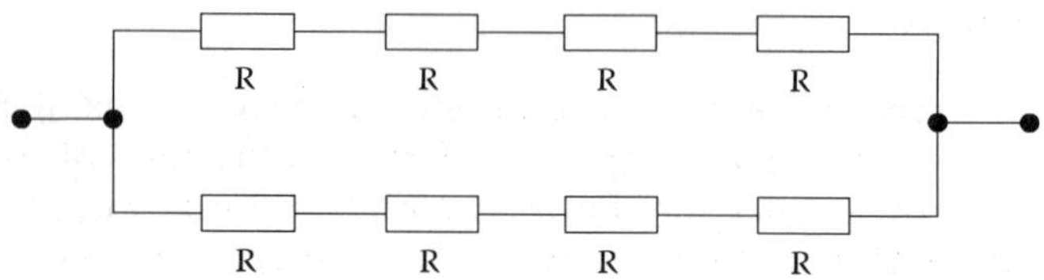

FIGURE 4.14(B)
Block diagram of subsystem 2 for Problem 4.6.

4.4 Using the data in Table 4.6, calculate the CAIDI index.

4.5 Using the data in Table 4.6, calculate the ASAI index.

4.6 For each of the reliability block diagrams shown in Figure 4.14(a) and Figure 4.14(b), given that they are based on the logic diagrams of each subsystem and that the reliability of each component is 0.9, calculate the reliability of the equivalent system.

4.7 State the definition and mathematical formulation used for computing the SAIFI, SAIDI, CAIFI, and ASUI distribution system reliability indices. Under what conditions are they used.

4.8 In a system there exists a very critical unit which requires spares to maintain a specified unit reliability of 99% over a period of 275 hours. The unit has a Mean Time Between Failure (MTBF) of 1,850 hours and exhibits a constant failure rate characteristic. How many spares would be required to achieve this 99% reliability if the faulty part is easily accessible and can be replaced almost immediately by inserting an identical spare when the functioning unit fails?

4.9 Consider that 2000 items are being tested for 500 hours. Early observations indicate that failures are occurring at a constant per-unit failure rate of $l = 2 \times 10^{-3} hr^{-1}$.

a. How many objects will survive the 500 hours?
b. What is the Mean Time To Failure (MTTF) for these items?

4.10 A transformer has a constant failure rate of $\lambda = 10^{-5}$ failure/hr.

a. What is its reliability for an operating period of 120 hrs?

b. If there are 100 such transformers in a given area, how many will fail in one hour?

c. What is the reliability for an operating time equal to MTTF?

d. What is the probability of service for an additional 120 hrs, given that a device has served for 120 hrs?

e. Design a maintenance table, for typical sub-station equipment (e.g. transformer, etc), for reporting failure based on possible reliability indices of choice.

5

Distribution Automation and Control Functions

5.1 Introduction

Electric power distribution is important in the delivery of energy to consumers from an electrical power system. The idea of distribution automation was motivated by the evolution of communication and information technology. Automated distribution uses these technologies to improve the operating performance of distributed systems, enhancing efficiency, reliability, and quality of service, as well as better management and control of the power distribution system. The principal objective can be summarized as energy conservation through reduction of losses, peak load, and energy consumption. IEEE defines a distribution automation system as one that enables an electric utility to remotely monitor, coordinate, and operate distribution components in a real-time mode from remote locations.

Distribution automation functions (DAFs) cover the following areas:

- Demand-side management
- Voltage regulation/VAr control
- Real-time pricing
- Dispersed generation and storage dispatch
- Fault diagnosis/location
- Power quality
- Reconfiguration
- Restoration

5.2 Demand-Side Management

Demand-side management (DSM) options provide an effective means of modifying consumer demand to cut operating expenses from costly generators while deferring capacity addition. DSM options also promote environmental conservation (reduced emission in fuel production) while sustaining industrialization at minimum cost and contributing to the reliability of generation systems. Demand-side management options have been categorized into:

- Peak shifting
- Valley filling
- Peak clipping
- Storage conservation

These options have an overall impact on the utility load curve. For DAFs, demand-side management is classified into three main categories:

1. *Direct control of load*: This uses a communication system such as power line carrier/radio to transmit control from the utility side to the customers. The aim is to directly control load, small generators, and storage.
2. *Local load control option*: This enables customers to self-adjust loads to limit peak demand, e.g., demand-activated breakers, load interlocks, timers, thermostats, occupancy sensors, cogeneration heating, cooling storage, etc.
3. *Distribution load control*: The utility controls the customer loads by sending real-time prices.

The cost benefits of direct-control options are numerous:

- Reduced peak load/capital investment
- Integrated least-cost planning
- Emergency control — system contingencies/overload
- Automatic control
- Voltage collapse
- Long-term stability
- Operating (spinning) reserve
- Distribution dispatch (normal conditions)
- Reduced loading on facilities
- Cold load pick-up

Some of the anticipated constraints of demand-side management options are:

1. *Technological constraints*: These depend on the use of advanced communication technologies for remote metering, billing and local controls, and power management.
2. *Economic constraints*: These involve determination of direct financial benefit, giving other investment commitments in network expansion and assets and, hence, providing an incentive for utilities to diversify the scope of their business operations.
3. *Social constraints*: The motivation is that demand-side management provides energy efficiency and meets environmental objectives. Demand-side management options are constrained by the utility's sincerity in cutting energy costs.
4. *Political and institutional constraints*: Demand-side management depends largely on the government, equipment manufacturing commitment, and institutional support.

5.2.1 Modeling Challenges and Methodology for Demand-Side Management

In the literature on demand-side management modeling, several researchers have discussed the concept of including customers as part of the planning options for new utilities. The current models for demand-side management contend that MW demand is no longer a fixed parameter; according to these models, MW demand is reduced at a certain cost, depending on demand-side management. The total system cost, including demand-side management cost, is minimized to obtain an optimal mix of supply-side generation and demand-side load reduction. Analysis of demand-side management is done using several techniques such as daily load curves or mathematical programming methods. Demand-side management has been carried out using the context of unit commitment studies, optimal power-flow studies, load-reduction forecasting methods, engineering features of the end-user equipment, interruptible load-management program, survey methods (data collection), and dynamic programming approach to optimize energy procurement and load management by utilities.

Some challenges in modeling DSM for distribution systems are as follows:

1. With the load-management options currently used, there is a reduction in reactive power demand along with the real power components. This can be overcome by using a full AC network representation for the modeling.
2. The accuracy of the demand-side management model depends on its ability to capture "time of use" aspects of AC cogenerators and

other utilities. This imposes a high computational burden due to the large number of controls at each demand node being monitored for demand-side management.

3. The computational burden of conventional optimal power flow and unit commitment reduces the chance for a high-efficiency computational algorithm unless further improvements are made.
4. The use of artificial neural networks (ANN), expert systems, and heuristics schemes, such as evolutionary programming in the area of unit commitment and optimal power flow (OPF), can be extended to demand-side management options in a distribution system. They include:
   - Dynamic OPF with network and dynamic constraints
   - Nonlinear programming for residential air conditioning load
   - Traditional optimization methods

5.2.2 Conceptual Overview of Methodology for DSM Studies

Demand-side management is carried out using the following four basic steps:

Step 1: identification of demand-side management and its characteristics as inputs to the automation process

Step 2: acquisition of a large number of surveys on life-cycle cost analysis, which includes the cost of saved energy as well as the cost of power (MW)

Step 3: identification of characteristics of demand-side management in terms of time-specific and technical control

Step 4: identification of utility requirements of fixed end users with ±8% margin on kW·h, ±3% °F temp, ±4 min on recorded time, days of storage, and installation cost for monitoring system

5.3 Voltage/VAr Control

Voltage control within a specified range of limits and capacitor switching are an effective means of minimizing loss and improving voltage profiles and deferred construction and maintenance costs in the end within the reliability and power-quality constraints of the system. Voltage/VAr control considers multiphase unbalanced distribution system operation, dispersed generation, and control equipment in the large system. In distribution automation, functions using voltage/VAr control options must maintain proper communication between planning problems like the decision to install capacitors, and recognizing the cost benefit analysis.

5.3.1 Methods of Voltage/VAr in Distribution Automation

Several methods have been used for voltage/VAr control over the years, including:

- Decoupled formulation with a linear voltage-regulator problem and capacitor-switching problem can be solved interactively using a linear and nonlinear optimization
- OPF-based reactive-power dispatching aimed at minimizing power losses by optimal placement of capacitors
- A proposed linear-programming-based method that minimizes losses by changing transformer tap setting and VAr injection
- A mixed-integer programming model that solves the problem of voltage/VAr control using the principle of recursive linear programming
- Several other techniques that use linear power flow to break the overall problem into a master-capacitor switching problem and slave-capacitor operation problem given the switching schedule (based on the Dantzig-Wolfe decomposition principle for a multi-area reactive-power-planning problem)
- A three-phase power-flow program that accounts for the distributed loads by approximating their effects on nodal voltage via equivalent lumped nodes; this method is also capable of accounting for dispersed generation, tap controls, and shunt capacitor switching
- A contingency-secured approach for voltage-profile improvements to bridge the gap between VAr planning and VAr dispatching
- A rule-based system combined with standard linear programming to solve the voltage/VAr control problem
- Other intelligent systems such as fuzzy logic, genetic algorithm, and their hybrids for voltage/VAr control that have been developed and tested successfully to keep the computational burden within possible limits

5.3.2 Evaluation of Methods Used for Voltage/VAr Control

1. Voltage/VAr control using optimization methods and intelligent systems and their hybrids have been applied to this complex problem.
2. Optimization methods include linear programming, nonlinear programming, and mixed-integer programming using the principles of

decomposition for the full AC OPF. Heuristic approaches have been used to get around the discrete variable used for selecting switching capacitors.

3. Rule-based evolutionary programming in combination with OPF based on efficient algorithms, such as interior-point OPF, is an option for designing a future efficient voltage/VAr control problem.

5.3.3 Modeling of Voltage/VAr Control Options

The VAr control problem is modeled as a large-scale, complex, nonlinear combinational problem. The decision on capacitor switching has to be modeled as a discrete variable. The sequence of switching in size and site has to be properly modeled for a given optimization method.

5.3.4 Formulation of Voltage/VAr

This includes the integrated voltage/VAr with the load-management problem to improve efficiency given the following four objectives:

1. Customer-outage cost

$$\text{Minimum outage cost} = \sum_{i} \left(X_{ik} \times PC_{ki} \times CC_k \right) \tag{5.1}$$

where

X_{ik} = level of curtailable load selection of type k at bus i (p.u. MW)

PC_{ki} = maximum curtailable MW of type k at the ith bus (p.u. MW)

CC_k = curtailment cost of customer type k ($/p.u. MW)

2. Loss minimization

The objective of the loss minimization function is given by

$$\text{Min } I^2 r = \sum_{ij} r_{ij} \frac{P_{ij}^2 + Q_{ij}^2}{|V_i|^2} \tag{5.2}$$

3. Load balancing

$$\text{Max} \frac{S_i}{S_i^{\max}} \tag{5.3}$$

where
P_{ij}, Q_{ij} = transfer power (branch *i-j* at p.u.)
V_i = voltage of bus *i* (p.u.)

4. Multiple objective function

$$Z = \text{Min}[a + b - c] \tag{5.4}$$

5.3.5 System Operating Constraints

These include

Branch flow equations:

$$Y_{i+1} = f_{i+1}(Y_i) \tag{5.5}$$

where

$$Y_i = \left[P_D, Q_D, \left|V\right|^2, X_k, Q_s, \delta_i \right] \tag{5.6}$$

Branch flow takes into account the recursive relationships between the successive nodes in the radial distribution system. The demand in P_D, Q_D also have an interruptible component, which is the load-management control options X. In addition, the reactive power equation has the capacity switching option Q_s.

Voltage limits/current limits:
The voltage and current limits are given as

$$\left|V_i^{\min}\right| \le \left|V_i\right| \le \left|V_i^{\max}\right| \tag{5.7}$$

$$\left|I_{ij}^{\min}\right| \le \left|I_{ij}\right| \le \left|I_{ij}^{\max}\right| \tag{5.8}$$

Capacitor control limits:

$$Qs_i^{\min} < Qs_i \le Qs_i^{\max} \tag{5.9}$$

Curtailable load-control limits:

$$P_i \times X \le Pc_{ki} \tag{5.10}$$

$$Q_i \times X_{ki} \le Qc_{ki} \tag{5.11}$$

5.3.6 Methodology

The mathematical optimization problem in Equations 5.5 to 5.11 falls into the class of nonlinear mixed-integer programming problems. We use an easier method to overcome the computational challenges. The hybrid artificial intelligence optimization scheme gets around the problem's computational challenges. They are used for:

- Off-line plan: use the feasible MIP (mixed-integer programming method) from Equations 5.5 to 5.11
- On-line execution

Use expert systems in real time to handle the optimization of:

- Discrete variables (selections of capacitor switching, load-management options)
- Operation development

The online artificial intelligence (AI) methodology uses the following steps:

1. Develop knowledge base from the off-line mode using the optimization model.
2. Perform real-time data acquisition on load, network, and topology (in real-time model).
3. Access the knowledge base to detect dispatch functions, load-management options, and capacitor switching under specific load conditions (in real time).
4. Invoke the rule base to ensure that the load management and capacitor switching are within limits.
5. Reform load flow to check violations of network constraints.

5.4 Fault Detection (Distribution Automation Function)

Fault detection and classification are of significance to both distribution engineers and consumers. This is due to the diversity of faults and their locations and to the limitations of the simulation program to generate fault data. The overall running cost is high if avoidance of supply interruption is not done in a timely manner.

Conventional fault studies are concerned with "what if" scenarios, i.e., on considering what happens after a fault occurs, identifying the location of the fault, and assessing the nature of the damage caused by the fault. In contrast, if potential faults could be identified by an early warning system before a catastrophic fault actually occurs, the chance of an interruption of service would be reduced. The decision-analysis functions use the relevant information from the detection technique to enable appropriate control actions. We summarize the commonly used methods.

5.4.1 Classical Approaches Used for Solving Detection Techniques

5.4.1.1 Harmonic Sequence Component Technique

This uses the third and fifth harmonics of a fault current after frequency decomposition of three-phase unbalanced faults. One can detect and classify high-impedance faults by measuring the degree of unbalance and comparing it with a threshold.

5.4.1.2 Amplitude Ratio Technique

The harmonic currents are very small in a system under normal conditions. When an arcing fault/high-impedance fault occurs, the harmonic currents increase. The amplitude ratio technique is used to compare the second harmonic to the fundamental current or compare the ratio between even and odd harmonic currents for the first seven harmonic ranges.

5.4.1.3 Phase Relationship Technique

The presence of a notch on the leading edge of each half-cycle of a high-impedance current waveform indicates that they must be rich in odd harmonics. This observation is used to develop ratios of the third harmonic with respect to the fundamental frequency current or voltage.

5.4.1.4 Energy Technique

This method utilizes the summation of squared sample values of the current over 60 cycles. Methods using high frequency up to 10-kHz current amplitude are used to detect high-impedance faults and burst-noise signals at frequencies near the fundamental, and low harmonics have been used for high-impedance fault-detection schemes.

5.4.1.5 Randomness Technique

This technique is based on the randomness of harmonic current and is developed using stochastic and dynamic behavior of the power system subject to high-impedance faults.

5.4.2 Modeling of Faults/Classification

Faults are classified as Single-Line-to-Ground (SLG), Double-Line-to-Ground (DLG), or three-phase (3ϕ) bolted or unbolted short circuit faults, or open circuit faults occurring on one or more lines. While 3-phase short circuit faults are the most severe case resulting in very high current levels, Single-Line-to-Ground (SLG) faults occur most frequently. The boundary conditions of phase and sequence voltages and currents before and after a network fault are discussed under the fault analysis section in Chapter 3.

A high-impedance fault is a short-circuit fault through high impedance to ground.

An arcing fault is caused by intermittent opening and closing of contacts with high-energy bursts.

The above formulations and definitions provide the framework for designing different fault detection and location strategies.

1. On-line fault detection indicators are developed as part of decision-analysis options.
2. The equipment status and difference network connectivity are available in new modern system engine interfaces but not in distributed systems.
3. A proper historic fault frequency data recorder should be installed at the utility and customer end to gather real-time data for analysis during a given fault event. User interfaces are for online interrogation.

5.5 Trouble Calls

The trouble-call distribution-automation option is a distribution management system in support of increased customer-focused service. It is built within the utility system to receive trouble calls from customers by phone, fax, or external communication services. This is a more cost effective method to reporting a fault event, as compared to physically going to the site location of the fault. Answering and logging of trouble calls are handled using advanced communication-support services.

Figure 5.1 shows the sequence of activities leading from reception of a trouble call to the dispatch of a crew. The trouble-call-handling scheme progresses through the following sequence. Local calls made to customer service and crews are immediately dispatched, or trouble-call information is processed via a customer call center to verify the problem type, confirm account activities, and proceed to authorization of a dispatch crew. A toll-free call can also be made directly to customer service to confirm account status and request service per trouble-call placement.

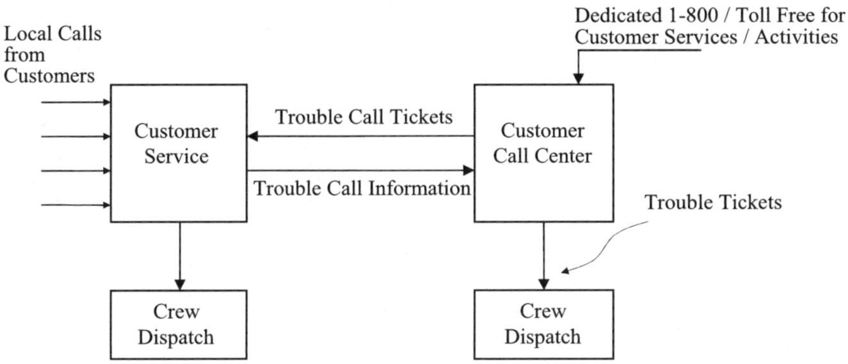

FIGURE 5.1
Trouble-call-handling sequence.

Trouble-call handling and alarm processing: New methods of remotely processing trouble (feeder problems, etc.) are done by using alarm processing indicating caller ID, telephone interface, loss-of-voltage indicator, and several other options.

Trouble-call placement: Overall system connectivity is checked locally if the network is available via geographic information systems (GIS) technology for network analysis, location of faulted distribution systems, and diagnostics. A repair crew may also be sent to the site for immediate repair or maintenance, depending on the nature of the problem. For instance, a switch gear may require maintenance, a capacitor bank may need to be switched, a pole may need replacement, etc. Feeder balancing and load balancing are other problems to be addressed during reconfiguration and restoration of the system. This aspect of distribution automation is referred to as trouble-call management, and performance of other management applications may be necessary, such as receiving calls, diagnosing and locating the fault, identifying all affected customers, and restoring the network in the shortest possible time.

The use of supervisory control and data acquisition (SCADA), energy management systems (EMS), customer information systems (CIS), and geographic information systems (GIS) interface is strongly recommended for a reliable and efficient trouble-call management. Future research in trouble-call analysis and alarm management for distribution systems is in development. The basic communication linkage between distribution automation and customer during trouble-call management is shown in Figure 5.2.

Customer-based trouble-call analysis under development includes the fault detection of the distributed generation network with the following problems: loss of voltage (leading to voltage collapse and instability), power factor correction, harmonics, etc.

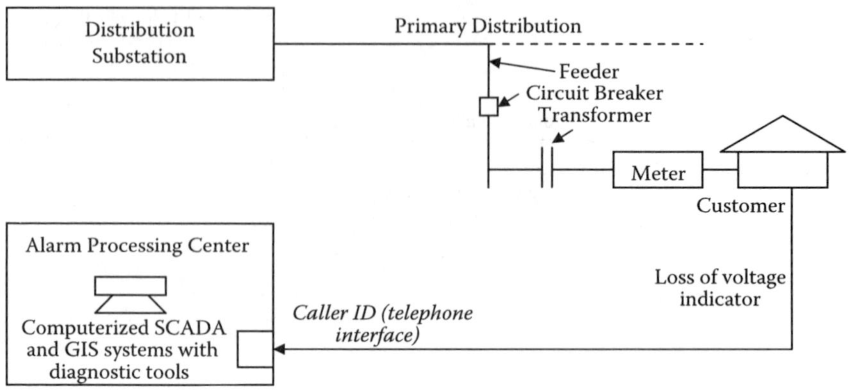

FIGURE 5.2
Basic communication schemes in a trouble-call reception system.

5.6 Restoration Functions

Planning a restoration service for a distribution system is a critical task for dispatchers in a power system control center. Restoration provides an ample amount of power to nonfaulty out-of-service areas for as many customers as possible while guaranteeing the safety and optimum reliability of the distribution systems. Several methods exist to solving restoration problems, ranging from the dispatcher's experience and to the operating values used in intelligent systems. The classical optimization technique is aimed at minimizing the number of unserved customers. (Use of sequential restoration schemes with analytical cold load pick up model to minimize the total restoration time.)

5.6.1 Evaluation of Methods

The competing methods for restoration based on optimization techniques are computationally intensive, but we provide the foundational work through formulation and appropriate selection of mathematical programming methods to solve the restoration problem. The approach utilized must account for:

1. Restoration time
2. Loss minimization
3. Optimal crew dispatch for service restoration
4. Voltage limits violation
5. Customer prioritization

6. Current/load unbalance alleviation
7. Protective devices and safe, efficient coordination
8. Equipment rating violation

5.6.2 Optimization Formulation

A case of optimization formulation and an associated method are discussed here.

Objective 1: Minimize an out-of-service area, min $f_1(\bar{x})$

where $\bar{x}$ is the switch-state vector such that

$$\bar{x} = [S_1, S_2, \ldots, S_{Ns}] \tag{5.12}$$

and N_s represents the total number of switches in the system under consideration. The state of each switch assumes a binary value such that

$$S_i = \begin{cases} 1, & \text{on} \\ 0, & \text{off} \end{cases} \tag{5.13}$$

where $f_1(\bar{x})$ denotes the number of nonfaulty out-of-service areas under the state $\bar{x}$.

Objective 2: Minimize the number of switching operations

$$\text{Min}\, f_2(\bar{x}) = \sum_{i=1}^{Ns} |S_i - S_{01}| \tag{5.14}$$

where $f_2(\bar{x})$ denotes the number of switching operations under the state $\bar{x}$, and S_{01} represents the original state of the ith switch (after the faults are isolated).

Objective 3: Minimize the deviations of the bus voltages

$$\text{Min}\, f_3(\bar{x}) = -\underset{i}{\text{Max}} |V_i - 1.00| \tag{5.15}$$

where

i = 1, 2, … , N_b

N_b = total number of buses in the distribution subsystem under consideration

V_i = bus voltage of the ith bus (p.u.)

$f_3(\bar{x})$ = maximal deviation of the bus voltage in the considered system

Objective 4: Minimize the line currents

$$\text{Max}\, f_4(\bar{x}) = \underset{i}{\text{Max}} \left\{ \frac{I_i^{\text{load}}}{I_i^{\text{rated}}} \right\} \tag{5.16}$$

where

i = 1, 2, … , N_L

N_L = total number of feeder lines in the distribution subsystem under consideration

I_i^{load} and I_i^{rated} represent the load current and rated current of the ith branch in the network, respectively

$f_4(\bar{x})$ = maximal normalized line current in the considered system

Objective 5: Minimize the loading of the transformer

$$\text{Max}\, f_5(\bar{x}) = \text{Max} \left\{ \frac{\text{tr}_i^{\text{load}}}{\text{tr}_i^{\text{rated}}} \right\} \tag{5.17}$$

where

i = 1, 2, … , N_t

N_t = total number of distribution transformers subsystem under consideration

$\text{tr}_i^{\text{load}}$ and $\text{tr}_i^{\text{rated}}$ represent the load currents and rated current of the transformer in the ith branch in the network, respectively

$f_5(\bar{x})$ = maximal normalized loading of the transformers

5.6.3 Optimization Constraints

To ensure that the radial distribution network remains radial after restoration, the switching operational sequence must be followed. For a restoration scheme, we impose the following constraints:

1. The switch to be opened is operated first (sectionalizing/isolating switchgear first).

2. If the radial structure is violated, after closing a switch, the switch cannot be selected as a backup switch; otherwise, it will cause a feeder with two suppliers from both sides and hence violate the radiality condition.
3. If interloops are still generated after the previous two steps, one switch in the loop must be arbitrarily opened.

The resulting multiple objective function is stated as follows:

$$\text{Min} f_1(\bar{x}) \tag{5.18}$$

where $i = 1, 2, \ldots, N$

subject to

$$g_j(\bar{x}) = 0 \text{ and } j = 1, 2, \ldots, N_c \tag{5.19}$$

where $f_1(\bar{x})$ is the number of distinct objective functions of decision vector $\bar{x}$, and $g_j(\bar{x}) = 0$ is the set of different constraints, some of which are listed above.

5.6.4 Methodology

The above multiple objective functions can be solved using a nonlinear optimal solution of the objective problems (where one objective function can be improved only at the expense of another). Using classical optimization techniques, the decision maker (such as a dispatcher) can make subjective decisions on which restoration plan is appropriate for the multiple objectives selected (weighted). New advances in intelligent systems, such as fuzzy logic (FL) and interactive fuzzy-satisfying methods, are used to solve this class of problem. We can also use genetic algorithms.

5.7 Reconfiguration of Distribution Systems

Distribution networks are generally configured in a radial structure. The configuration can be varied with manual or automatic switching operations so that all the loads are supplied with minimum losses and increased reliability, power quality, and security. The automatic switching sequence is an important subject in distribution automation. Switching operations are performed to ensure that the radiality of the network is maintained while preventing the distribution system from out-of-service conditions, overloads, or unbalanced conditions. The greater the number of switches, the greater are the possibilities for reconfiguration and ease of application.

To evaluate every possible configuration of the network feeders results in too many combinations to select for an optimal or near-optimal solution. Several techniques from heuristics, optimization, and intelligent systems have been proposed. The principal aim of reconfiguration is to satisfy the following objectives:

1. Minimize distribution losses
2. Optimize voltage profile
3. Relieve overload requirements while maintaining the radial structure of the network

5.7.1 Methods Used for Reconfiguration

These include:

1. *Loss minimization*. This has been an active research area. It was first developed by Merlin and Black using the branch-and-bound-type optimization method to determine the minimum loss configuration based on a schedule-switching pattern that corresponds to the loss.
2. *Heuristic algorithm*. This is an extension of the previous method that involves introducing an improved load flow and closing all switches, which are then opened one after each other so as to establish an optimum power-flow pattern.
3. Other variants of the first two methods are developed to improve the load estimation, to facilitate effective determination of system configurations, and to enhance modules for computing cost-benefit analyses of the reconfigured structure.

Other *noncombinational heuristic search* methods, binary integer programming techniques, optimization techniques, and annealing methods have been used to determine minimum energy losses for a given period. Additionally, AI techniques have been proposed for minimum loss using artificial neural networks (ANN), rule-based systems, genetic algorithm (GA), fuzzy logic (FL), and other evolutionary programming algorithms. The schemes for network configuration do not explicitly take into account the radiality aspects due to modeling issues in mathematical programming techniques.

5.7.2 Formulation of Modeling of Reconfiguration

As in restoration, a multiple-objective problem of reconfiguration, subject to operational constraints, is considered.

5.7.2.1 Method of Load Balancing 1

Minimize power loss

$$\Delta P_{\text{Loss}} = \left|I_i - I_j\right|^2 r_i - \left|I_i\right|^2 r_i \tag{5.20}$$

where r_i is the resistance of branch i. Therefore,

$$\Delta P_{\text{Loss}} = \left|I_i\right|^2 r_i - 2\left|I_i\right|\left|I_j\right| r_i + \left|I_j\right|^2 r_i - \left|I_i\right|^2 r_i \tag{5.21}$$

$$\Delta P_{\text{Loss}} = \left|I_j\right|^2 r_i - 2\left|I_i\right|\left|I_j\right| r_i \tag{5.22}$$

Hence for N branches in the loop, we have

$$\Delta P_{\text{Loss}} = \sum_{i=1}^{N} \left|I_j\right|^2 r_i - 2\left|I_i\right|\left|I_j\right| r_i \tag{5.23}$$

Using this index, the branch from the one that has the lowest increase in losses for each time or close sequence is used to maintain the optimal configuration. An algorithm for loss minimization is shown in Figure 5.3.

5.7.2.2 Method of Load Balancing 2

Using the loss-minimization method without reactor compensation (similar to that above), the goal is to minimize a load balance index, L_i, given by

$$L_i = \frac{\left|I_i\right|}{I_i^{\max}} \tag{5.24}$$

such that

$$\Delta L_i = \frac{\left|I_i - I_j\right|}{I_i^{\max}} - \frac{\left|I_i\right|}{I_i^{\max}} \tag{5.25}$$

and for the loop

$$\Delta L_{\text{Loop}} = \sum_{i=1}^{n} \left(\frac{\left|I_i - I_j\right|}{I_i^{\max}} - \frac{\left|I_i\right|}{I_i^{\max}} \right) \tag{5.26}$$

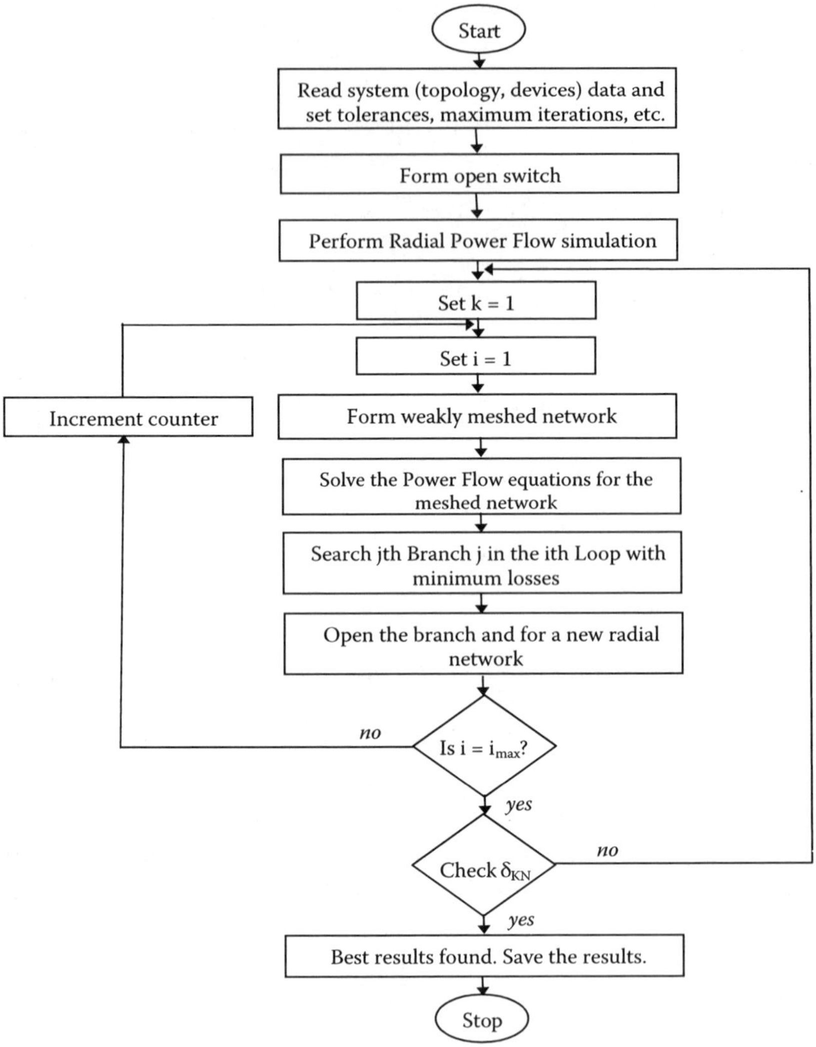

FIGURE 5.3
Algorithm for minimizing network loss.

The branch opening in the minimum change in the load-average load-balance index is obtained from the equations above. The procedure is repeated for loss minimization moving from loop to loop. To avoid repetitions, the first in the open list is chosen for loss minimization. (See algorithm in Figure 5.3.) Other modules used in the specified capacity-constraint model are given as

$$\frac{P_i}{P_i^{\max}} < L_i^{\max} \tag{5.27}$$

where

P_i^{max} = current capacity limit for the *i*th line

L_i^{max} = maximum load balance in the *i*th line

Each of these objectives is subject to the line and other topology constraints.

5.7.2.3 Method of Minimizing Voltage Deviation

Let us define

$$y = \left| V_{ij} - V_s \right|$$

for

$$i = 1, 2, 3, \ldots, NC \quad \text{and} \quad j = 1, 2, 3, \ldots, NB$$

where N_B is the total number of branches of the system, V_s is the voltage of the substation (in p.u.), and V_{ij} is the voltage of the *j*th node corresponding to the opening of the *i*th branch in the loop (in p.u.).

In general, once a normally open switch is selected and closed, a combinational search is used to determine the branch whose opening results in minimum voltage deviation. To reduce the time involved in the search, a heuristic search technique is utilized. We apply Kirchoff's voltage law (KVL) around the loop to ensure that the voltage around the loop is summed to zero, thus giving the lowest voltage drop where $\Delta V = 0$ (p.u.) and $\frac{dV}{dx} = 0$.

5.7.2.4 Algorithm for Single-Loop Voltage Minimization

The generalized algorithm constructed for minimizing losses and also capable of handling the load-balancing optimization process is derived based on the following steps:

1. Read the system data.
2. Run the load-flow program for radial distribution networks.
3. Compute the voltage difference across the open tie switches, i.e., $\Delta V_{\text{tie},i}$ for all ties in the set $\{i\text{: } 1, \ldots, i_{\text{tie,max}}\}$.
4. Identify the open switch across which the voltage difference is maximum and its code *k* with $\Delta V_{\text{tie,max}} = \Delta V_{\text{tie},k}$.
5. If $V_{\text{tie,max}} > \varepsilon$, then go to step 10 to print results and stop. Otherwise, continue.
6. Select the tie switch *k* and identify the total number of loops (N_k), including the tie branch where the switch is closed.

7. Open one branch at a time in a loop and evaluate the resulting objective function using classical optimization, artificial neural networks (ANN), fuzzy logic (FL), or a deterministic scheme. Select D_{ki} (the minimum load balancing, minimum voltage deviation, branch currents, etc.).
8. Obtain the optimal solution for the operation of the switch k such that $\Delta S_k = \max\{D_{ki}\}$.
9. Rearrange coding for the remainder of the switches and go to step 2.
10. Print output results and stop.

The implementation flowchart is displayed in Figure 5.4.

From optimization methods and others used, it is worth noting that global or near-global optimum results depend on the minimum or maximum limiting value of each objective function and the value of the threshold specified. It is possible to have a local optimum result if these are not properly selected. The proper choice the of minimum and maximum limits value of the objective functions used and the value of threshold is very important for obtaining the global or near-global optimum solution.

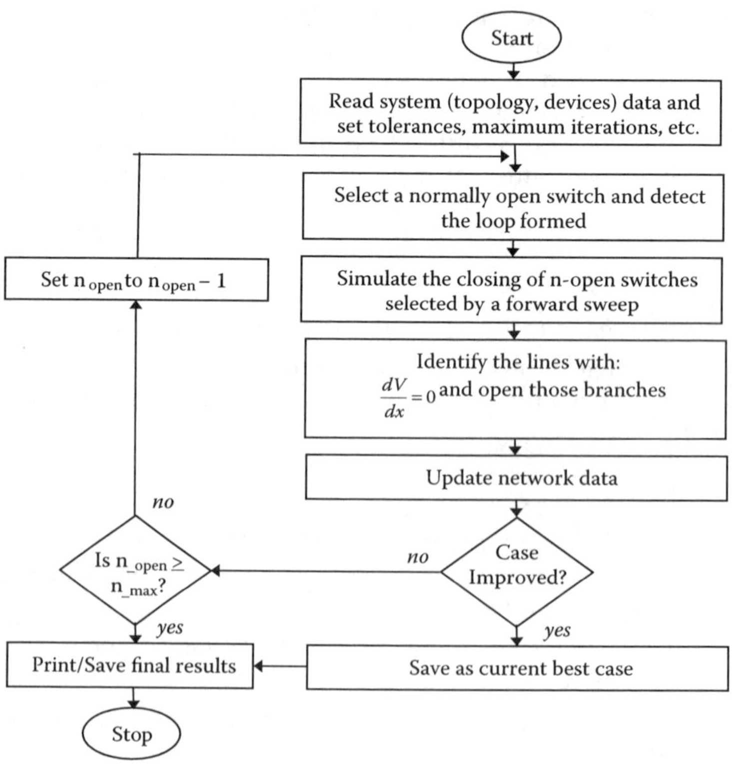

FIGURE 5.4
Single-loop voltage-minimization algorithm.

5.8 Power Quality

Power quality has been an area of research investigation and continues to be of interest as new devices are connected to the distribution system. Power quality becomes a difficult term to define because the measures depend on the need of the utility, on the equipment manufacturer, and on the nature of the supplies, which are different in most cases. Simply, power quality has a large number of anomalies related to voltage, current, and frequency deviation that result in failure or abnormal operation of customer/utility equipment. The related events affecting power quality are defined as follows:

- Outage: a complete loss of voltage, usually covering a time period varying from 30 cycles up to several hours or even days. Outage is caused by the fault-induced operation of circuit breakers or fuses and can be temporary or permanent.
- Surge: another important anomaly caused by transient voltage or current that can have extremely short duration and high magnitude. It is caused by lightning at the switching operation of customer loads or capacitors. This type of anomaly requires attention in recent years due to the use of electronic equipment such as VCRs and PCs.
- Undervoltage: another anomaly experienced when voltage is less than the proper (or contractual) nominal voltage. It can be caused by overload, poor wiring, or poor connection to utility system.
- Harmonics: these are nonfundamental components of a distorted 60-Hz waveform. They have frequencies that are integral multiples of the fundamental frequency of 60 Hz. Harmonics are produced by customers' equipment. Industrial, commercial, and domestic nonlinear loads also generate distortions that can propagate through the system and affect the customer.

Among all these anomalies, harmonics are the most important. They can be detected by using robust pattern-recognition automation control and security assessment in distribution systems. The differences in harmonics are nonlinearity and randomlike behavior of load, sensitivity of distribution systems loads to both frequency and voltage, and the influence on network configuration.

5.8.1 Techniques for Modeling Harmonics in Power-Quality-Assessment Methodology

Since power quality is a quality-of-service issue for the customer and the utility power company, it covers a wide variety of electromagnetic phenomena in power systems. For assessment, several techniques are proposed. They are classified into two categories: the time-domain analysis and the linear-frequency-domain analysis (as well as their combinations). The various techniques for measuring or assessing the degree of power quality are given as follows:

Time domain, which is given in terms of

- Crest factor
- Peak and RMS values
- Total harmonic distortion (THD)
- Distortion index
- K-factor and telephone factor, (TF), etc.

These time-domain methods are modeled through the integration of different equations that assume the same initial conditions. Once the harmonics system is modeled, the current and voltage waveforms at steady state are extracted, and fast Fourier transforms (FFT) are used to generate the harmonic spectra.

Time-domain analysis for harmonic assessment in power-quality work is an accurate method; however, it is computationally time consuming. We present here the formulation of different power-quality indices in the time domain.

Method 1: The most commonly used power-quality index is the total harmonic distortion (THD) index. It is defined as

$$(\mathrm{THD})_v = \frac{1}{V_1}\sqrt{\sum_{k=2}^{\infty}|V_k|^2} \tag{5.28}$$

This index depends on the Fourier coefficients of a periodic signal computed from the time/frequency domain in the harmonic analysis based on Parseval's theorem. In general, THD assesses the relative amount of harmonic content associated with a periodic signal.

Method 2: The VT product finds use as a voltage distortion index, as it integrates the voltage amplitude. It is defined as

$$V \cdot T = \sqrt{\sum_{i=1}^{\infty}\left(w_i V_i^2\right)} \tag{5.29}$$

Method 3: Power factor is the ratio of the actual power (kW) and the apparent power (kVA) delivered by a utility. This is a good indicator of how effectively current is being converted to useful work.

$$\mathrm{PF} = \frac{P_{\mathrm{tot}}}{|V_{\mathrm{rms}}||I_{\mathrm{rms}}|} \tag{5.30}$$

Power factor has the limitation of failing to indicate undesirable and potentially harmful effects of high-frequency harmonics.

Method 4: For the purpose of transformer derating, which is conducted on transformers that are carrying nonsinusoidal load current, the *K*-factor index is preferred. It is defined as

$$K = \frac{\left(\sum_{h=1}^{\infty} h^2 I_h^2\right)}{\sum_{h=1}^{\infty} I_h^2} \tag{5.31}$$

Method 5: Flicker factor is applied for bus voltage regulation and in determining the sufficiency of short-circuit capacity.

$$F = \frac{\Delta V}{|V|} \tag{5.32}$$

Method 6: Crest factor is applied in calculations of dielectric stress, as it determines whether breakdown will occur. The crest value is closely linked to the voltage across the dielectric and is related to the area under the current waveform and thus the charge, *Q*.

$$\mathrm{crest} = \frac{V_{\mathrm{peak}}}{V_{\mathrm{rms}}} \tag{5.33}$$

Method 7: Unbalanced factor is valued when considering a three-phase circuit.

Method 8: Linear-frequency-domain technique. In the linear-frequency method, it is assumed that the harmonic currents are independent of voltages and of harmonic impedance of the system. Therefore, the admittance matrix equations are used to model the system. The method is efficient computationally and is very useful where there are many harmonic devices.

5.8.2 New Approaches of Power Quality

Given the limitations of classical techniques, research approaches using state estimation (SE) theory are encouraging. The SE method is based on a fast

Fourier transform (FFT) and combinations of state estimation to generate and analyze the harmonic spectra of the system. Heydt has proposed using a weighted least-squares harmonic estimator; the use of a quadratic criterion based on a decomposition technique has also been proposed. Both methods are reliable, accurate, and can identify many nonlinear loads. Other new advances use time-frequency methods for a more robust power-quality assessment.

Other new techniques include wavelet transforms, artificial neural networks (ANN), and genetic algorithm (GA) methods. New signal-processing techniques such as wavelet transforms are used to improve feature extraction. The wavelet transform has the ability to handle nonstationary harmonic distortions in power distribution systems, and the results obtained by applying wavelet transforms provide a better assessment scheme for power-quality study for broadband signals that may not be periodic — a case for power transients. The ability of waveform transforms to dilate or contract transient signals while varying the frequency allows for the representation of power disturbances in a three-dimensional space.

5.9 Optimization Techniques

Optimization is a mathematical process in which a search is activated that aims at a best value of an objective function that is optimal (extrema). This can either be a maximum or a minimum, depending on the choice of the function. In a power system, the objective function could be a multivalued function involving cost, losses, security, and stability at the same time. In DAF, it is defined as losses, optimum switching, voltage deviation, or costs. The combination of multiple objective functions is done using individual judgment and weights. The optimization, in a general sense, provides or searches for the best value of an objective function where equalities and inequalities are satisfied as part of the optimization problem.

5.9.1 Objectives

Objective functions derived from the practical definition are usually continuous but not necessarily convex. Sometimes the objectives are discrete. For distribution power systems, the objective function can be converted or approximated by a convex function. Linear objective functions, quadratic forms or approximation to a nonlinear quadratic form, and piecewise linear functions are employed. In all cases, the objective function is a scalar, e.g., cost, loss, voltage and power or other deviations, allocations/scheduling.

In standard form, the objective function can be written as $F(x)$ or $F(x,u)$ and can be maximized or minimized. F is a scalar function, and x is the vector of variables, including both state variables and control variables. State variables are typically voltage and angles, flows; and control variables are nodal voltage magnitude at generator buses, tap changer positions, output of reactive sources. The minimization of function is obtained from a solution that fulfills the first derivative of the objective function with respect to all variables to be zero. This is normally called the unconstrained optimization process.

5.9.2 Constraints

There typically are two types of constraints:

- Equality constraints
- Inequality constraints

For the power system, the equality constraints are typically written as $g(x) = 0$ or $g(x, u) = 0$, where x, u, and g are vector functions. An example of a set of equality constraints is the set of nonlinear power-flow equations.

For the inequality constraints, the introduction of limits to state and control variables and functions leads to inequality constraints. For distribution automation, the inequality can be a continuous function or a discrete or continuous variable. The constraints satisfy both less than and equal to or greater than the value of each of the control variables. For example, $h(x) \leq 0$ or $h(x, u) \leq 0$, where x, u, and h are vectors or vector functions. The inequality can also be rewritten as equality constraints with the addition of slack variables. It should be mentioned that inactive constraints can be eliminated from the problem if they do not contribute to the solution process. The limits imposed on the constraints are called hard limits. No solution exceeding the limits will be tolerated. However, from the present viewpoint we can introduce some engineering judgment to soften the constraints such as varying or relaxing limits. Thus a standard optimization problem can be given as

$$\text{Minimize } F(x, u)$$

subject to

$$g(x, u) = 0$$

$$h(x, u) = 0$$

where all vectors and vector functions can be continuous, discrete, or a combination. In distribution automation, the optimization objectives and constraints presented in Table 5.1 are possible.

TABLE 5.1

Optimization Background Review for Solving the Proposed Problem

Objective Function	Subject to
1. Minimize an out-of-service area 2. Minimize the number of switching operations 3. Minimize the deviations of the bus voltages 4. Minimize the line currents 5. Minimize the loading of the transformer	a. The switch to be opened is operated first b. If the radial structure is violated, after closing a switch, the switch cannot be selected as a backup switch; otherwise, it will cause a feeder with two suppliers from both sides and hence violate the radiality condition c. If interloops are still generated after the previous two steps, one switch in the loop must be arbitrarily opened

5.9.3 Classical Solution

The techniques for solving optimization problems are specialized for different problems given the type of objective (linear, quadratic, or nonlinear) or constraints (linear, quadratic, or nonlinear) and continuous or discrete variable form.

A method that plays a central role in power-system optimization is Newton's method. It is similar to solving power-flow problems, except it has constraints. Simply put, it consists of a solution method for $g(x) = 0$ as a nonlinear function of x to obtain a value of x for $g(x) = 0$.

Using a Taylor series

$$g(x) = g(x_0) = \frac{\partial g \Delta x}{\partial x} + \frac{1}{2!}\frac{\partial^2 g}{\partial x^2}\Delta x^2 \tag{5.34}$$

gives

$$\Delta x = -\left[\frac{\partial g}{\partial x}\right]^{-1} g(x_0) \tag{5.35}$$

in matrix form

$$\Delta x = [J]^{-1}[g(x_0)] \tag{5.36}$$

where J is the Jacobian matrix.

This method is therefore used in solving optimization with equality constraints, such as

$$\text{Given } F(x) \Rightarrow \min$$

$$\text{subject to } g(x) = 0$$

We minimize

$$L = F(x) + \lambda^T g(x) \tag{5.37}$$

where L is the Lagrange function. The condition for optimum is found as follows from

$$\frac{\partial L}{\partial x} = \frac{\partial F}{\partial x} + \left[\frac{\partial g}{\partial x}\right]^T \lambda = 0 \tag{5.38}$$

$$\frac{\partial L}{\partial \lambda} = g(x) = 0 \tag{5.39}$$

With these conditions given, we must now determine which method of optimization to use. The following outline is used to find the method of solution:

1. If the objective function is quadratic and the constraints are linear, we use nonlinear optimization methods.
2. If the objective function is linear and the constraints are linear, we use a simplex-like linear programming method that does not depend on Lagrange and optimality conditions. Rather, it is based on a simplex linear algebra rule given as follows:

From

$$\text{Min } F(x) = C^T X \tag{5.40}$$

$$\text{subject to } \begin{array}{l} Ax = b \\ x \geq 0 \end{array} \tag{5.41}$$

with $Ax = b$ already including the state variables that convert inequality to equality constraints. There is no difference between inequalities due to state and control variables. Here, the requirement is that all components of $x \geq 0$ must be nonnegative for linear programming. This constraint is relaxed, hence its extension to linear-integer programming, which finds its use in distribution automation functions for restoration and reconfiguration.

5.9.4 Linear Programming

$$\text{Min } F(x) = C^T x \tag{5.42}$$

$$\text{subject to} \quad \begin{array}{l} Ax = b \\ x \geq 0 \end{array} \tag{5.43}$$

where

$$A = |B, D|$$

$$x = |X_B, X_D|$$

$$C = |C_B, C_D|$$

where B refers to basic variables, D to nonbasic variables, and the standard linear programming problem becomes

$$\begin{aligned} F(x) &= C_B^T X_B + C_D^T X_D \\ BX_B + DX_D &= b \end{aligned} \tag{5.44}$$

$$X_B \geq 0,\ X_D \geq 0 \tag{5.45}$$

$$\begin{aligned} X_B &= B^{-1}b - B^{-1}DX_D \\ F(x) &= C_B B^{-1}b + \left(C_D - C_B B^{-1}D\right)X_D \end{aligned} \tag{5.46}$$

which gives the current value of the sensitivity due to nonbasic variables. The component showing the sensitivity is considered as the relative cost vector

$$r = C_D - C_B B^{-1}D \tag{5.47}$$

$$\begin{vmatrix} I & B^{-1}D & B^{-1}b \\ 0 & C_D - C_B B^{-1}D & -C_B B^{-1}b \end{vmatrix} \tag{5.48}$$

The matrix-cost vector in the last row gives an indication as to which nonbasic variable is to become a basic variable, and when all components have become positive, the minimum is reached. The right lower corner shows the negative value of the cost function.

5.9.5 Mixed-Integer Programming

Integer and mixed-integer programming problems are special classes of linear programming, where all or some of the decision variables are restricted to integer values. There are many practical examples where the "divisibility" assumption in linear programming needs to be dropped and some of the variables can take only discrete values. However, even greater importance can be attributed to problems where the discrete values are restricted to zero and one only, i.e., "yes" or "no" decisions (binary decision variables). In fact, in many instances, mixed-integer programming problems can be reformulated to have only binary decision variables, which are easier to handle. The occurrence of binary variables can be due to a variety of decision requirements, the most common of which are:

1. *ON/OFF decisions*: The most common type of binary decision falls into this category for engineering optimization problems. This decision variable can also have alternative representation of GO/NO GO, BUILD/NOT BUILD, or SCHEDULE/NOT SCHEDULE, and so on, depending on the specific application under consideration in short, medium, and long-term planning contexts.
2. *Logical EITHER-OR/AND constraints*: Binary variables can also indirectly handle mutually inclusive or exclusive restrictions. For example, there might be cases where a choice can be made between two constraints, so that only one can hold. There could also be cases where process B must be selected if process A has already been selected.
3. *K out of N constraints must hold*: Consider the case where the overall model includes a set of N possible constraints such that only some K of these constraints must hold (assuming $k < N$). Part of the optimization task is to choose which combination of K constraints permits the objective function to reach its best possible value. In fact, this is nothing but a generalization of the either/or constraints, and can handle a variety of problems.
4. *Function with N-possible values*: In many real-life problems, the functions do not have smooth, continuous properties, but can take up only a few discrete values. For example, consider the following case:

$$f(x_1, \dots, x_n) = d_1 \quad \text{or} \quad d_2, \dots, d_n \tag{5.49}$$

The equivalent integer programming formulation would be

$$f(x_1, \dots, x_n) = \sum_{i=1}^{N} d_I y_I \tag{5.50}$$

$$\sum_{i=1}^{N} y_i = 1 \tag{5.51}$$

and

$$y_i = \text{binary (0 or 1)} \quad \text{for} \quad i = 1, 2, \ldots, N \tag{5.52}$$

5. *The fixed-charge problem*: In most problems, it is common to incur a fixed-cost/setup charge when undertaking a new activity. In a process-engineering context, it might be related to the setup cost for the production facility to initiate a run. A typical power system example is the startup cost of a thermal-generating unit. This fixed charge is often independent of the length or level of the activity and, hence, cannot be approximated by allocating it to the (continuous) level of activity variables.

   Mathematically, the total cost comprising fixed and variable charges can be expressed as

$$f_i(x_j) = \begin{cases} K_j + C_j x_j, & \text{if } x_j > 0 \\ 0, & \text{if } x_j = 0 \end{cases} \tag{5.53}$$

   The mixed-integer programming (MIP) transformation would look like

$$\text{Min } Z = \sum_{J=1}^{N} (C_j x_j + K_j y_j) \tag{5.54}$$

   where

$$y_j = \begin{cases} 1, & \text{if } x_j > 0 \\ 0, & \text{if } x_j = 0 \end{cases} \tag{5.55}$$

   Pure integer or mixed-integer programming problems pose a great computational challenge. While there are highly efficient linear-programming (LP) techniques to enumerate the basic LP problem at each possible combination of the discrete variables (nodes), the problem lies in the astronomically large number of combinations to be enumerated. If there are N discrete variables, the total number of combinations becomes 2^N! The simplest procedure one can think of for solving an integer or mixed-integer programming problem is to solve the linear-programming relaxation of the problem (i.e., allowing the discrete variables to take continuous values so that the mixed-integer programming reduces to nonlinear programming) and then rounding the noninteger values to the closest integer solution. There are, however, major pitfalls:

- The resulting integer solution may not be feasible in the first place.
- Even if the rounding leads to a feasible solution, it may, in fact, be far from the optimal solution.

Algorithmic development for handling large-scale integer or mixed-integer programming problems continues to be an area of active research. There were exciting algorithmic advances during the middle and late 1980s. The most popular method to date has been the branch-and-bound technique and related ideas to implicitly enumerate the feasible integer solutions.

5.9.6 Interior-Point Linear Programming

For solving a large system, another linear programming technique developed by Karmarker based on the interior-point method is employed. It can also be extended to quadratic nonlinear programming as well as the integer branch-and-bound technique.

The algorithm for simple *LP* is

$$\text{Min } P = C^T X \tag{5.56}$$

$$\text{subject to } AX = b \tag{5.57}$$

$$\text{with } Xi \geq 0,\ i = 1, \ldots \text{n}$$

An interior point is given on page 100 of the textbook.

A drawback of modeling the problem by a linear objective function is that the solution will be found at a limit or a combination of limits. To ensure that the system objective is close to reality, we use a linear approximation called quadratic programming (QP), with several variations. The standard formulation goal is again given as

$$\text{Min } F(X) = C^T X + \frac{1}{2} X^T Q X \tag{5.58}$$

$$\text{subject to } A_1 X \leq b_1 \tag{5.59}$$

$$A_2 \leq b_2 \tag{5.60}$$

Convert to dual quadratic programming in standard form

$$\text{Min } F(X) = P^T X + \frac{1}{2} X^T Q X \tag{5.61}$$

subject to

$$\text{subject to } DX - b_1 = 0 \tag{5.62}$$

$$DX - b_2 = 0 \tag{5.63}$$

Using the Lagrange method,

$$\min L = F(X) + \lambda^T(DX - b_1) + \mu^T(AX - b_2) \tag{5.64}$$

Optimality condition is performed using the following:

$$\frac{\partial L}{\partial \lambda} = 0 \Rightarrow QX + D^T\lambda = -P \tag{5.65}$$

$$\frac{\partial C}{\partial \lambda} = 0 \Rightarrow DX = -b_1 \tag{5.66}$$

$$\frac{\partial L}{\partial \mu} = 0 \Rightarrow AX = -b_2 \tag{5.67}$$

$$\begin{bmatrix} Q & A_1^T & A_2^T \\ A_1 & 0 & 0 \\ & & \end{bmatrix} \begin{bmatrix} X \\ \lambda \\ \mu \end{bmatrix} = \begin{bmatrix} -P \\ 0 \\ \end{bmatrix} \tag{5.68}$$

$$A_2X - b_2 \leq 0 \tag{5.69}$$

$$\text{Diag } (\mu)\ A_2X - b_2 = 0 \tag{5.70}$$

$$\mu \geq 0 \tag{5.71}$$

To solve

$$C^T X$$

$$\text{Min } AX = b \tag{5.71}$$

$$L \leq X \leq u$$

Step 1: Initialization set $x = 0$, set $x^0 \geq 0$
to get

$$Ax^0 = b \tag{5.73}$$

Step 2: Obtain

$$W_k = \left(AX_k^2A^T\right)^{-1} AX_k^2c \tag{5.74}$$

where x_k is a diagonal matrix of elements x_k.
Step 3:

$$r^k = c - A^TW^k \tag{5.75}$$

Step 4:

$$r^k \geq 0 \tag{5.76}$$

$$c^TX^Cr^k \leq \varepsilon \tag{5.77}$$

Stop if okay; otherwise, go to step 5.
Step 5:

$$dy^k = -x_kr^k \tag{5.78}$$

Step 6: Check

$$dy^k \geq 0 \tag{5.79}$$

Stop.
Step 7:

$$\alpha_k = \min\left\{\frac{\alpha}{-\left(dy^k\right)_i} \middle| \left(dy^k\right)_i \leq 0\right\} \tag{5.80}$$

$$0 \leq \alpha \leq 1 \tag{5.81}$$

Step 8:

$$X_R^{k+1} = X^k + \alpha_kX_kdy^k \tag{5.82}$$

5.9.7 Sequential Quadratic Programming

This algorithm is an extension of the quasi-Newton method for constrained optimization. The method solves the original problem by repeatedly solving a quadratic programming approximation. A quadratic programming problem is a special case of a Non-linear Programming (NLP) problem, wherein the objective function is quadratic and the constraints are linear. Both the quadratic approximation of the objective and the linear approximation of the constraints are based on Taylor series expansion of nonlinear functions around the current iterate X^k.

The objective function *f(X)* is replaced by a quadratic approximation; thus

$$q^k(D) = \nabla f(X^k)D + \frac{1}{2} D^T \nabla^2 L(X^k, \lambda^k) D \tag{5.83}$$

The step D^{k-} is calculated by solving the following quadratic programming subproblem:

$$\text{Min } q^k\ (D) \tag{5.84}$$

subject to

$$G(X^k) + \mathbf{J}(X^k)D = 0 \tag{5.85}$$

$$H(X^k) + \mathbf{I}(X^k)D \leq 0 \tag{5.86}$$

where **J** and **I** are the Jacobian matrices corresponding to the constraint vectors G and H, respectively.

The Hessian of the Lagrangian $\nabla^2 L(X^k, \lambda^k)$ that appears in the objective function, Equation 5.83, is computed using a quasi-Newton approximation. Once D^{k-} is computed by solving Equations 5.84 to 5.86, X is updated using

$$X^{k+1} = X^k + \alpha^k D^k \tag{5.87}$$

where α^k is the step length.

Finding α^k is more complicated in the constrained case. This is because α^k must be chosen to minimize constraint violations in addition to minimizing the objective in the chosen direction D^k. These two criteria are often conflicting, and thus a merit function is employed to reflect the relative importance of these two aims. There are several ways to choose a merit function, with one choice being

$$P_1(X, v) = f(X) + \sum_{i=1}^{a} v_i |g_i| + \sum_{j=1}^{b} v_{a+j} \max\left[h_j, 0\right] \tag{5.88}$$

where $v \, \varepsilon \, R^{a+b}$ is the vector of positive penalty parameters, and g_i and h_j are elements of the constraint vectors $G(X)$ and $H(X)$, respectively.

For the merit function $P_1(X, v)$, as defined in Equation 5.88, the choice of v is defined by the following criterion:

$$v_i \geq |\lambda_i| \text{ , } i = 1, 2, \ldots, a, a+1, \ldots, b \tag{5.89}$$

where λ_i represents Lagrange multipliers from the solution of the quadratic programming subproblem of Equations 5.84 to 5.86 that define D^k. Furthermore, the step length α^{k-} is chosen so as to approximately minimize the function given by

$$P_1(X^k + \alpha D^k, v) \tag{5.90}$$

A different merit function that can be used is known as the augmented Lagrangian merit function

$$L_A(X,\lambda,v) = f(X) - \sum_{i=1}^{a} \lambda_k g_i + \frac{1}{2}\sum_{j=1}^{b} v_i g_i^{\,2} + \sum_{j=a}^{b} \Phi_{j-a}(X, v_j, \lambda_j^{\,2}) \tag{5.91}$$

where

$$\Phi_{j-a}(X, v_j, \lambda_j^{\,2}) = \frac{1}{v_j}\left[\max(0, (\lambda_j + v_j h_{j-a})^2) - \lambda_j^{\,2}\right] \tag{5.92}$$

and g_i and h_j are elements of the constraint functions $G(X)$ and $H(X)$, respectively; v is the vector of the positive penalty parameters; and λ_i represents Lagrange multipliers from the solution of the quadratic programming subproblem given by Equations 5.84 to 5.86 that define D^k.

If Equation 5.84 is used as the merit function, the step length is chosen to approximately minimize the function

$$L_A(X^k + \alpha D^k, \lambda^k +, \quad \alpha(\lambda^{k+1} - \lambda^k), v) \tag{5.93}$$

where D^{k-} is the solution of the quadratic programming subproblem given by Equations 5.84 to 5.86 and defines λ^{k+1} as the associated Lagrange multiplier.

5.10 Illustrative Examples

5.10.1 Example 1

Maximize

$$Z = -(2x_1 - 5)^2 - (2x_2 - 1)^2$$

subject to

$$x_1 + 2x_2 \leq 2$$

and

$$x_1, x_2 \geq 0$$

$$\frac{\partial z}{\partial x_1} = -4(2x_1 - 5) = 0$$

$$\frac{\partial z}{\partial x_2} = -4(2x_2 - 1) = 0$$

Therefore,

$$(x_1, x_2) = \left(\frac{5}{2}, \frac{1}{2}\right)$$

$$L(x_1, x_2) = -(2x_1 - 5)^2 - (2x_2 - 1)^2 - \lambda(x_1 + 2x_2)$$

$$\frac{\partial L}{\partial x_1} = -4(2x_1 - 5) - \lambda = 0$$

$$\frac{\partial L}{\partial x_2} = -4(2x_2 - 1) = 0$$

$$\frac{\partial L}{\partial \lambda} = -x_1 = 0$$

$$\therefore x_1, x_2 = \left(0, \frac{1}{2}\right)$$

$$\therefore Z = -25$$

5.11 Summary

This chapter provides a working definition for different distribution automation functions (DAFs). Formulations using mathematical programming techniques are given, as well as procedures for solving DAFs using classical optimization techniques and their potential benefits.

Existing tools for DAFs developed by researchers are presented. References to outstanding works for possible case studies are available as references for the researcher. Examples of research products as candidate DAFs are provided.

Problem Set 5

5.1

1. Define the term distribution automation.
2. Construct a detailed mathematical formulation for reconfiguration, restoration, load balance, and remedial control. Clearly define all terminologies.
3. Use a simple optimization process to construct the algorithm for implementation.
4. Choose two of the DAFs for the implementation using your simple example and carry out the calculation by hand.

5.2 Use the integer programming method to solve the following optimization problem:

$$Z = 3x_1 + x_2 + 2x_3$$

subject to

$$-x_1 + 2x_2 + x_3 \le 4$$

$$4x_2 - 3x_3 \le 2$$

$$x_1 - 3x_2 + 2x_3 \le 3$$

where x_1, x_2, and x_3 are nonnegative numbers.

5.3 Solve for

$$f(x) = 5x_1 + 3x_2$$

given constraints

$$g_1(x) = x_1 + 2x_2 - x_3 - 6 = 0$$

and

$$g_2(x) = 3x_1 + x_2 + x_4 - 9 = 0$$

by

1. Linear programming method
2. Jacobian method
3. Lagrange method

5.4 Maximize

$$Z = 6x_1 + 3x_2 - 4x_1x_2 - 2x_1^2 - 3x_2^2$$

subject to

$$x_1 + x_2 \leq 1$$

$$2x_1 + 3x_2 \leq 4$$

and

$$x_1, x_2 \leq 0$$

Show that it is strictly concave and solve by quadratic programming.

5.5 Discuss the importance of the Demand Side Management (DSM) and construct and algorithm or flowchart for implementing a typical DSM.

5.6 List the 4 classical approaches used to solve the problem of fault detection in a distribution system. What are the merits and demerits of the selected approaches.

5.7 Define the term "Trouble Call Analysis" and construct its corresponding sequence diagram.

5.8 Identify the various techniques for measuring Power Quality and give a brief mathematical description of each technique.

5.9 VAr optimization is an important aspect of supporting nodal voltages within contractual limits throughout the distribution network.

a. Formulate a typical Voltage/VAr optimization problem (i.e., construct a detailed mathematical formulation for voltage/VAr control problem).

b. Develop an implementation algorithm for solving the problem in (a).

5.10 Discuss briefly the optimization techniques applicable to solve linear and nonlinear mathematical problem (type of objective function, constraints, and variables). Solve or determine the feasibility of the following optimization problems via:

a. Linear Programming (LP)
$Max\ f(x_1, x_2) = 2x_1 + 3x_2$ subject to $x_1 + x_2 \leq 2$, $x_2 - x_1 \leq 3$ and $x_1, x_2 \geq 0$

b. Quadratic Programming (QP) (hint: use a calculus technique)
$Min\ f(x_1, x_2) = (2x_1 - 4)^2 + (3x_2 - 2)^2$ subject to $2x_1 + x_2 \leq 2$, $x_1, x_2 \geq 0$

6

Intelligent Systems in Distribution Automation

6.1 Introduction

The previous chapter on distribution automation functions (DAFs) presents the different functions for monitoring, reconfiguration, restoration and power quality and customer-based support systems. The formulation and solution strategies are given using optimization techniques or power-flow evaluation algorithms. Several important notable successes have been achieved using the numerical methods, which assume off-line studies in designing an automated distribution system. There remain a large number of problems to be addressed in power systems, especially in distribution system automation, which requires heuristic or intelligence computing.

Recent works in this field have included integrated numerical methods with intelligent systems such as artificial neural networks (ANN) and genetic algorithms (GA), which are used to achieve an efficient and reliable distribution system. This chapter reviews the directions of research in this field and the application of intelligent systems to distribution automation functions. Next, we will identify the common trends and emerging trends in intelligent systems as they cut across several domains of power system automation. Finally, we will outline research themes of significant importance for the future evolution of intelligent systems in distribution automation functions.

The following features frequently characterize these problems:

1. Inadequate model of the real world
2. Complexity and size of the problems, which prohibit timely computation
3. Solution method employed by the human incapable of being expressed in an algorithm or mathematical form; usually involves many rules of thumb
4. Operator decision making based on fuzzy linguistics description

These drawbacks have motivated the power system community to seek alternative solutions techniques through the use of artificial intelligence (AI) systems and variants of its applications.

In this chapter, we present a brief summary of such techniques, including expert systems, artificial neural networks, fuzzy logic systems, and GA. These techniques have been employed for solving various power system operations and planning problems and especially for the different types of control measures for given power system abnormalities.

6.2 Distribution Automation Function

The problems of distribution automation functions have been discussed earlier. They include the following:

Reconfiguration: principal aim of reconfiguration is to minimize distribution losses, optimize voltage profiles, and relieve overload requirements while maintaining the radial structure of the network

Restoration: provides an ample amount of power to nonfaulty, out-of-service areas for as many customers as possible while guaranteeing the safety and optimum reliability of the distribution systems

Power quality: refers to a large number of anomalies related to voltage, current, and frequency deviation that result in failure or abnormal operation of customer/utility equipment

Fault analysis: involves considering what happens after a fault occurs, identifying the location of the fault, and assessing the nature of the damage caused by the fault

Some commonality exists among intelligent systems approaches. The system requirements for developing or assessing intelligent systems approaches are as follows:

1. Ability to identify system state
2. Selectivity of controls
3. Learning ability to update knowledge
4. Coordination of tasks
5. Flexibility
6. Ability to handle uncertainty

These factors must all be taken into consideration when attempting to solve distribution automation problems by applying Intelligent systems.

6.3 Artificial Intelligence Methods

AI is a subfield of computer science that investigates how the thought and action of human beings can be modeled or mimicked by machine. The symbolic computation involved in AI is numeric and nonnumeric. The mimicking of intelligence includes not only the ability to make rational decisions, but also to deal with missing data, to adapt to existing situations, and to improve itself over a long time horizon based on accumulated experience. In general, it is conceived as a computer program that possesses an algorithm that attempts to model and emulate, thus automating an engineering task that was previously carried out by a human. In this section we provide an overview of four major families of AI techniques that are applicable to distribution systems, namely:

- Expert system techniques (ES)
- Artificial intelligence neural networks (ANN)
- Fuzzy logic systems (FL)
- Genetic algorithms (GA)

6.3.1 Expert System Techniques

An expert system (ES), also referred to as a knowledge-based system, embodies human expertise in a narrow field or domain in a machine-implementation form. It utilizes elements of human knowledge to provide decision support at a level comparable with a human expert and is capable of justifying its reasoning. It separates the inference mechanism from the knowledge and uses one or more knowledge structures such as production rules frames, semantic nets, predicate calculus, and objects to represent knowledge.

An expert system is an artificial intelligence (AI) program incorporating a knowledge-and-inference system. The expert system software includes heuristic rules based on the expert's experience. In such a system, the knowledge takes the form of so-called production rules written in the form of if/then syntax (knowledge base). The system includes facts, data that generally describe the domain and the state of the system contained in the so-called database.

It is an inference engine that can be data driven or goal driven; it uses facts, rules, and data/goals to deduce new facts, which allow the firing of other rules. The knowledge base is a collection of domain-specific knowledge, and the inference system is the logic component for processing the knowledge base to solve the problem. The search process continues until the base of facts is saturated and a conclusion has been attained. An explanation facility is provided for some advanced expert systems to come up with a recommended conclusion or selection.

Five special features distinguish expert systems from traditional power system analysis techniques:

1. *Flexibility*: The expert system allows flexible manipulation of domain-specific knowledge without having to watch for the impact of changes or the way we are reasoning it.
2. *Manipulation of symbolic information*: The expert system is concerned with manipulating symbolic information rather than direct manipulation of numerical information.
3. *Ability to handle imprecise knowledge*: The expert system addresses problems where knowledge may be deterministic and more imprecise and allows for handling of uncertainty in reasoning.
4. *Ease of modification*: The integrity of the knowledge base depends on how accurate and up to date it is. In domains where rapid changes take place, it is important to provide a quick and easy way to modify the knowledge base.
5. *Portability*: An expert system is designed to operate in one particular environment; expert system software for distribution automation should be transportable and adaptable to different system configurations and environments. Expert systems in most cases should also be able to adapt to different learning scenarios. They must be able to learn from experience.

The general framework of an expert system architecture is illustrated in Figure 6.1.

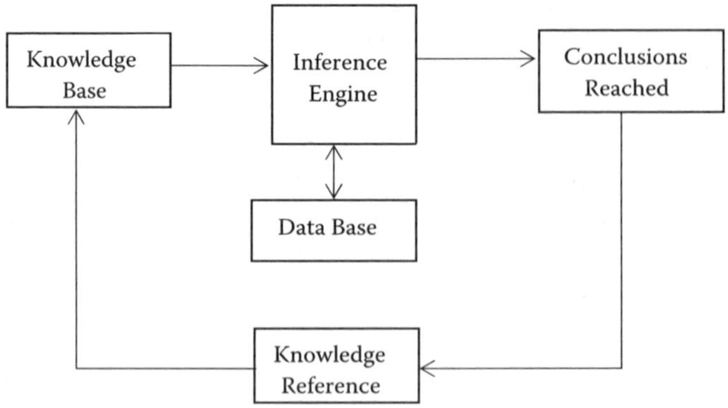

FIGURE 6.1
Architecture of an expert system.

6.3.2 Artificial Neural Networks

The ANNs are very different from expert systems, as they do not need a knowledge base to work. Instead, they have to be trained with numerous actual cases. An ANN consists of interconnected processing elements known as neurons or nodes. It acts as a directed graph in which each node performs a transfer function f_i of the form

$$y_i = f\left(\sum_{j=1}^{n}\left(w_{ij}x_j - \theta_i\right)\right) \tag{6.1}$$

or for high-order networks of multiple input

$$y_i = f_i\left(\sum_{j=1}^{n}\left(w_{ij}x_j x_m - \theta_i\right)\right) \tag{6.2}$$

where

y_i is the output of node i
x_j is the jth input to the node
w_{ij} is the connection weight between nodes i and j
θ_i is the threshold (bias) of the node

Usually f_i is nonlinear, and it is represented as a heavy-side, sigmoid, Gaussian, or exponential function.

ANN techniques are attractive because they do not require tedious knowledge-acquisition, representation, and writing (if/then) stages and can therefore be used for tasks not previously described in advance. ANN learns from a response based on given inputs and a required output by adjusting the node weights and biases accordingly. ANNs can be divided into two general classes — feed-forward and recurrent classes — described as follows:

Feed-forward ANN: a method that numbers all nodes in the network such that there is no connection from a node with a larger number to a node with a smaller number; all connections are from nodes with smaller numbers to nodes with larger numbers

Recurrent-net ANN: that does not have such a numbering method does not exist

The architecture of an ANN is determined by its topological structure. The overall connectivity and transfer function of each node in the network is illustrated in Figure 6.2.

The speed of processing of an ANN allows for real-time application, hence its many applications in distribution automation functions. ANN has the ability to generalize; there is no exact guide for choosing the number of

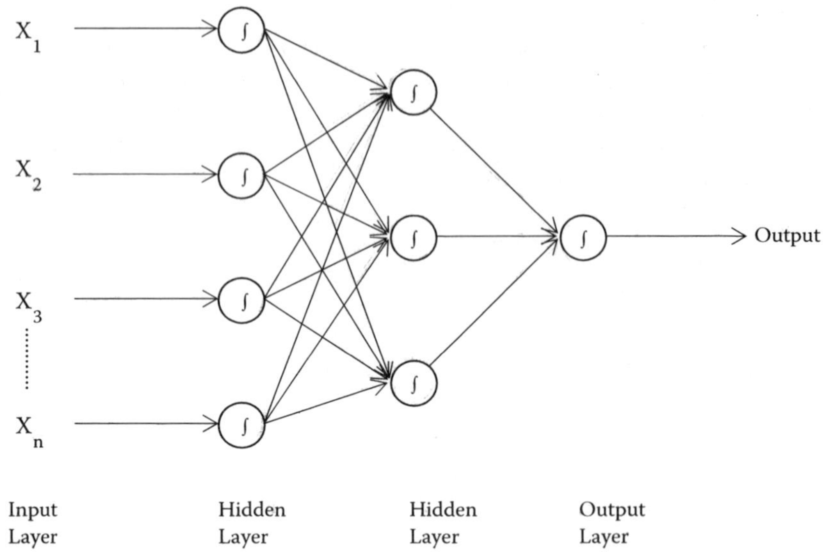

FIGURE 6.2
Architecture of an ANN.

hidden layers. Several works describe applications demonstrating that pattern classification and associated memory can learn to distinguish between inputs, which explains the technique's ability for decision making and classification.

6.3.2.1 *Evolution of Connection Weights*

Weight training in an ANN is usually formulated as minimization of an error function such as $\Sigma = \int w[x_1^t - x_1^{act}]$, the mean square error between target and actual output averaged over all examples, by iteratively adjusting connection weights. Most training algorithms such as back propagation (BP) and conjugate gradient are based on gradient descent. Because the BP method is based on gradient descent, it has a drawback on convergence, leading to its inability to find a global optimum if the error function is a multimodal or nondifferentiable function.

6.3.3 Fuzzy Logic

"Fuzzy set" is a term coined by Professor Zadeh to argue that human reason cannot be represented in terms of discrete symbols and numbers but in fuzzy sets. Fuzzy set are functions that map a value that might be a member of the set to a number between zero and one, indicating the actual degree of membership.

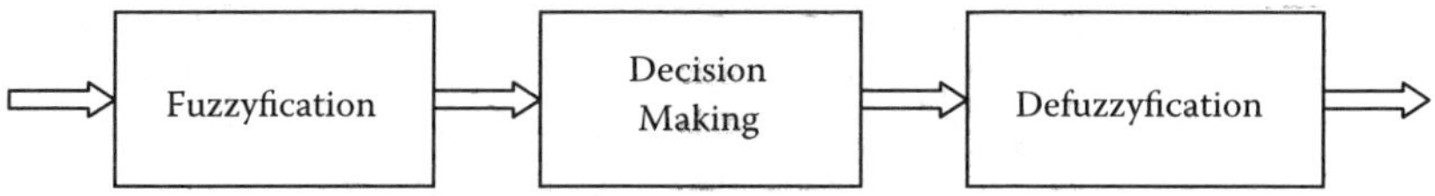

FIGURE 6.3
Simplified block diagram of fuzzy logic approach.

The criteria signals are fuzzified to account for dynamic errors of measured signals. The thresholds are represented in fuzzy numbers to account for the lack of precision in decision making. Then the fuzzy signals are compared with fuzzy settings, which revolve in the form of Boolean algebra levels of true and false.

A simplified block diagram of fuzzy logic approach is illustrated in Figure 6.3.

6.3.3.1 Fuzzy Sets and Systems

In 1965, Zadeh laid the foundation of fuzzy set theory as a method to deal with the imprecision of practical systems. Bellman and Zadeh wrote: "Much decision making in the real world takes place in an environment in which the goals, the constraints and the consequences of possible actions are not known precisely." This "imprecision" or fuzziness is the core of fuzzy sets or fuzzy logic. Fuzzy sets were proposed as a generalization of conventional set theory. Partially as a result of this fact, fuzzy logic remained the purview of highly specialized and mathematical technical journals for many years. This changed abruptly in the late 1980s.

6.3.3.2 Fuzzy Sets

In a conventional (nonfuzzy, hard, or crisp) set, an element of the universe either belongs or does not belong to the set. That is, the membership of an element is crisp — it is either yes (in the set) or no (not in the set). A fuzzy set is a generalization of an ordinary set in that it allows the degree of membership for each element to range over the unit interval [0,1]. Thus, the membership function of a fuzzy set maps each element of the universe of discourse to its range space, which, in most cases, is assumed to be the unit interval.

One major difference between crisp and fuzzy sets is that crisp sets always have unique membership functions, whereas every fuzzy set has an infinite number of possible membership functions that may represent it. This enables fuzzy systems to be adjusted for maximum utility to a given situation.

6.3.3.3 Fuzzy Systems, Complexity, and Ambiguity

Zadeh's principle of incompatibility was given in 1973 to explain why there is a need for a fuzzy systems theory. The principle states, in essence, that as the complexity of a system increases, our ability to make precise and yet significant statements about its behavior diminishes until a threshold is reached beyond which precision and significance (or relevance) become almost mutually exclusive characteristics. This suggests that complexity and ambiguity (imprecision) are correlated: "The closer one looks at a real-world problem, the fuzzier becomes its solution."

It is a characteristic of the way a human thinks to treat problems involving complexity and ambiguity in a subjective manner. Complexity generally stems from uncertainty in the form of ambiguity; these are features of most social, technical, and economic situations experienced on a daily basis. In considering a complex system, humans reason approximately about its behavior (a capability that computers do not have) and thus maintain only a generic understanding of the problem. This generality and ambiguity are adequate for a human to perceive and understand complex systems.

As one learns more and more about a system, its complexity decreases, and understanding increases. As complexity decreases, the precision afforded by computational methods becomes more useful in modeling the system. For less complex systems, thus involving little uncertainty, closed-form mathematical expressions offer precise descriptions of the system's behavior. For systems that are slightly more complex but for which significant data exist, model-free methods, such as computational neural networks, provide powerful and effective means to reduce some uncertainty through learning based on patterns in the available data.

Basic statistical analysis is founded on probability theory or stationary random processes, whereas most experimental results contain both random (typically noise) and nonrandom processes. One class of random processes or stationary processes exhibits the following three characteristics:

1. The sample space on which the processes are defined cannot change from one experiment to another, i.e., the outcome space cannot change.
2. The frequency of occurrence, or probability, of an event within that sample space is constant and cannot change from trial to trial or experiment to experiment.
3. The outcomes must be repeatable from experiment to experiment. The outcome of one trial does not influence the outcome of a previous or future trial.

However, fuzzy sets are not governed by these characteristics.

6.3.4 Genetic Algorithms (GA)

Evolution algorithms have become very popular tools for search, optimization, and machine learning algorithms. There are many different types of evolutionary algorithms. Genetic algorithms and evolution strategies are two of the most basic forms of evolutionary algorithms. Genetic algorithms were

TABLE 6.1

Comparison of Classical AI Techniques

	Artificial Intelligence (AI) Techniques			
Features	**ES**	**ANN**	**FL**	**GA**
Knowledge used	expert knowledge in forms of rules	information from training sets	expert knowledge in developing fault criteria	information data search
Modify results	inference engine rules can be changed	internal signal cannot be changed	easy to change internal signal	cannot be changed internally
Self-learning	possible	natural	possible	natural
Robustness	noncritical, easy to ensure	difficult to ensure	not critical, easy to ensure	difficult to ensure
Diagnose fault	convenient	large number of simulations required	convenient knowledge and simulation are used	large number of simulation required
Computations	extensive	dedicated hardware	moderate	extensive

developed by Holand (1973) and his students, who worked on GA from the 1970s through the mid 1980s.

Genetic algorithms emphasize the use of a gene type that is decoded and evaluated. These gene types are often simple data structures. The chromosomes are bit strings that can be recombined in a simple form of several reproductions and can be mutated by simple bit flips. These algorithms can be described as function optimizers. GA algorithms find competitive solutions, but GA is also useful as a search process rather than strictly as an optimization process. As such, competition of selection of the fittest is the key aspect of a GA search.

The intelligent systems (IS) — ES, ANN, FL, and GA — have their own advantages and limitations. The IS systems have the following features in common for comparison, as seen in Table 6.1: what knowledge is used, how to modify the results, self-learning ability, robustness, fault diagnosis, and computations required. These attributes will be compared for different automation functions, for example fault analysis/diagnosis.

6.4 Intelligent Systems in Distribution Automation

6.4.1 DSM and AI

There have been successful implementations of ANN, expert systems, and other heuristic schemes and generic algorithms in the area of load dispatching, optimal power flow (OPF), and unit commitment, but they have not been fully utilized in the area of load management until recently. It is worthwhile to explore the possibility and potential of using intelligent systems in managing loads in distribution systems.

The limitations of existing demand-side management (DSM) are as follows:

1. The chronological load impact of DSM can be taken care of, but often at the cost of ignoring the network aspect. Thus a full AC network that accounts for VAr/MW is necessary.
2. An accurate model of DSM options and characteristics produces a high computational burden due to the large number of control variables at each element node.
3. The requirement of full AC modulation calls for a highly efficient algorithm.

The mathematical programming models that are used in DSM fall into the category of a mixed-integer nonlinear programming (MINLP) problem, and this poses a different computational challenge. In the presence of highly nonlinear power-flow constraints and a large number of other variables and constraints, the real-time DSM dispatching IS is beyond the use of mathematical programming techniques. This forms a void where such techniques as ANN or ES could be useful.

The following aspects suggest an appropriate network for IS applications:

1. *Computational speed requirement*: The predispatch mode of optimal DSM is done off-line. The real-time dispatch requires very fast solution, and ANN and ES have the requisite capabilities.
2. *Operator's judgment*: The operator's personal judgment and experience can form important inputs, rather than relying solely on the results from a mathematical (optimal) model. An expert system allows us to incorporate such information through heuristic rules. For example, an operator may have a priori knowledge of the value and impact of a given load interruption in a particular demand mode.
3. *Satisfying rather than optimal*: In a mathematical programming practice, it is difficult to assess whether a feasible solution is satisfactory if it does not happen to be the optimal solution. We may experience local minima for a MINLP problem in the case of a nonconverse power system network. AI methods, on the other hand, can lead to a "satisfying" solution with reasonable certainty even though it may not be the optimal solution. However, given the notion of a DSM dispatch problem and the computational complexity, the system operator may be interested in getting a satisfying solution rather than the theoretical optimal solution in the DSM problem, thereby reducing the computative burden.

4. *Special structure of DSM problem*: There are special structures and power system properties that can be utilized in an AI framework to simplify the problem within reasonable tolerance of the probability of the solution. Unit commitments (UC) are derived as network constraints when scheduling between hourly OPF over time. Many of the unit commitment constraints have useful fuzzy representation rather than the hard-constraint approach used in a mathematical programming model.

Based on the above observations, an AI approach could be developed that would provide a compromise between mathematical programming rigor and computational practicality. For example, the following is one way of achieving optimal DSM scheduling using AI:

1. Taking the output of optimal power flows for different hours from the predispatch stage to train a neural network to yield a set of power-flow solutions in real time given the real-time data on nodal demand and DSM resource availability
2. Fuzzifying the DSM characteristics and some of the unit-commitment-related constraints, like start-up/back-down constraints and minimum uptime/downtime constraints, to reflect the degree of satisfaction of these constraints
3. Applying a rule base compiled with a fast-decouple load flow to check the constraints of the network solution and analyze DSM impacts

6.5 Voltage/VAr Control

Voltage control (VAr control) within a specified range of tolerance and capacitor/LTC (load tap changer) is an effective means of minimizing loss while improving voltage profile. This also improves reliability and defers the need for future construction of additional capacity. The problem has been solved initially by using such traditional optimization algorithms as linear programming, nonlinear programming, and mixed-integer programming. These solution schemes have been studied relative to their computational burden as well as a cost-benefit analysis relative to the software and hardware requirements for work in the budding area of voltage/VAr control.

Recent efforts to improve the mathematical techniques include the use of artificial intelligence (AI). One shortcoming of existing AI methods is the treatment of voltage regulation and capacitor switching in an isolated manner. The use of optimization techniques such as linear programming (LP), decoupled AC optimal power flow (OPF), and mixed-integer programming have also provided limited useful results.

Some forms of heuristics have been proposed to handle the problem of discrete variables used for selecting/switching capacitors, changes of modeling the constraints of variables in distribution in mathematical programming techniques for distribution voltage or VAr control.

These challenges suggest that future research to facilitate development of a regulator-capacitor switching scheme and a network for optimal VAr control could focus on the following areas:

1. A well-designed rule-based heuristic combination with linearized OPF has potential as an optimization scheme. The heuristic rule base can help the operator in selecting the discrete variable for the capacitor switching scheme, while the linearized OPF can be used to check the network feasibility. This involves a good mix of operator judgment and heuristic rules.
2. Genetic algorithms (GA) are potential candidates for use in distribution automation like control of voltage/VAr. GAs, which provide multiple search paths and simplified computation, can be utilized as a preprocessor for the optimization algorithm. In this scenario, the GA preprocessor selects only the subset of switches that are critical for opening and closing rather than having millions of possible combinations of switches to be evaluated for loss minimization.

6.6 Network Reconfiguration via AI

Network reconfiguration refers to balancing the load distribution in a power system during or after a disturbance while accounting for power-loss-minimization voltage, thermal-generation constraints, and power-outage costs. ANN, ES, and fuzzy logic have found applications in the problem of network reconfiguration. Recent methods have used ANN to reconfigure distribution systems by determining the optimal system topology that reduces the power loss according to the variation of the load pattern. It is based on a two-stage ANN. The first ANN estimates the load levels of each zone, while the second ANN determines the appropriate system topology on/off status of switches from the input of the first-stage ANN.

Expert systems have also been applied to network reconfiguration. For example, ES can be used to solve for transformer and feeder overloads. The heuristic rule is used to locate the appropriate feeders to which loads must be transformed and the amount of loads to transfer. Another method of ES by Mondon integrates planning-knowledge sources containing restoration expertise. The scheme uses a qualitative simulation model to predict the results of the considered action and then uses a quantitative model to verify the correctness of plans in terms of numerical constraints.

Fuzzy logic has also been used to solve the load-balancing problem coupled with a min/max optimization problem. The fuzzy membership function used in the analysis represents the degree of satisfaction for (a) load balancing of transformers and feeders and (b) the number of switching operations that minimize the power loss during load transfer.

6.6.1 Further Research Work in Network Reconfiguration Using Artificial Intelligence

The use of AI-OPF models for network reconfiguration is desirable for real-time applications. It consists of the following features:

- Enhancement of the generic power flow to efficiently handle radial distribution systems by using rule-based techniques to handle priority of loads and feeder measurements
- Enhancement of optimization algorithms such as interior-point method to handle the mixed-integer variables via a mixed linear-integer programming
- Inclusion of the postoptimal sensitivity calculations for the interior point (IP) algorithm to compare different possible network configurations based on the associated cost of holding binding constraints
- Off-line training of ANN for use in reconfiguring the network; hybrid AI-OPF can also be used to achieve an optimally reconfigured system

6.7 Fault Detection, Classification, and Location in Distribution Systems

Fault detection and classification remain one of the problems that continues to challenge distribution engineers. Classical methods have been used for fault detection and classification. The most common techniques are based on harmonic-sequence components. This involves the detection and classification of high-impedance faults by measuring the degree of imbalance and comparing it with a given threshold. Researchers have also developed an amplitude ratio technique, a method that senses the ratio of the second harmonic to the fundamental current. Other techniques compare the even and odd harmonic currents at the first and seventh harmonics. A third classical method uses the phase relationship, which monitors the third harmonic phase with respect to the fundamental frequency.

Energy techniques, on the other hand, use an approach based on the increase of high-impedance faults. A randomness technique is another method that uses the randomness of harmonic currents as a characteristic of high-impedance faults. All of these techniques have their drawbacks and strengths.

6.7.1 Use of AI Techniques for Fault Analysis

With the progress made in AI, the classical approach used to analyze fault problems has changed. AI schemes are being used to solve the most complex problems such that we can achieve speed, accuracy, and reliability of the results and low cost for the detection schemes. Many AI-based methods have been considered by researchers. Some of the outstanding work needs to address the following:

1. Development of a rule-based technique for classifying different types of faults combined with the use of a fuzzy logic system for appropriate fault location
2. The use of ANN to classify faults that are hard to model, such as arcing and high-impedance faults
3. A hybrid of ANN and expert systems to combine the special features of diagnosis and location of different faults

6.8 Summary

This chapter continues the evaluation of distribution automation functions (DAFs) using intelligent systems. First, an overview of intelligent systems was given for commonly used methods such as expert systems, fuzzy logic, and artificial neural networks (ANNs). The procedures for utilizing performance of DAFs were also given. References to case studies of AI-based DAFs and of working examples of candidate AI-based DAFs are available in the literature.

Case Study of Voltage/VAr Control Using Artificial Intelligence

Voltage/VAr control involves adjusting a system's voltage profile by controlling the reactive power using various methods and components such as capacitor banks, load-tap-changer transformers, and line regulators, to name a few.

Voltage control within specified limits and capacitor switching is an effective means of minimizing loss and improving the voltage profile and reliability of a power network. The real-time application of AI to solving the voltage/VAr-control problem must take into consideration multiphase unbalanced operation of the distribution system, dispersed generation, multiphase multimode control equipment, and large system configurations.

The steps for applying AI methods to the voltage/VAr problem comprise the following:

1. Developing the knowledge base in the off-line mode using an optimization model
2. Real-time data acquisition of network information
3. Accessing the knowledge base to select the planned dispatch functions (load-management options, capacitor switching) under the specific load condition that has occurred in real time
4. Invoking the rule base to ensure that the load-management and VAr-control options are realistic (checking the limits on capacitors, maximum curtailable loads, etc.)
5. Performing load flow to check violations of the network constraints; at this point, the operator uses judgment to choose among conflicting objectives

The main transformer is equipped with a load tap changer (LTC) to keep the secondary bus (11.4 kV) voltage close to the preset value (Figure 6.4). In addition, a shunt capacitor is installed at the 11.4-kV bus to compensate the reactive power flow through the main transformer. The current practice is to switch on/off the shunt capacitor according to system reactive power demand such that the reactive power flow over the main transformer is minimized. Thus, reactive power/voltage control in a distribution substation can be achieved either by changing the on/off status of the capacitor or by adjusting the tap position of the LTC. Coordination between capacitor

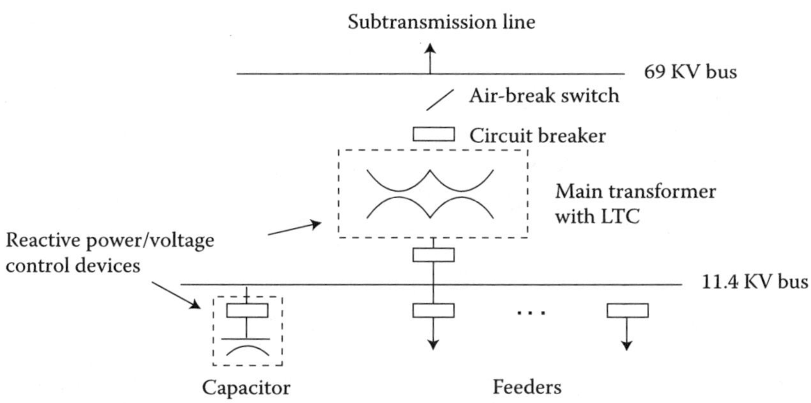

FIGURE 6.4
Problem Set 6: a system that is part of a 69/11.4-kV distribution substation.

switching and tap movements is necessary to achieve satisfactory control of reactive power and voltage.

We want to find the proper on/off status for capacitor and LTC tap position for the 24 h in the next day; therefore, we do the following:

Let us define

$$X = 1, \text{ when capacitor is on at hour } i$$

$$X = 0, \text{ when capacitor is off at hour } i \ (i = 1, 2, \ldots, 24)$$

In addition, let us define

$$\text{TAP} = \text{LTC tap position at hour } i \ (i = 1, 2, \ldots, 24)$$

$$= \text{integer between } -8 \text{ and } 8$$

Problem Set 6

6.1 Consider a voltage/VAr-control problem that has been solved using alternative methods. Using the above case study for voltage/VAr control, formulate a method of implementing voltage/VAr control using artificial intelligence.

6.2 Apply the algorithm developed in Problem 6.1 and show how artificial intelligence can be used to solve the given voltage/VAr-control problem.

6.3 Define the following terms as they pertain to the distribution automation:

   a. Reconfiguration
   b. Restoration
   c. Power Quality
   d. Fault Analysis

6.4 Construct the architecture of the following Artificial Intelligence (AI) techniques, highlighting their main features, layers and blocks:

   a. Expert Systems
   b. Artificial Neural Networks
   c. Fuzzy Logic

6.6 Using one of the Intelligent Systems (IS) search techniques in problem 6.4, develop a flowchart and a pseudo-code for solving the

Power Quality monitoring and correction problem in distributions systems.

6.7 By selecting an appropriate IS search technique repeat problem 6.6 for the Distribution System Reconfiguration function.

7

Renewable Energy Options and Technology

7.1 Introduction

Increasing demand for electric power in the 21st century and the need for more environmentally benign electric power systems are of critical concern to both government and stakeholders (industry and end users). Electricity shortages, power quality, rotating outages, and increasing oil prices have motivated many utilities and consumers to look for alternative forms of highly reliable energy. The traditional utility ties have been deregulated, yielding room for new market structures and players. The regulatory commissions in different parts of the world have unbundled the vertical utility industry into separate business units that can be categorized into three broad categories: generation companies, which include utility and nonutility companies; transmission companies, which may be under state ownership; and distribution companies, which are privately owned business units.

7.2 Distributed Generation

The subcategories for renewable energy sources are assigned as broad categories of utility and nonutility, as outlined in Figure 7.1.

The distributed generation is classified into two main categories: utility-owned and nonutility-owned. The nonutility generation is further classified as qualifying facilities, which are privately owned, with specific regulation on interconnection standards, location, operating efficiency, and tariffs. These qualifying facilities are further classified as renewable (photovoltaic [PV], biomass, wind, etc.) and cogeneration, which has the capability to generate electricity and heat.

The second subcategory nonutility is the IPPs (independent power providers), which are nonregulated but are authorized under operating standards to generate power. They are built in different sizes and options. The

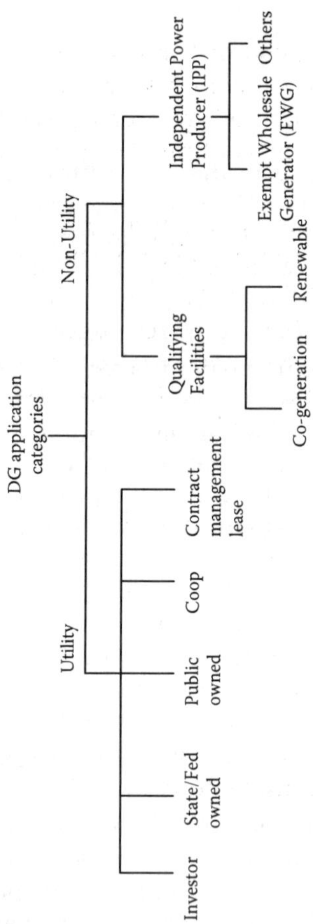

FIGURE 7.1
Distributed generation subcategories.

exempt wholesale generator (EWG) and other individuals (self-generating units) exist to merchandise aggregate power. The nonutility subsystem, on the other hand, consists of generation business that is owned as an investment company, state or federal ownership, as it is in the vertical regulated industry. Publicly owned cooperatives are another form of generation company for delivering power in small-to-medium/large quantities.

Regardless of subcategory, distributed generation (DG) has become a household name in the power sector. The different categories are proposed as:

1. On the generator side: It is called distributed energy resources. It is a premium power with the capability to produce backup power during frequency variation and voltage drops, peak power shaving, low-cost energy (base load), and continued cogeneration of heat and power.
2. On the demand side: It is a distributed resource that uses electricity efficiently, such as in heat pumps, solar heating/cooling devices, efficient regulation, load shifting, and other energy-swing schemes.
3. At the grid level: It involves distributed resources such as embedded generation, sited as storage systems, distributed power factor correction, and schemes to achieve reduction in losses and improve overall grid performance and grid capacity.

This chapter concentrates on distributed energy resources. First we define DGs as they are commonly used in nonutility systems. The term DG refers to small-scale generation of electric power by a unit close to the load being served. DG technologies range in size from 5 to 30 MW. They involve the use of such technologies as microturbines, sterling engines, fuel cells, and renewable energies such as photovoltaic, wind, and biomass systems.

As stated in the literature, DGs can meet the needs of a wide range of users, ranging from residential to commercial and industrial sections. Given the different definitions of DGs in the literature, we provide a general frame that encompasses some of the salient features of DG.

7.3 Working Definition and Classification of Renewable Energy

Renewable energy is derived from natural sources that replenish themselves over a short period of time. These resources include sun, wind, hydropower, organic plant and waste material (biomass), and earth heat (geothermal). Whereas renewable resources can generate both electricity and heat, the term "green power" is used in a narrow sense to mean electricity products that are generated from renewable sources that are environmentally and socially acceptable.

In all cases, distributed generation can be on-site generation, at the end user's facility. As stated earlier, they help stockholders to meet environmental and human health standards. Furthermore, distributed generation is considered in terms of its added value to financial and cost competitiveness, its ability to serve as backup power to the central station without the emissions produced by fossil fuels, its ability to stimulate improvement in the robustness of energy system supply chains, and its provision of a natural secondary/public good by reducing dependence on a centralized power infrastructure.

An alternative working definition assumes a distributed generation as a unit with one of the following:

- Modular electric generation or storage located close to the point of use
- Small-generation resources interconnected at the distribution level
- Small power plants, defined as less than 10 MW, typically less than 25 to 250 kW

DG is typically defined at the customer side of meter, but it can also be installed by the distribution utility.

7.4 Renewable Energy Options

We describe each of the commonly used renewable-energy options.

7.4.1 Solar

Photovoltaic (PV) cells and modules are configurable from 1 to 5 MW. Figure 7.2 shows a typical modern PV system. In 1839, French physicist Edmund Becquerel was the first to discover that certain materials exposed to light produce current. Refinements at Bell Laboratory in 1954 led to the development of silicon-based PV cells producing electricity conversion with over 4% efficiency. Following the energy crisis in recent years, the use of solar power has become more widespread. Commonly known as solar panels, PV modules are commercially available, provide no emissions, are an alternative to other energy sources, are reliable, and require minimum maintenance to operate. However, they are expensive compared with other renewable energy options and up to four to six times bigger than complicated alternative technologies. Today, advances in material science have led to the engineering and fabrication of solar panels with about 30% efficiency. In the

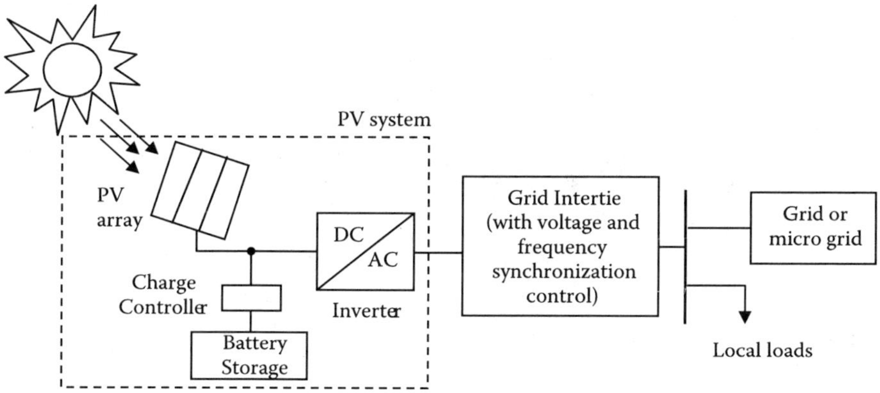

FIGURE 7.2
PV system.

presence of sunlight, solar panels comprising discrete cells radiate DC electricity that, after appropriate conversion to AC, is connected to power the load (lamp) or grid.

Insulation: a term used in PV systems to describe the available solar energy for conversion to electricity. Insulation levels are affected by the operating temperature of PV cells and the intensity of light (which is dependent on location). A third factor is the position of the solar panel. It must be positioned to maximize the power tracking from the panel while maximizing perpendicular incident light rays.

Emission: PV systems produce zero emissions.

Application: Photo cells can be part of a building, duplicating other building materials, and can have a wide range of application as distributed generation ranging from residential and commercial users to remote power consumers (structures in school, homes, community facilities, and commercial buildings). PV's greatest potential is as a green power because it does not emit pollutants or CO_2. However, it is a poor fit as a peaking power application source, since the unit outputs are not easy to control. PV systems produce better power during daylight periods of maximum available solar radiation and require battery storage for backup. It is not a good fit as premium power due to the unpredictable nature of power from solar cells.

Cost implication: Manufacturers continue to reduce the cost of installation while increasing efficiency as new technology is developed for manufacturing materials and other operational and management costs.

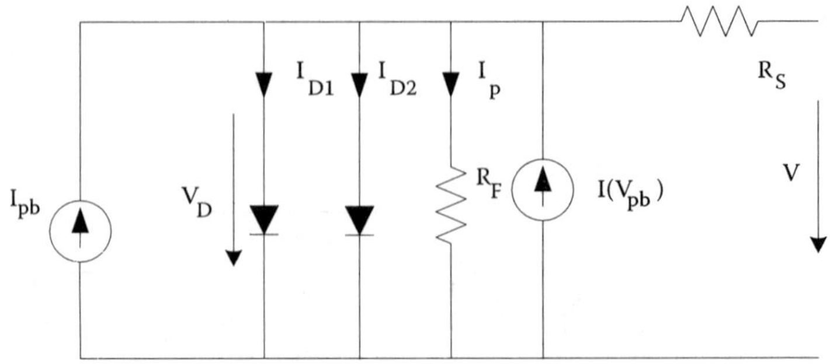

FIGURE 7.3
PV equivalent circuit.

7.4.1.1 Modeling

PV systems need to be developed and verified to optimize the output of the system at the design stage for maximum energy production and peak-sharing applications. Three models of a PV equivalent circuit are given in Figure 7.3.

The equivalent circuit consists of a diode and source, which are in parallel. A simplified model has the following $V - I$ equations:

$$I = I_p - I_d = I_p - I_s\left(\exp\frac{e(V + IR_s)}{mKT} - 1\right) \tag{7.1}$$

or

$$I = I_{sc} - I_d = I_{sc} - I(e^{qV/KT} - 1) \tag{7.2}$$

where

I_p = photon current
I_d = diode current
I_s = diode ideal factor
K = Boltzmann constant
T = absolute temperature
e = charge of an electron
I_{sc} = solar insulation

The open circuit voltage, V_{oc}, of the PV, is given as

$$V_{oc} = 0.0257\ln\left(\frac{I_{sc}}{I_d} + 1\right) \tag{7.3}$$

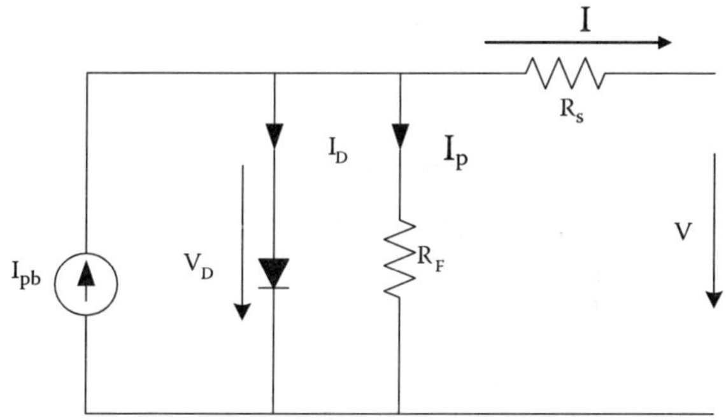

FIGURE 7.4
Modified equivalent circuit.

At 25°C, Equation 7.2 becomes

$$I = I_{sc} - I(e^{38.9} - 1) \tag{7.4}$$

Modified equivalent circuit

M_{ss} modified equivalent circuit (Figure 7.4) accounts for a real solar cell with external contacts, and voltage loss (drop) is accounted for through leakage currents and resistances. Using the V-I and Kirchoff's law

$$I_p = \frac{V_D}{R_P} = \frac{V + IR_S}{R_P} \tag{7.5}$$

$$I_{pb} - I_D - I_p - I = 0 \tag{7.6}$$

$$I_{ph} - I_S\left[\exp\left(\frac{V + IR_S}{mV_+}\right) - 1\right] - \frac{V + IR_S}{R_P} - I = 0 \tag{7.7}$$

Simpler way (parallel connection)

For parallel resistance

$$I = I_{sc} - I_d - \frac{V}{R_p} \tag{7.8}$$

For series connection

$$I = I_{sc} - I_D = I_{sc} - I_s\left(e^{\frac{qV_d}{KT}} - 1\right) \tag{7.9}$$

$$V_D = V + IR_s \tag{7.10}$$

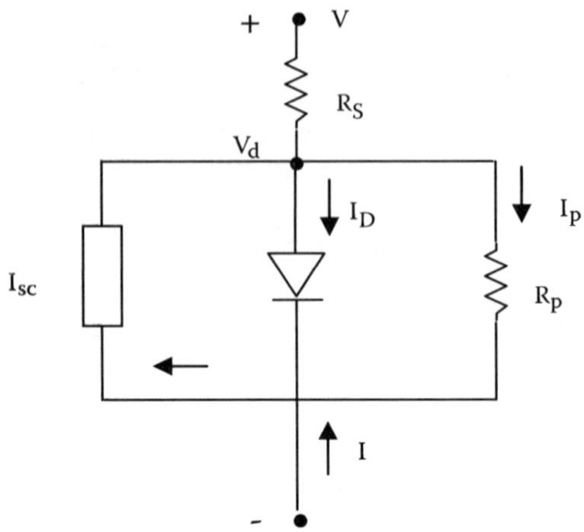

FIGURE 7.5
PV equivalent circuit for both series and parallel resistances.

$$I = I_{sc} - I_d\left\{\exp^{\left[\frac{q(V+IR_s)}{KT}\right]} - 1\right\} \tag{7.11}$$

Model 3

Combine PV to series and parallel resistances R_s and R_p, respectively (Figure 7.5)

Generalized PV equivalent circuit for both series and parallel resistances

$$I = I_{sc} - I_d\left\{\exp^{\left[\frac{q(V+IR_s)}{KT}\right]} - 1\right\} - \left(\frac{V + IR_s}{R_p}\right) \tag{7.12}$$

$$I = I_{sc} - I_d\left\{\exp^{38.9(V+IR_s)} - 1\right\} - \frac{1}{R_p}\left(V + IR_s\right) \tag{7.13}$$

$$I_{sc} = I + I_d + I_p \tag{7.14}$$

$$I = I_{sc} - I_d(e^{38.9V_d} - 1) - \frac{V_d}{R_p} \tag{7.15}$$

Note that

$$V = V_d - IR_s \tag{7.16}$$

To obtain from cell to module, we use

$$V_{\text{module}} = n(V_d - IR_s) \tag{7.17}$$

where n is the number of cell modules.

7.4.1.2 PV Systems

A PV system can be connected to a grid (utility system), can stand alone, or can be integrated.

- A PV system connected to a grid sends DC power to a power-condition unit converter, which converts DC to AC power.
- A stand-alone PV system off the grid is equipped with battery storage and a generator for backup power.
- An integrated PV system is directly coupled to its loads without battery or major power-conducting equipment.

Different types of loads affect the PV system, as shown from the performance characteristics of the V-I curves.

7.4.1.3 V-I Characteristics

PV systems connected to different loads exhibit different V-I characteristics, as seen in Figure 7.6.

a. Simple Resistive Load

$$V = IR \text{ or } I = \left(\frac{1}{R}\right)V \tag{7.18}$$

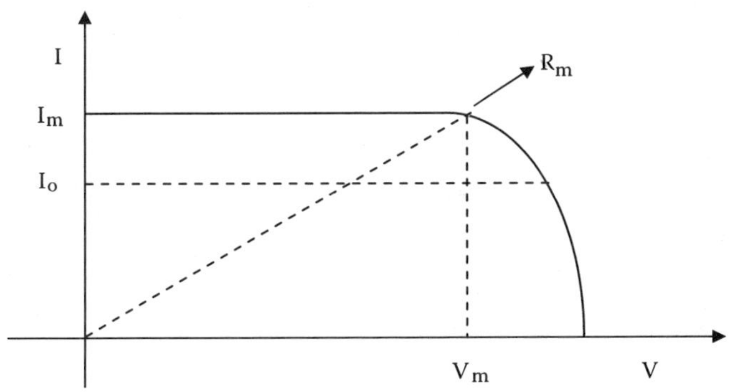

FIGURE 7.6
V-I characteristics: PV connected to resistive load.

$$R_m = \frac{V_m}{I_m} \tag{7.19}$$

To achieve R_{max}, we can use a maximum power tracker (MPT), which keeps a PV system operating at its highest efficiency point at all times.

b. DC Motor I-V curve (Figure 7.7)

$$V = IR_a + kW \tag{7.20}$$

Again, MPT is used to achieve the maximum operating point.

c. If battery is connected to the PV

$$V = V_B + R_i I \tag{7.21}$$

Maximum power point tracker (MPPT) is a feature of charge controllers which finds use in PV applications. It maximizes power transfer from the PV. The maximum power point is that point along the I-V curve corresponding to the maximum output power possible.

7.4.2 Wind Turbine Systems

Windmills have been used for many years to harness wind energy for mechanical work such as pumping water in farms and ranches. Today, it is one of the fastest-growing sources of energy. After the energy crisis of the 1970s, wind energy was considered to be the most economically viable choice in the portfolio of available renewable-energy options. Wind turbines can produce electricity at affordable cost without additional investments in infrastructure such as transmission lines. Wind turbines basically include the rotor, generator, turbine blades, and driver or coupling device.

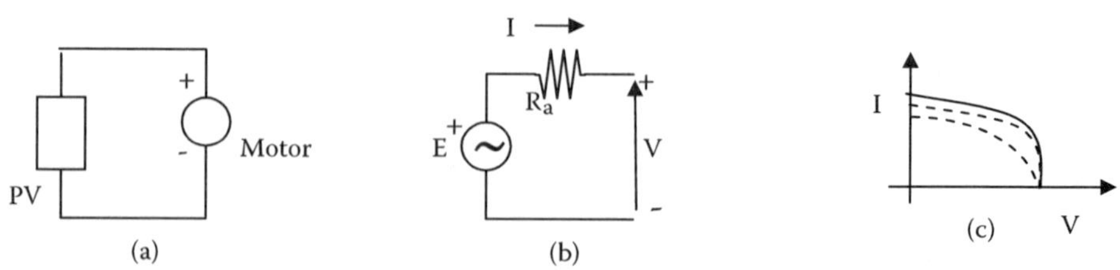

FIGURE 7.7
DC motor I-V curve.

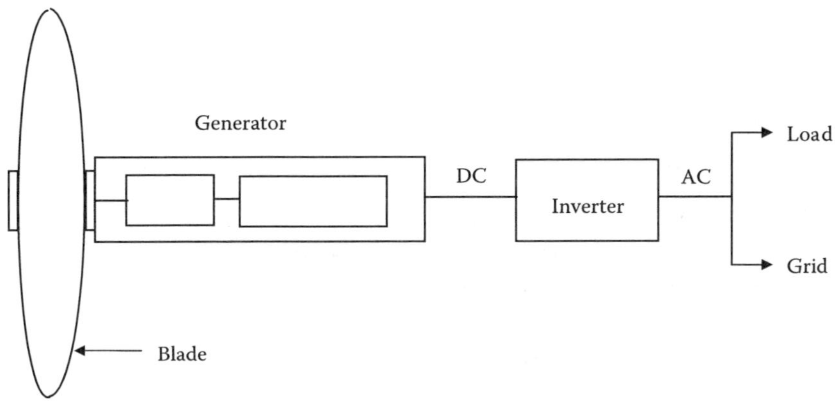

FIGURE 7.8
Basic components of a wind turbine system.

The operation is simple. The wind blows through the blades, with pressure exerted on the cross-sectional area of the blade. Aerodynamic force causes the blades to turn the rotor. The gearbox and the generator shown in Figure 7.8 are all in a single unit behind the machine blades. The output of the generator is passed through a unit for appropriate conversion from DC to AC.

The windmill comes in different configurations, normally horizontal or vertical. The wind speed and the height of a pole-mounted windmill above the ground contribute to the power output of the wind turbine system. The location of the system is equally important in sizing the output of windmill.

Wind turbines produce no emissions. They have a variety of sizes and applications and are classified into utility scale and individual scale. For large-scale utility projects, they can range from 1.5 to 5 MW loads. A small system can be as simple as a single pole and a blade.

7.4.2.1 Modeling

The mathematical model for wind power stems from aerodynamic power, given in

$$P = \frac{1}{2} p \pi R^2 v^3 C_p \tag{7.22}$$

where

p = air density
R = turbine radius
v = speed
C_p = turbine power coefficient, which represents the power conversion efficiency of a wind turbine
λ = ratio of the tip speed of the machine turbine blades to wind speed

$$\lambda = \frac{R\Omega}{v} \tag{7.23}$$

where

Ω = wind speed

C_p is maximum at $\lambda_{optimal}$

The wind turbine system uses induction generators that are independent of torque variation while speed varies between 1 to 2%.

7.4.2.2 Impact of Tower Height on Wind Power

A taller tower is expected to provide higher-speed winds to the turbine. Surface winds can also easily be affected by the irregularities or roughness of the earth's surface or forests/buildings. It is given as follows:

$$\text{Let}\left(\frac{v}{v_0}\right) = \left(\frac{H}{H_0}\right)^{\alpha} \tag{7.24}$$

where

v = wind speed at height H

v_0 = reference speed at ref height H_0

α = friction coefficient

In the U.S.,

$$\alpha = \frac{1}{T} \tag{7.25}$$

whereas in Europe,

$$\left(\frac{v}{v_0}\right) = \frac{\ln\left(H/2\right)}{\ln\left(H_0/2\right)} \tag{7.26}$$

7.4.2.3 Emission Control Technologies

Wind turbines produce no emissions. As a green-power application, the efficiency of wind turbines is superb, since they do not emit CO_2 or pollutants. However, wind power has an obvious disadvantage. Because its output cannot be controlled, it is mostly suited for peaking applications, producing power only when there is sufficient wind. Wind power cannot serve as premium power because of its unpredictability. Moreover, it is unsuitable for combined heat and power (CHP) applications.

Despite these drawbacks, windmills remain a subject of continuing research, with the principal focus on:

1. Lowering the minimum wind speed of operation
2. Developing voltage regulators to improve the turbine's ability to recharge the batteries while simultaneously producing electricity

Wind turbines are a relatively inexpensive way to produce electricity compared with PV, the most competitive green power to date.

7.4.3 Biomass-Bioenergy

Bioenergy is the energy derived from biomass organic matter such as corn, wheat, soybeans, wood, and residues that can produce chemicals and materials that we normally get petroleum. Biopower is also obtained from a process called biomass gasification, which converts biomass to a gas that can be used to power a turbine and generate electricity. The biomass gasification process is shown in Figure 7.9. The energy conversion process for biomass also utilizes the concept of pyrolysis oil, whereby biomass is converted directly into fluid fuels. The most common fuels are ethanol alcohol or biodiesel derived from corn ethanol.

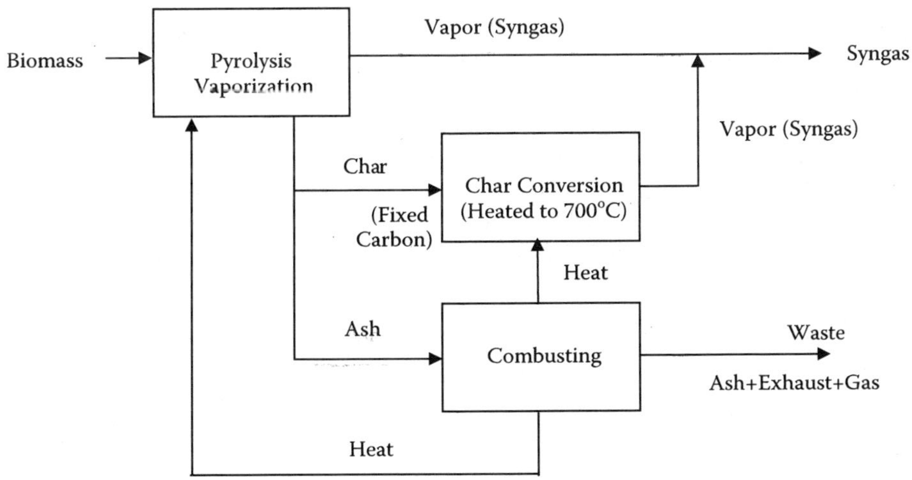

FIGURE 7.9
Biomass gasification process.

Biomass power plants are commercially available in the U.S. for up to 11 GW of installed capacity. Biomass power ranges from 0.5 to 3.0 GW using landfill gas and forest products, respectively. Biomass has traditionally been used for domestic cooking and heating, and such use is still widely practiced in developing countries.

7.4.3.1 Advantage and Disadvantages of Biomass Power

This source of power is viable only when a sufficient quantity of bioproducts is available and a conversion process is done. Truly continuous applications are likely for biomass systems, and it appears to be a good fit for CHP application.

Since the output of these units cannot be controlled, they are not suited for peaking applications. Ideally, biomass power is not a premium power due to the limited availability of bioproducts. It is also not an ideal green power due to the emission of CO_2 and other pollutants.

Several research efforts are underway to improve the quality of biomass power and reduce its environmental impacts. The form of energy input is very inexpensive. However, the efficiency of biomass power is low (typically less than 20%), and it is a relatively expensive way to produce electricity compared with PV. Several advances of technology are being used to reduce the emission of CO_2 and improve the green-power nature of biomass fuels. The success of biogas energy depends on the continuity of fuel supplies.

7.4.4 Small and Micro Hydropower

Hydropower is by far the oldest renewable source of power/energy. Small hydrosystems vary from 100 kW to 30 MW, while micro hydropower plants are smaller than 100 kW.

Small hydropower generators work at variable speed because the water upon which they depend flows at variable speeds. Induction motors are effectively used to provide a generator for a turbine system. The hydraulic turbine converts the water energy to mechanical rotational energy. The power available (P) from the flow of water (Q) is given as

$$P_{\text{avail}} = QH \tag{7.27}$$

where

Q = discharge, m^3/sec

H = net head

γ = specific weight of water, kN/m

The induction motor and stator quantities are modeled as a motion equation as follows:

$$V_{qs} = R_s i_{qs} - \frac{d}{dt}\Phi_{qs} + \omega\Phi_{ds} \tag{7.28}$$

$$V_{ds} = R_s i_{ds} + \frac{d}{dt}\Phi_{ds} - \omega\Phi_{qs} \tag{7.29}$$

$$V'_{qr} = R'_r\, i'_{qr} + \frac{d}{dt}\Phi'_{qr} + (\omega - \omega_1)\Phi'_{dr} \tag{7.30}$$

$$V'_{dr} = R'_r\, i'_{dr} + \frac{d}{dt}\Phi'_{dr} + (\omega - \omega_1)\Phi'_{qr} \tag{7.31}$$

$$Te = \frac{3}{2}p(\Phi_{ds} i_{qs} - \Phi_{qs} i_{ds}) \tag{7.32}$$

$$\frac{d}{dt}\omega_m = \frac{1}{2\pi}(T_e - F\omega_m - T_m) \tag{7.33}$$

where

F	combined rotor and viscous coefficients of friction
Φ_{qs} and Φ_{ds}	stator q and d axes, respectively
Φ'_{dr} and Φ'_{qr}	direct and quadrature flux, respectively
p	number of poles pairs
ω_m	angle of the rotor
ω_r	electrical angular velocity
H	rotor inertia

7.5 Other Nonrenewable Energy Sources

We present here some background and development issues on fuel cells, microturbines, and sterling engines, which are commonly used as a form of distributed generation. Later we will summarize all of the energy sources as renewable and nonrenewable energy for the purpose of comparison in terms of their functions, size, cost, and efficiency.

7.5.1 Fuel Cell

The fuel cell was first developed by Sir George Grove in 1839 and put to practical use in the 1960s by NASA to generate fuel for electricity needed by the spacecrafts *Apollo* and *Gemini*. Fuel cells are quiet, clean, and highly efficient on-site generators of electricity that use the electrochemical process to convert fuel into electricity. In addition to generating electricity, fuel cells can also serve as a thermal energy source for water and space heating or for cooling absorption.

Fuel cells can run using hydrogen, natural gas, methanol, or gasoline. The efficiency for conversion of fuel to electricity can be as high as 65%, as it does not depend on Carnot limits. This efficiency is what makes fuel cells environmentally friendly. Fuel cells come in a variety of different forms, all of which are under development. Examples include phosphoric acid fuel cells (PAFC), proton-exchange membrane (PEM), solid-polymer molten carbonate fuel cells (MCFC), solid oxide fuel cells (SOFC), alkaline (a direct methanol) fuel cells, regenerative fuel cells, and botanical ceramic fuel cells (BCFC). Fuel cells produce virtually no emissions of air pollutants or greenhouse gases. However, their costs are significantly high compared with those of conventional technologies.

7.5.1.1 Operation of Fuel Cells

Although fuel cells use different types of fuels, they operate using the same basic principle. A fuel cell consists of two electrodes: an anode and a cathode, separated by an electrolyte, as seen in Figure 7.10. Through the hydrogen catalyst, atoms split into a proton H^+ and an electron, and the proton passes through the electrolyte to the positive cathode. The resulting current is a DC current. Using a converter, we can easily generate an AC current. The combined hydrogen and oxygen at the cathode produce water and heat.

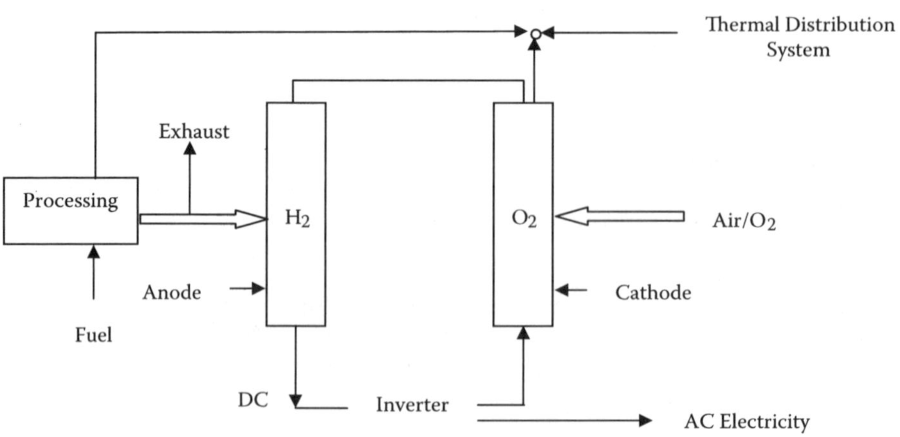

FIGURE 7.10
Schematic diagram of fuel cell system.

Details on the differences between fuel cells are based on materials and manufacturing costs, operating temperature, efficiency, and power-to-volume (weight) ratio. Additional models of fuel cells and their distinguishing features are available in the literature.

The topology of a fuel cell is defined as a stack that consists of the part of the fuel cell that holds the electrodes and the electrolytic material. Hydrogen is extracted from gasoline, propane, and natural gas refineries to operate commercial fuel cells. Emissions from fuel cells are very low, and so they have minimal environmental impacts. Their high efficiency leads to lower fuel costs and minimal maintenance due to a lack of moving parts. They have virtually no pollutant emissions, and CO_2 is rather low. Fuel cells are a good fit for green power and premium power. In addition, they provide a moderately high thermal quantity output and hence are ideal for CHP applications. However, they perform poorly as peaking power due to extremely high capital cost.

Developmental research work to refine the effectiveness of fuel cell is receiving greater attention.

Fuel cell efficiency ranges from 40 to 80%. Two commonly used fuel cell types are (a) phosphoric acid fuel cells (PAFC), which operate at relatively high temperature and use an external water-cooling system to cool the stack and (b) proton-exchange membrane fuel cells, which operate at a lower temperature than most of the other fuel cells and contain no chemicals such as liquid acids or molten bases that would cause concerns about the materials of construction.

7.5.1.2 Sample Calculation

Figure 7.11 shows a system for sample calculation for a fuel cell.

$$\text{Power: } P = VI \tag{7.34}$$

$$\text{Energy: } E = VI \cdot t \tag{7.35}$$

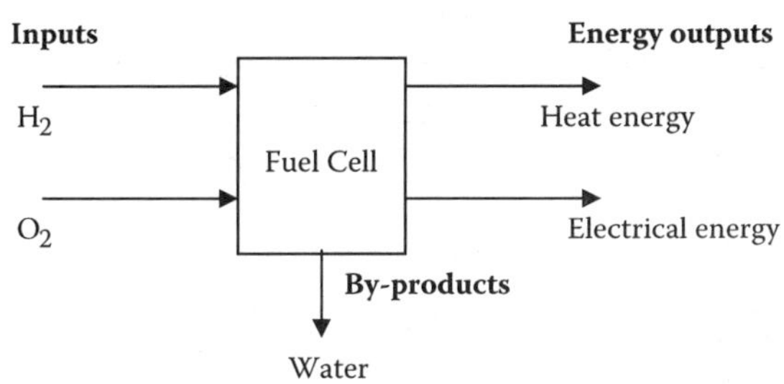

FIGURE 7.11
System for sample calculation for fuel cell.

The energy of chemical input and output, i.e., of H_2, O_2, and H_2O, is defined as a chemical energy given as enthalpy, Helmholtz function, or Gibbs free energy. Fuel cell energy can also be expressed in terms of calorific value. However, Gibbs free energy (G_f) is the preferred measure of fuel cell energy, defined as the energy to do external work, neglecting any work done by changes in pressure or volume. Gibbs energy represents the external work involved in moving electrons around an external circuit.

Gibbs free energy is used to represent the zero-energy point, and a change in G_f is given by

$$\Delta G_f = G_{f_{product}} - G_{f_{reactant}} \tag{7.36}$$

Consider $2H_2 + O_2 \rightarrow 2H_2O$, which is equivalent to the following (after chemical interaction) as

$$H_2 + \frac{1}{2}O_2 \rightarrow H_2O$$

where the new product is 1 mole of H_2O.

Reactants are 1 mole of H_2 and $^1/_2$ mole of O_2, which gives a balance of energy

$$\Delta \bar{g}_f = \bar{g}_f \text{ of products} - \bar{g}_f \text{ of reactants} \tag{7.37}$$

i.e.,

$$\Delta \bar{g}_f = (\bar{g}_f)_{H_2O} - (\bar{g}_f)_{H_2} \frac{1}{2}(\bar{g}_f)_{O_2} \tag{7.38}$$

However, the Gibbs free energy of the function is not constant. It changes with temperature and the state of the liquid or gas.

Continuing the calculation in terms of electrical work, we get $2N$ electrons around the equivalent fuel cell circuit. N is the Avogadro number, i.e., the charge on one electron.

$$-2Ne = -2F \tag{7.39}$$

where F is the Faraday constant or charge on 1 mole of electrons.

$$\text{Electrical work done} = \text{charge} \times \text{voltage}$$

$$= -2F \cdot E \text{ joules} \tag{7.40}$$

$$\therefore \Delta\bar{G}_f = -2F \cdot E \tag{7.41}$$

$$E = -\frac{\Delta\bar{G}_f}{2F} \tag{7.42}$$

This equation gives the fundamental EMF = reversible open-circuit voltage of fuel cell.

Example 1: for fuel cell at 200°C

$$\Delta\bar{G}_f = -250 \text{ kJ}$$

$$E = \frac{250,000}{2 \times 96,485} = 1.30 \text{ V}$$

The open-circuit voltage can also be used for other electrical power sources such as battery.

Efficiency is defined as

$$\eta_{max} = \frac{\text{electrical energy produced}}{\text{Gibbs free energy charge}} = \frac{\Delta \bar{g}_f}{\Delta \bar{h}_f} \times 100\% \tag{7.43}$$

Δg_f = Gibbs free energy at maximum electrical energy produced
Δh_f = negative when energy is released (between higher heating value [HHV] and the lower heating value [LHV])

$$\text{Efficiency of fuel cell (hydrogen fuel cell)} = \eta = p_f \frac{V}{1.48} \times 100\% \tag{7.44}$$

where
p_f = fuel cell utilization, typically 0.95
V = voltage of a single cell with a fuel cell stack

7.5.2 Ocean Energy

Ocean energy can be considered as another type of renewable energy. Briefly, the ocean can produce (a) thermal energy from the sun and (b) heat and mechanical energy from the tides and waves.

Ocean thermal energy has a variety of applications involving electricity generation. It uses a simple conversion from warm surface water or boiled sea water to turn a turbine that activates a generator. The conversion of ocean power to electricity involves heavy use of a mechanical turbine. A dam is usually used to convert sea power energy to electricity.

Active research is under development in the Pacific West of the U.S. and in Europe.

7.5.3 Geothermal Heat Pumps

Another minor source of renewable energy is the geothermal heat pump. This form of power is based on accessing underground steam or hot water from wells several miles into the earth. The conversion is done by pumping hot water to drive conventional steam turbines, which drive the electrical generator to produce electrical power. The water is then recycled back to earth to recharge the reservoir for a continuous energy cycle.

There are several types of geothermal power plants, namely dry steam, flash steam, and binary cycle. Dry-steam plants draw water from the reservoirs of steam, while both flash-steam and binary-cycle plants draw their energy from the recycled hot water reservoir.

Geothermal power is currently under development in the U.S., and some reasonable levels of power have been produced in California, Utah, Nevada, and Hawaii. Various applications of geothermal power exist, such as heat pumps, agricultural applications, fishing farms, food processing, etc.

Geothermal projects force significant upfront capital investment for exploration, drilling wells, and capital equipment cost. Exploration risk and environmental impacts are also considered in geothermal power plant projects.

7.5.4 Microturbine and Sterling Engine

Microturbines are a new generation of gas turbines that are small in size, typically producing between 25 and 500 kW of power. The technology is derived from the auxiliary power systems used in aircraft, diesel engines, turbochargers, and automotive designs. It consists of a compressor, combustor, turbine, and generator, as shown in Figure 7.12.

7.5.4.1 Description

Incoming air is compressed to about 3 atm of pressure and sent to a heat exchanger called a recuperator, where the temperature is raised by hot exhaust gases. The heated steam is mixed with fuel and burnt with enough energy to drive the turbine, which subsequently powers the electrical generator. The turbine has only one moving part, which drives the generator at 96,000 rpm on the air bearings and hence does not require lubrication and cuts down on operational/maintenance costs.

The generator creates AC current but can be rectified for DC output as needed. It can easily be operated for 60 Hz and reverted to a 50-Hz supply.

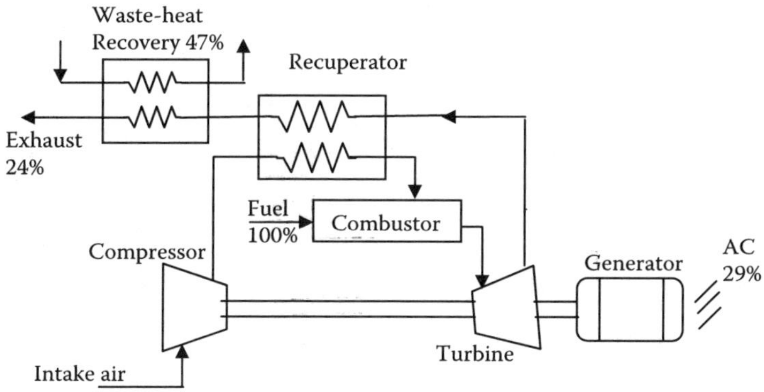

FIGURE 7.12
Schematic diagram of microturbine.

They can also easily be started in parallel for increased power output up to 30 to 60 kW.

In general, microturbine emissions are comparable with those of large turbines. NO_x levels are based on field tests and projections by manufacturers. Emission control to achieve an acceptable standard is focused on combustion design and flame control.

Microturbines are moderately applicable for peaking power but can be used as a stand-alone in limited areas. Its inverter-based generators offer a high premium of power quality. If efficiency degrades as temperature increases, it leads to CO_2 emission, which further degrades its efficiency. It is a moderate fit for CHP applications.

Further development on microturbines to lower costs due to electronics, power conditioning, and grid connection are concerns. The application of fuel diversity (low Btu) and digester gas are under development. Microturbine efficiency is up to 30 to 40% for nonrecuperated units and 20 to 30% for recuperated units. Microturbine efficiency is mostly impacted by the available natural gas pressure level. Further work that hybridizes microturbines with fuel cells will facilitate the generation of additional electricity of up to 60% efficiency.

There are many models of microturbines produced by a number of manufacturers, e.g., Eliot Energy System, Ingersoll Rand Co. Ltd., Honeywell, Capstone Turbine Corporation, etc. They are rated in terms of rated power, fuel input, heat rate, efficiency, emissions (NO_x, CO_2), turbine rotation, weight, noise, and size.

7.5.4.2 Sterling Engine

Sterling engines are classified as external combustion engines in which energy is supplied to the working fluid inside the engine from a source outside of the engine. They are scaled systems with an inert working fluid, either helium or hydrogen. They are generally found in small sizes of 1 to

TABLE 7.1

Comparisons of Renewable/Nonrenewable Energy Options

Technology	Size Ranges (kW)	Efficiency (%) Electric	Efficiency (%) Overall	Generation Electricity ($/MWh)	Environment Issues/Emission Control	Reliability Drawbacks on Fuels Source?
Reciprocating Diesel	30–5000	26–43	85–90	7.1–14.2	Controls required for NOx and COx	Yes
Turbines/Micro turbines	5–10	20–30	60–75	11.9–18.9	Low impact	Yes
Fuel cell PEM*	1–250	27–40	40–75	21.9–31.3	Nearly zero emissions	Yes
Photovoltaic (PV)	5–5,000	9–14	8–35	18.0–36.3	Zero direct emissions	No
Wind	5–1,000	20–40	20–26	0.2–28.5	Zero direct emissions	No
Biomass	20,000–50,000	5–10	5–20	0.05–0.09	Indirect emission	No
Geothermal	5,000–10,000	5–15	5–25	0.03–0.05	Low emission, only excess steam	No
Ocean	100–10,000	5–45	5–60	5–7	Zero direct emissions	No
Small hydro	100–1,000	50–55	60–90	0.03–0.25	Zero direct emissions	No

* PEM: Polymer Electrolyte Membrane.

25 kW. Their efficiency is typically less than 30%. Potential applications include use as small-scale portable power for battery chargers and as cogenerators for electricity and thermal/cooling. Sterling engines have low emissions when natural gas is used.

7.5.5 Comparison

Table 7.1 provides a comparison of renewable and nonrenewable energy options.

7.6 Distributed Generation Concepts and Benefits

Distributed generation (DG) is a relatively new approach to describe the new wave of generation at the customer side, which is less than that of the typical control power station in a competitive electricity market. DG has been given a variety of definitions relative to its rating, its power delivery area, its environmental impact, and its penetration level and point of connection. Although these criteria are necessary, they are not sufficient. We provide here a practical definition for DG as "an electric power source connected directly to the power network, preferably at the customer side of the meter, sufficiently smaller than the controlling generating plant." This definition does not define the rating of the generation source, since the maximum rating depends on the voltage level of distribution on the subtransmission network.

7.6.1 Categories of DG

The different voltages at various DG levels are described as follows:

Micro DG (between 1 W and 5 kW)
Small DG (between 3 kW and 5 MW)
Medium DG (between 5 and 50 MW)
Large DG (between 50 and 300 MW)

Note that the DG definition does not specify the technology options. Hence, DG can include any of the renewable sources, combined heat and power (CHP) applications are modular applications.

7.6.2 Criteria for DG Concepts

To further narrow the definition of DG, the following criteria should be observed:

1. Since voltage between transmission and distribution varies and is country dependent, DG should be close to the load and not the voltage level, which would limit DG to the distribution network.
2. Generation capacity is important in classifying DG, but there is no universal agreement on maximum generation.
3. Generator units should be able to satisfy reactive supply, but some DGs cannot and, hence, this may not be a sufficient criterion for definition.
4. Some DGs use renewable energy sources; some do not.
5. The technology for DG varies and hence cannot be used to narrow the definition. Similarly, generation mode, area, or ownership do not define DG. For example, both IPPs (independent power producers) and traditional generators can own a DG.

7.6.3 DG Benefits

Distributed generation (DG) can help to reduce investments in transmission and distribution capacity. From a planning point of view, DG can be placed close to load centers, thus minimizing loss in the networks from an operations point of view and reducing the costs for operations and controls. DG is especially favored to help reduce losses in distribution networks and to serve as stand-alone or back-up generation.

Grid support: DG can contribute to provision of ancillary services needed to maintain stable operation of the grid.

Environmental concerns: DG can help to improve or enforce environmental regulations.

DG combined generation and heat capacity: DG has provided the so-called cogeneration, trigeneration to provide electricity, heat, and steam for different applications.

Fuel diversity: DG uses different fuels at optimized prices, depending on the technology.

- Liberalization of electricity markets to an environment within a competitive market and due to various technologies makes it hard to generalize.
- Due to fuel diversity, any shortfall in fuel DG that is nonrenewable is considered to be risky and costly. As such, it may not help in alleviating blackouts and invariably degrades the security of the power system, hence emphasizing the need for increased regulating (backup) power.

Power quality and system frequency: Policies are necessary to ensure that GD systems adhere to some quality of supply and be able to maintain system frequency. High voltage levels approved for DG connections relative to the utility company must also be known and properly controlled to achieve voltage security to respond to changing market conditions. This flexibility in construction of lines and centralized generation has made the DG market economically attractive.

The major policy issues surrounding DG potential include:

- High cost of implementing various DG technologies, a concern of policy makers and stakeholders
- Less choice between more costly DG with expensive fuel supply, a fact that has discouraged the selection of DG when compared with the cost of central generation
- A policy to ensure economic efficiency of DG selection to liberalize the market

There are other attributes that factor into DG selection. These attributes exist on the part of the customer as well as the utility system, and the implementation depends on a variety of issues, including technical, economic, infrastructural, environmental, and regulatory issues. These may present themselves as barriers to the process of installing and utilizing renewable energy sources. For interconnections, there are technical requirements such as voltage and frequency matching.

The decision to utilize DGs is sometimes linked to purchasing green power. The success of such an endeavor requires a methodical approach, as it involves research and planning. An energy consumption audit should first be conducted by the potential DG user to determine the need. Factors such as the monthly usage, areas where energy could be saved, the need for green power, and the environmental impacts of the user's current electricity consumption should be considered. The aim of this step is to minimize the need for DGs where necessary while also minimizing the user's environmental impact.

With the analysis of the energy needs complete, the next step is an evaluation of the power options available. This raises questions such as:

Should power be generated on site, or should power or a renewable energy certificate (REC) be purchased from outside vendors?

Which type of green power is suitable?

The answers to these questions would be individual to the user's circumstance and location. On-site generation brings with it the need for an up-front investment but also a long-term reduction in the consumption of conventional energy and increased reliability of the power supply. Limitations to options available to the consumer stem from factors such as the electricity market structure, which varies by state as well as the availability and quality of resources such as solar, wind, or biomass fuel.

Comparatively, RECs and renewable-energy purchases, though not requiring up-front investment, result in savings for the duration of the contract only. In this case, it is necessary that a dependable supplier be identified. Some of the criteria by which suppliers should be evaluated include their reputation, financial strength, location, range of products, suppliers, and commitment to the society in which they operate. For a particular product, in addition to its cost and the length of the contract, factors such as the ratio of renewable energy to resource mix, the percentage of the product that stems from renewable energy, the location of the point of generation, and the certification or verification of third parties should also be considered. Examples of such third-party certifiers are Environmental Resources Trust and Green-e. A fact that should not be ignored is that the most suitable solution may be a hybrid of several of these sources, e.g., using both on-site generation as well as purchasing RECs to meet demand.

The next step in the process of implementing a renewable-energy-powered system is the calculation of a cost-benefit ratio for the user. When it is determined that green power is a viable option, the development of a procurement plan is the next step. This is a document that promotes the idea of utilizing such a power supply to the decision makers of the company or facility. It would indicate the pros and cons of utilizing the green power and the accumulated information that supports the purchase decision by addressing the scope of the project, the expected benefits, financial considerations, financial budget, and possible incentives where available.

Returning to the idea of utilizing on-site generation as the viable option, this would be followed by the planning and execution of an on-site renewable-generation project. In some cases, this would be more involved than simply purchasing the power RECs and would require the acquisition of assistance from experts in this field. The decision of a suitable and readily available fuel would be followed by the selection of technology that would be cost-effective. It should be determined whether the installation should be done incrementally or in one job, and whether the on-site system should be run only in conjunction with the grid supply or as a stand-alone system. The execution of such a project would also include a procurement strategy, where the necessary resources would be acquired. This would include contractors, vendors, and an energy-services company (ESCO) that would attend to the design, installation, and maintenance of the project.

It is important to note that the utilization of DG may often be integrated in an updated energy portfolio that includes efficiency upgrades and load management, which could have a holistic impact on the user's energy consumption.

7.7 Illustrative Examples

7.7.1 Example 1

Assume that a PV module is made up of 336 identical cells, all wired in series with the sun insulation (1 kW/m^2). Each cell has short-circuit current $I_{sc,}$ = 3.4 A, and at 25°C its reverse saturation current is $I_0 = 6 \times 10^{-10}$. The parallel resistance $R_p = 6.6\ \Omega$, and series resistance $R_s = 0.005\ \Omega$. Find the voltage, current, and power delivered when the junction voltage of each is 0.05 V; compute for different values of V_d increments of 1 to 10%.

Solution

Use $V_d = 0.50$ with data given.

$$I = I_{sc} - I_d\left(e^{38.9V_d - 1} - 1\right) - \frac{V_d}{R_p}$$

$$= 3.4 - 6\times 10^{-10}\left(e^{38.9\times 0.50 - 1} - 1\right) - \frac{0.50}{6.6}$$

$$= 3.3242 \text{ A}$$

$$V_{\text{module}} = n(V_{\text{d}} - IR_{\text{s}})$$

$$= 36(0.50 - 3.3242 \times 0.005)$$

$$= 17.40 \text{ V}$$

$$\text{Power (watts)} = V_{\text{module}} I$$

$$= 17.40 \times 3.3242$$

$$= 57.84$$

7.7.2 Example 2

Compute the energy at 15°C, 1 atm pressure, contained in 1 m^2 of a given wind turbine for (a) 100 h of 6-m/sec winds (13 mph) and (b) 50 h at 2 m/sec plus 50 h at 10 m/sec (average wind speed of 6 m/sec).

Solution

(a) for 100 h of steady 6-m/sec winds

$$\text{Energy (6 m/sec)} = \frac{1}{2} pAv^3 \Delta t$$

$$= \frac{1}{2}(1.225 \text{ kg/m}^3)(1 \text{ m}^2)(6 \text{ m/sec})^3(20 \text{ h})$$

$$= 13{,}320 \text{ W·h}$$

(b) for 50 h at 2 m/sec plus 50 h at 10 m/sec

$$\text{Energy (2 m/sec)} = \frac{1}{2}(1.225 \text{ kg/m}^3)(1 \text{ m}^2)(2 \text{ m/sec})^3(50 \text{ h})$$

$$= 245 \text{ W·h}$$

for 50 h at 10 m/sec

$$\text{Energy (10 m/sec)} = \frac{1}{2}(1.225\ \text{kg}/\text{m}^3)(1\ \text{m}^2)(10\ \text{m}/\text{sec})^3(50\ \text{h})$$

$$= 30{,}625\ \text{W·h}$$

$$\text{Total energy} = 245 + 30{,}625\ \text{W·h}$$

$$= 30{,}870\ \text{W·h}$$

The result shows the inaccuracy of using average wind speed. Here, the average wind speed produces less energy than the average of 2 m/sec and 10 m/sec, which equals 6 m/sec, and gives 133% more energy than winds blowing a steady 6 m/sec. Because of the difference in power as a function of wind velocity and time, the probability distribution of wind speed is used in formal work. It gives twice the amount used in average values.

Other features affecting wind turbine are

Temperature correction for air density

Altitude correction for air density, given as a function of molecular weight (MW) of gas

$$P = \frac{\rho \text{MW} \times 10^{-3}}{RT} \tag{7.45}$$

where

- P = pressure
- ρ = air density
- MW = molecular weight of gas
- R = ideal gas constant $\approx 8.2056 \times 10^{-5}\text{m}^3\text{atm/k.mol}$
- T = temperature

For example

For warmer air, density of air at 1 atm and 30°C (86°F)

$$P = \frac{(1\ \text{atm}) \cdot (28.97\ \text{g}/\text{mol}) \cdot (10^{-3}\ \text{kg/g})}{(8.2056 \times 10^{-5}\ \text{m}^3\text{atm}/\text{k.mol}) \cdot (273.15 + 30)\text{K}}$$

$$= 1.165\ \text{kg/m}^3$$

Altitude correction for density is computed simply as

$$P = 1.225\ K_T K_A$$

where K_T and K_A are temperature and altitude correction factors given in tables.

7.7.3 Example 3

Find the air density at

(a) 15°C (288.15 K) at an elevation of 2000 m (6562 ft)
(b) 5°C at 2000 m
(c) repeat for combined temperature and altitude correction

Solution

(a) Without combined temperature and altitude
Let

$$P = 1\text{ atm} \cdot e^{-1.185\times 10^{-4}\times 2000} = 0.789\text{ atm}$$

$$\therefore\ P = \frac{\rho.MW.10^{-3}}{R.T}$$

$$= \frac{(0.789\text{ atm})\times(28.97\text{ g/ml})\times(10^{-3}\text{ kg/g})}{(8.2056\times 10^{-5}\text{m}^3.\text{atm.k}^{-1}.\text{mol}^{-1})\times(288.15\text{ K})}$$

$$= 0.967\text{ kg/m}^3$$

(b) At 5°C and 2000 m, the air density

$$P = \frac{(0.789\text{ atm})\times(28.97\text{ g/ml})\times(10^{-3}\text{ kg/g})}{(8.2056\times 10^{-5}\text{ m}^3.\text{atm.k}^{-1}.\text{mol}^{-1})\times(273.15+5)\text{K}}$$

$$= 1.00\text{ kg/m}^3$$

(c) Combining temperature and altitude
For wind speed of 10 m/sec and elevation of 2000 m at temperature 5°C

$$P = 1.225\ K_T K_A$$

$$= 1.225 \times 1.04 \times 0.789 = 1.00\text{ kg/m}^3$$

From tables, K_A = 0.789 at altitude of 2000 m, and K_T = 1.04 at 5°C. Thus power density is

$$\frac{P}{A} = \frac{1}{2}\rho v^3$$

$$= \frac{1}{2} \cdot 100 \times 10^3$$

$$= 500\ \text{W/m}^3$$

7.7.4 Example 4

An anemometer mounted at a height of 10 m above a surface with crops, hedges, and shrubs shows a wind of 5 m/sec. Compute the speed and specific power in the wind at a height of 100 m. Assume 15°C and 1 atm of pressure.

Solution

Assume α for the hedges and shrubs is 0.20

From 15°C, 1 atm condition

$$P = 1.225\ \text{kg/m}^3 \text{ given}$$

∴ wind speed at 50 m

$$v_{50} = 5\left(\frac{100}{10}\right)^{0.20} = 7.92\ \text{m/sec}$$

Power at 100 m

$$P_{50} = \frac{1}{2} P v^3 = 0.5 \times 1.225 \times 7.92^3$$

$$= 304.3\ \text{W/m}^2$$

This power is more than 2.5 times as much power as the 76.5 W/m^2 available at 10 m. From equation $\left(\frac{v}{v_0}\right) = \left(\frac{H}{H_0}\right)^{\alpha}$, the relative power of the wind at height H versus the power at reference height H_0 is

$$\left(\frac{P}{P_0}\right) = \frac{\frac{1}{2} PAV^3}{\frac{1}{2} PAV_0^3} = \left(\frac{v}{v_0}\right)^3 = \left(\frac{H}{H_0}\right)^{3\alpha} \tag{7.46}$$

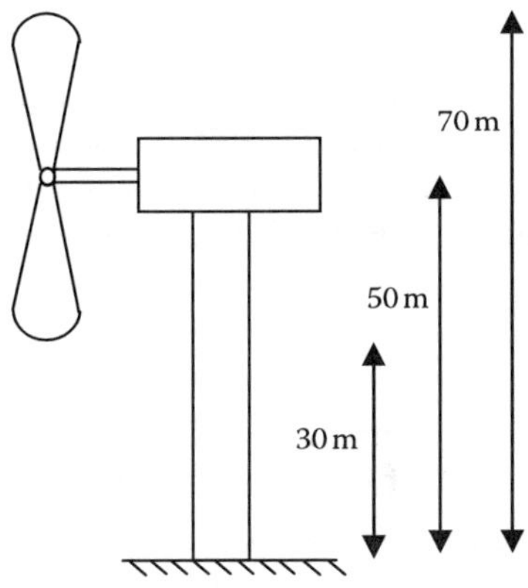

FIGURE 7.13
System for illustrative Example 5.

Thus the ratio indicates that the P at the elevation of 100 m compared with that of 10 m shows the dramatic impact of a cubic relationship between the wind speed and power.

7.7.5 Example 5

A wind turbine with a 40-m rotor diameter is mounted with its hub at 50 m above a ground surface characterized by shrubs and hedges. Estimate the ratio of specific power in the wind at the highest point that a rotor blade tip reaches to that at the lowest point the blade tip falls to (Figure 7.13).

Solution

$$\alpha = 0.2$$

$$\frac{P}{P_0} = \left(\frac{H}{H_0}\right)^{3\alpha}$$

$$= \left(\frac{70}{30}\right)^{3\times0.2} = 1.66$$

Power at the top tip of the rotor is $^2/_3$ more than the lower tip of the blade.

Other specific wind turbine performance calculations, including using the probability distribution of wind speed (Rayleigh or Weibull distribution functions) and capacity factor to estimate energy produced by the wind turbine, are done using the topology of the blades and the overall windmill characteristics.

7.7.6 Example 6

A microturbine rated at 100 A at its full 105-kW output burns 1.24×10^6 Btu/h of natural gas. Waste heat supplies water and space heating in an apartment house. The design calls for boiler temperatures ranging from 120 to 145°F, with the water being returned to the boiler. The system operates for 8000 h/yr.

a. Compute the water flow rate if 47% of the fuel energy is transferred.
b. If the boiler is 75% efficient and gas fuel cost is \$6/MBtu, how much money will the microturbine save in displaced boiler fuel?
c. If the utility electricity cost is \$0.08/kW·h, how much will the microturbine save in avoiding the use of utility electricity?
d. If operation and management is 1000/yr, what is saved in a year when using a microturbine?
e. If a microturbine costs \$220,000, what is the ratio of annual savings to the initial investment (ROI)?

Solution

a. The heat required to raise a substance given a specific heat rate, mass flow rate $\dot{m}$, and temperature change ΔT.

   $Q = \dot{m}C\Delta T$ 1 Btu will raise 1 lb of water by 1°F

   One gallon of water = 8.34 lb

$$\dot{m} = \text{water flow rate}$$

$$= \frac{(0.47) \times (1.24 \times 10^6 \text{ Btu/h})}{(1 \text{ Btu/lb°F}) \times (20°\text{F}) \times (8.34 \text{ lb/gal}) \times (60 \text{ min/h})}$$

$$= 58 \text{ gpm}$$

b. If $\eta = 75\%$ (of boiler)

$$\text{Fuel saving} = \frac{(0.47) \times (1.24 \times 10^6 \text{ Btu/h}) \times \left(\dfrac{\$6.0}{10^6 \text{ Btu}}\right) \times (8000 \text{ h/yr})}{0.75}$$

$$= \$37{,}300/\text{yr}$$

c. Utility electricity

Electricity utility savings = 105 kW × 8000 h/yr × (0.08/kW) = $67,200/yr

Cost of fuel for operating turbine, which is microturbine fuel cost,

$$= 1.24 \times 10^6 \text{ Btu/h} \times \left(\frac{\$6.0}{10^6 \text{ Btu}} \right) \times (8000 \text{ h/yr})$$

$$= \$59{,}520/\text{yr}$$

d. Microturbine saving = ($37,300 + $67,200) – $59,520 – $1,500

$$= \$43{,}480/\text{yr}$$

e. Initial rate of reaction $= \dfrac{\text{annual savings}}{\text{initial investments}}$

$$= \frac{43{,}480\,/\,\text{yr}}{220{,}000} = 0.198$$

$$= 19.8\%$$

7.8 Summary

The chapter provides a summary of renewable energy and a working definition of distributed generation resources. Models and characteristics of distributed generation (DG) and renewable energy sources are discussed. Furthermore, the potential benefits and other economic, environmental, and institutional barriers are discussed.

Problem Set 7

7.1 Identify and discuss the various forms of renewable energy alternatives and efficient methods of implementing them.

7.2 Given that a PV module is made up of 275 identical cells, all wired in series with the sun insulation (1 kW/m^2). Each cell has short-circuit current I_{sc} = 6.4 A, and at 80°F its reverse saturation current is $I_0 = 4 \times 10^{-10}$. A parallel resistance R_p = 5.8 Ω and series resistance R_s = 0.003 Ω. Find

a. Find the voltage, current, and power delivered when the junction voltage of each is 0.05 V.

b. Compute for different values of V_d increments of 1 to 10%. Use $V_d = 0.50$ with data given.

7.3 A wind turbine with a 80-m rotor diameter is mounted with its hub at 30 m above ground surface characterized by shrubs and hedges. Estimate the ratio of specific power in the wind at the highest point that a rotor blade tip reaches to that at the lowest point the blade tip falls to (Figure 7.14).

7.4 Discuss an implementation strategy for applying a distributed generation source using any renewable energy form of your choice.

7.5 Compute the energy at 31°C, 7 atm. contained in a given wind turbine of average cross sectional area 2.5m^2 over a time of:

a. 375 hours with wind speed of 4m/sec.

b. 95 hours with wind speed of 5m/sec.

7.6 Calculate the air density given that a wind turbine has a temperature of 25°C, at an elevation of 1500m. Repeat for combined temperature and altitude correction.

7.7 For a small hydro system, at 98 A and 200kW at full output, 2.5×10^6 Btu/h is burnt. The temperature range is given as 95 –150°C and the system operates for 7500h/yr. Compute the water flow rate if 28% of the fuel energy is transferred.

7.8 Corresponding to problem 3 stated above:

a. If operation and management is 800h/yr, what is saved in a year when using a micro turbine?

b. If the utility electricity cost is \$0.05/kWh, how much would the micro turbine save in avoiding the use of utility electricity?

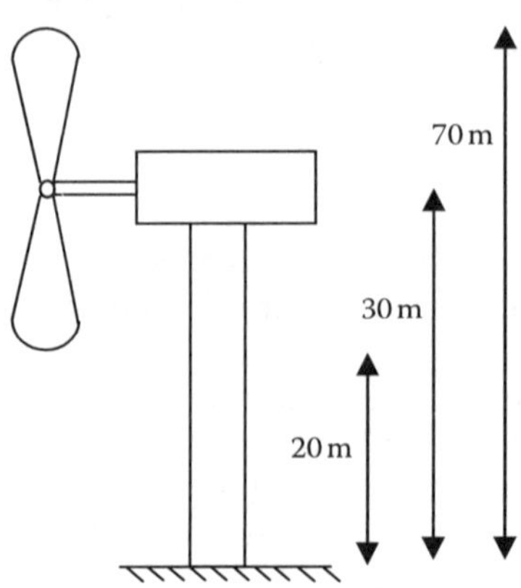

FIGURE 7.14
System for problem 7.3.

7.9 Develop a brief review of commonly used Distributed Generation (DG) technologies that include Renewable and Non-Renewable Energy Sources.

a. Discuss comparatively pertinent features of each type energy options based in capacity, efficiency, generation cost, environmental impacts, reliability and stability interconnection issues, and portability of the technology.

b. Select 2 DGs sources and implement its penetration on a power system topology of choice that has generation adequacy problems, and perform impact analysis to help justify DG feasibility.

c. What are some of the Cost Benefit Analysis (CBA) issues associated with the implementation of DGs at the distribution or subtransmission level?

8

Distribution Management Systems

8.1 Introduction to EMS

An energy management system (EMS) balances the sources of energy and consumption of energy to achieve the lowest cost. The energy sources can be electricity, water, gas, oil, steam, or renewable energy in the form of distributed generators (DG). The consumption of energy can be industrial, commercial, manufacture, or residential.

An EMS generally puts the user in control of energy consumption through monitoring, billing, and cost allocation. The integrated management software consists of power flow, security assessment, system stability, and system reliability. Furthermore, EMS represents a large collaboration of power distribution control products that connect state-of-the-art devices for communication control. It interfaces with communication and intelligent devices such as switchgear and intelligent switching controllers that are connected through an Ethernet network to computer systems equipped with software for collecting and displaying data from the network.

8.1.1 DMS and EMS

A DMS (distribution management system) and an EMS are similar in many ways:

1. Both collect measurements of the state of the system and its power devices remotely at the data collection terminals equipped with remote terminal units (RTU).
2. Both processes present information to operators through a user interface on a video display.
3. Both store information for later retrieval and analysis of historic events.
4. Both contain analytical functions to help operators interpret the information and analyze future situations.

5. Both are typically connected to other computer systems for data sharing and analyzing results.

However, there are fundamental differences between distribution and transmission systems; hence there are also differences between DMS and EMS:

- Distribution systems are typically radial; transmission systems are typically network connected.
- Distribution system devices are located along the length of distribution circuits, often on pole tops: transmission system devices are generally located only at substations.
- The number of locations requiring RTUs in a distribution system is at least an order of magnitude greater than the number of locations in the associated transmission system.
- On a distribution system, most field devices are manually operated; on a transmission system, most fields devices can be remotely controlled.
- The amount of data at a given distribution system device location is about an order of magnitude less than that at a transmission system substation.

EMS can provide:

1. Cost allocation of generator, etc.
2. Demand prediction of loads
3. Guide for optimal load shedding
4. Online metering of power flows and voltage and angle of different buses, lines and interregional transfers
5. Meter system parameters such as power quality, power factor and system voltage sag conditions
6. Protection control and relay settings

EMS has been developed over the years as a branch of control center, which comes on a scaled platform. EMS further encompasses supervision and control of power plants and high to medium voltage levels.

8.2 Functions of EMS

The complete chain of secondary monitoring functions of EMS includes assessment topology, state estimation, steady-state condition, time constant, and error from the network model, followed by operational enhancement functions. EMS also specifies coloring options such as node coloring and status island coloring functions, which serve as a guide and can be extended to DMS applications.

EMS has two main objective functions: production cost minimization and loss minimization. Production cost minimization is usually an objective for an EMS to minimize the cost of overall operations subject to scheduled controls in the system being monitored. The following controls for regulating MW are available:

1. Generator MW control output: This consists of generator cost curves that are modeled appropriately for each local or area-wide power plants.
2. Regulating phase shifter MW: This consists of phase regulation, which is normally done at the transmission level.

The EMS therefore provides more controls to minimize the objective function of loss reduction by improving voltage profile via the use of phase-shifter or regulating transformers as controls and unit commitment as controls at the optimum level. The conditions of Automatic Gain Control (AGC) are also suitable as controls, including load balancing.

In summary, the structure of a SCADA (supervisory control and data acquisition)-based EMS is organized to achieve the following:

- Reserve allocation (local/global region)
- Quality and security assessment and state estimation
- Load management
- Equipment protection

To achieve these multiple goals of the EMS system, a variety of software applications have been developed. Communication networks using TCP/IP and other protocols over copper, fiber, Modbus, or wireless are utilized for effective communication and control. Network devices such as monitors, Ethernet gateways, and Modbus Internet are all needed for adequate connectivity to sensors and for data acquisition and processing. The equipment is integrated with repeater signal conditioners, communication cables, junction boxes, and power supplies to make up the architecture of the SCADA system.

8.3 SCADA (Supervisory Control and Data Acquisition)

SCADA is a platform with basic functionality to classify or handle events, alarm processing, monitoring, and the limits of measurable power qualities.

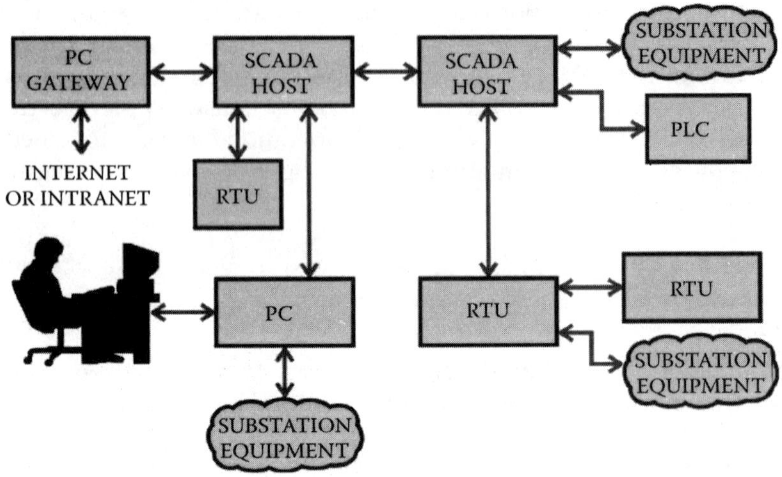

FIGURE 8.1
General SCADA architecture.

It consists of a process database, a man–machine interface (MMI) (a PC with a graphical user interface [GUI]), and application software. The MMI accesses the data from the process database and presents it in the form of single-line-diagram tabular displays and reports. MMI is based on a client-server architecture and the display device, which can be a workstation or a PC with a standard GUI. The general architecture of a SCADA system is shown in Figure 8.1.

The overall SCADA functions include:

- Data acquisition from the transmission system equipment and subsequent processing of the data received for further uses
- Provision of state estimation data based on data collected at the substation level
- Control of the transmission system equipment and alarms to notify operators when an abnormal event occurs
- Event and data logging to record all interactions between the operator and the system
- Man–machine interface that provides an interactive channel for the operator
- Voltage control for automatically controlling the voltage of any specific point in the transmission system

8.4 RTU (Remote Terminal Units)

RTUs are installed in distribution substations at various feeders and other pieces of equipment to facilitate automation of the distribution network. They are also used as a digital communication interface with computer-based substation control systems. They are designed in modular form for use in pole-top, single-node configurations as well as large multimode configurations in substations.

Distributed architecture is used to connect an RTU to the control node. The control node in turn connects with the DMS master using DNP3.0 protocol. IEC 810-5-101 communication protocol is also possible. Input/output (I/O) nodes include digital signal processing, which enables AC input from potential and current transformers. This information is used to compute real and reactive power flows; to calculate harmonic contents and other power quality indicators, such as voltage sag and swells; and to detect and collect distributed data, including the sequence of events. RTU also supports the definition and execution of programmable large functions, such as closed-loop voltage control of transformer taps.

The substation RTU communicates with the DMS master over an existing digital microwave link or over leased lines and a time-division multiple-access radio system. RTU communication generally uses 9600K links, and these are polled for data every 2 sec for status changes and every 10 sec for analog changes. The RTU analog points are typically configured with 1% deadband for reporting changes. RTUs are generally equipped to report data as specified in the protocol arrangement.

8.5 Distribution Management System (DMS)

A DMS is designed for supervisory control of residential or commercial loads at low distribution voltage levels. At these voltage levels, the number of lines, transformers, and switches may increase when compared with the transmission network. Very few switches are remotely controlled by the DMS. DMS acts as a system monitor and remotely controls switches and substations in the distribution system. A modern DMS consists of a distributed computer system (DCS), which has the capability of handling any application software, hence its usefulness in the design of DMS.

DMS, like EMS, is designed to be incorporated into the main control server, which provides support for transmission options, monitoring, and control. The off-line server provides outage information, maintenance simulation, and network planning and reconfiguration. As in EMS, the off-line server has I/O devices and is designed to accommodate database integration from

SCADA, AM/FM (automated mapping/facilities management), GIS (geographic information system), and real-time modes of application. DMS is coupled to other application programs that are interconnected with a user-interface software/programs with GUI capabilities.

8.5.1 System Hardware for DMS Station

DMS hardware consists of multiple computer processor nodes in an open distributed architecture with T-LAN (local area) networks to connect the computer nodes. Control rooms are also built to accommodate DMS consoles, large screens, and a large number of PCs for a variety of data maintenance and other functions. They are used for remote data collection or remote assessment of system functions.

8.5.2 SCADA System Functions for DMS

Each DMS has full high-performance SCADA functionality. This provides all typical data acquisition, alarming, supervisory control, historical data collection, and other functions expected in a modern control center (Figure 8.2). It is characterized by the following attributes:

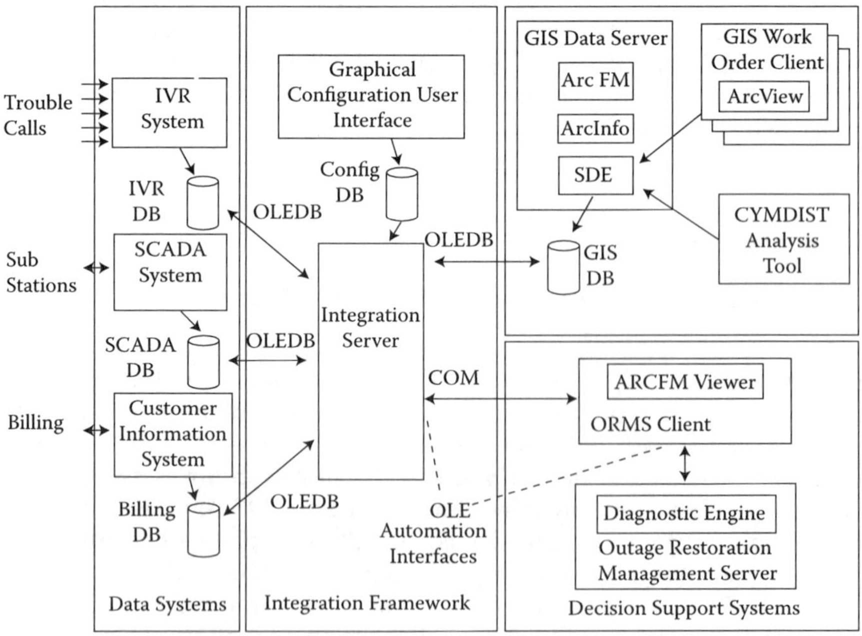

FIGURE 8.2
Architecture of integrated distribution management system.

- *Flexibility*: The architecture should be capable of providing sealable application and support a diverse set of distribution applications.
- *Expandability*: New functions can be integrated into the existing program easily without affecting other functions.
- *Maintainability/portability:* If changes to the database scheme for a power system model are required, the effect is limited to data-access routines; no application code should be affected.
- *Data integrity*: Data integrity must be easily accomplished and be independent from any application.

8.5.3 DMS Functions

At the heart of the DMS are the application functions that provide network model analysis and capability. The functions of a DMS application system are grouped into layers, as illustrated in Figure 8.3:

1. Substation and feeder SCADA (SFS)
2. Substation automation (SA)
3. Feeder automation (FA)
4. Distribution system analysis (DSA)
5. Application based on geographic information system (GIS), such as automated mapping (AM) and facilities management (FM)
6. Trouble-call analysis management (TCM)
7. Automatic meter reading (AMR) and distribution system analysis

These functions are displayed with selected application areas used in distribution automation functions, as seen in Table 8.1.

8.5.4 Substation and Feeder SCADA

A SCADA system coupled with RTUs serves as supporting hardware to the DMS in monitoring: the distribution equipment of substations (circuit breakers and other switching devices); the status of reclosers, cutoff switches, and load tap changers; voltage regulator positions; capacitor banks; bus phase voltages; transformer temperatures; relay settings; real and reactive power flows; harmonics; and voltage sag and swells. DMS monitors equipment located on pole taps and other locations along feeders, including line recloser sections, and other measurable quantities such as capacitor bank status, phase voltages and magnitudes, switchgear status, etc. The SCADA system at the substation provides sequence-of-events recording data collection and event logging, and it generates reports on system stations.

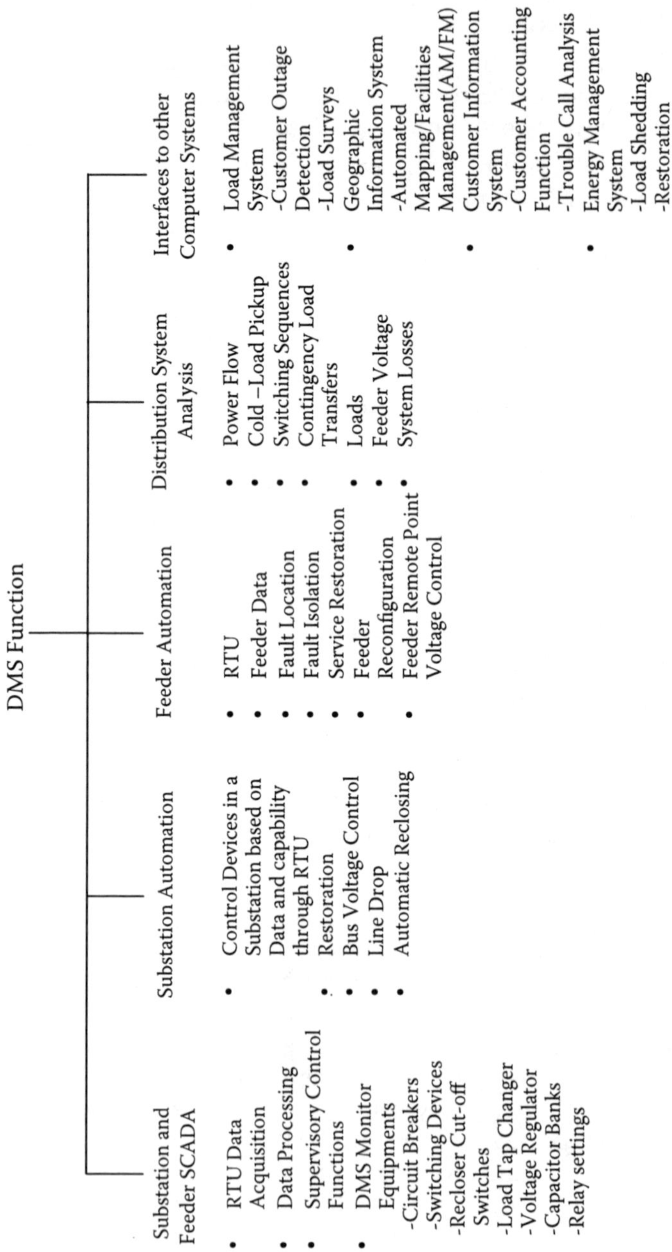

FIGURE 8.3
DMS function layers.

TABLE 8.1

Distribution Automation Functions

Substation Automation Functions	Feeder Automation Functions	Customer Interface Automation Functions
Data acquisition from: · Circuit breakers · Load tap changers · Capacitor banks · Transformers Supervisory control of: · Circuit breakers · Load tap changers Fault location Fault isolation Service restoration Substation reactive power control	Data acquisition from: · Line reclosers · Voltage regulators · Capacitor banks · Sectionalizers · Line switches · Fault indicators Supervisory control of: · Line reclosers · Voltage regulators · Capacitor banks · Sectionalizers · Line switches Fault location Fault isolation Service restoration Feeder reconfiguration Feeder reactive power control	Automated meter reading Remote reprogramming of time-of-use (TOU) meters Remote service connect/disconnect Automated customer claims analysis

8.5.5 Feeder Automation

The DMS system supports feeder automation functions, which include:

- Feeder automatic sectionalizing for fault location, isolation, and restoration (FLIR)
- Service restoration of feeder
- Feeder reconfiguration
- Voltage/VAr control
- Substation reactive power control
- Substation transformer load balancing
- Cold pickup and automatic reclosing

We discuss each of these functions briefly in the following subsections.

8.5.5.1 Fault Location, Isolation, and Restoration (FLIR)

The system application agent minimizes the duration of outages to customers caused by network faults. FLIR automatically assists the dispatcher in locating network faults that cause feeder breakers to trip and quickly determining the switching actions that will isolate faulted sections and then restore the power to unfaulted feeder sections both upstream and downstream of the faulted sections. The FLIR action depends on feeder RTU stations, which check the pressure of fault overcurrents. Restoration is done by using the switching actions command by FLIR.

8.5.5.2 Voltage/VAr Control

The voltage/VAr control application function uses telemetered measurements of VAr flows in the substation to establish the value of VAr flows and determine the power factor application to determine when to automatically switch capacitors. The objective of the function is to keep the power flow through the distant network within preset voltage limits.

8.5.5.3 Voltage Control

The objective of the voltage-control application is to achieve temporary small reductions in system load by reducing the voltage at the secondary side of HV/MV transformers. The voltage-control application system ensures that voltages at the customer level are within the contractual or obligatory level.

8.5.5.4 Substation Automation (SA)

The substation automation layer includes control devices, and data collected from these devices are processed via RTU or DMS application software that does digital signal processing (DSP). The data information from substation automation are used to perform system restoration based on voltage control, optimal reclosing, and switching.

8.5.5.5 Trouble-Call and Outage Management (TCOM)

The TCOM application system is used to identify and respond to network outages that are undetected by SCADA telemetry but are reported as a customer trouble call to the utility. The TCOM provides computer facilities that keep track of each customer trouble call. This is done by mapping customer calls to the electric model and tracing them to a common open fuse or other protective device. The system software module is able to determine the number of customers affected by each outage and assists the dispatcher in prioritizing outages according to their size and in dispatching trouble crews to deal with the outage. The distance of the outage location and the station service quality, determines the expected reliability indices.

8.5.5.6 Reconfiguration Function

DMS supports a system-reconfiguration function that provides computer support for switching plans that reduce the workload for the dispatcher. Automatic tracking of proposed switching action is done by using a reconfiguration program module.

8.5.6 Distribution System Analysis (DSA)

The distribution power flow is the key analysis tool in a DMS system. It models system components and is used to determine the steady-state criteria of the system voltage and to compute balanced or unbalanced system conditions. It is also developed to handle radial, loop, or mixed configurations. Distribution power flow is capable of handling single-phase, double-phase, or three-phase systems. Chapter 5 provides computational algorithms for distribution power flow.

Fast analysis techniques for voltage control, distribution losses, cold-load pickup, system restoration, and contingency analysis are available. Off-line stand-alone applications based on telemetered or static-mode data and state-estimation techniques can be used to detect erroneous measurements in the distribution system. The distribution power flow, in general, serves the same function as in the EMS counterpart used in the transmission system environment.

8.5.7 Load Management System (LMS)

These are special interconnection systems designed for direct load control. They are equipped with a load management system accessible through a communication system. Large load management systems employ power-line communication (PLC) or some other communication technology using the distribution feeder as a communication path. The load management system is used for different automation functions in a distribution management system (DMS). It provides interface automation for automatic meter reading (AMR), direct load control, customer-outage detection, customer and load management, as well as trouble-call analysis for a given distribution system.

8.5.8 Geographic Information System (GIS)

The geographic information system (GIS) is an automated mapping/facilities management (AM/FM) system that was developed in the 1980s in the U.S. It links automated digital maps of utility infrastructure to databases containing nonspatial facility-management data. The GIS is easily interfaced with a distribution-automation operator and other customer-based information systems. A GIS database could be used as the source for distribution model data supporting the distribution system analysis formulation. GIS can also provide automatic data transfer on the status of monitored switches and operator entry of manual switches, trouble-call management, and update information to the customer during an outage.

Geographic information systems (GIS) are now developed with Web-supported servers. This enables updating of maps more efficiently and accurately compared with the manual-based GIS services done daily based on field marking from crew workers.

8.5.9 Customer Information System (CIS)

The customer information system (CIS) was developed to solve the customer-accounting function and the trouble-call analysis function. The CIS is often used in connection with a trouble call that connects the troubleshooter to the location of the suspect device. The DMS in the distribution system also provides support to the trouble-call analysis function.

8.6 Automatic Meter Reading (AMR)

Automated meter reading (AMR) of advanced meters has become a necessity for most energy suppliers, especially as the utilities are becoming deregulated for open-market competition. Metering technologies and advances in communication have enabled the development of new electronic meters and the subsequent development of AMR systems.

Automated meter reading has many applications, including outage management, since most utilities rely on a trouble call from customers reporting an outage to the utility. Automated meters can be used to determine system power usage at any time and facilitate communication of information needed by customers to determine appropriate off-peak pricing and flexible billing options. AMR can also be used to analyze, manage, and forecast energy usage.

The meters encourage more efficient use of electricity. Utility companies like to measure the power factor of the load and the time of electricity consumption. The capability of course requires such advanced technology as sensors, electronics, and specialized integrated circuits (IC) that can handle different load models and communicate efficiently to provide real-time processing.

The following list briefly enumerates the advantages of using electronic-based meters for automatic reading.

Reliability: Rugged electronic meters made from solid-state components can be designed to withstand a high level of mechanical stress. They are also smaller in size due to the electrical components used. They can withstand different environmental changes and, as such, are very reliable.

Improved accuracy: Meters are classified by the accuracy of their measurements. They are standardized by the American National Standards Institute (ANSI) to meet up to 0.3 to 0.8% accuracy specification of full-scale deflection.

Energy meters: Electronic meters are able to measure real and reactive power factor instantaneously.

Ease of calibration: Electronic meters are capable of being adjusted to cope with variation in temperatures.

Antitampering and protection: Electronic meters are designed to be able to detect tampering and theft. Several methods are in place to detect attacks and vandalism.

Security: Advances in electronic meters, such as automated meter reading, require communication technology to ensure the security of information and the integrity of the data, both of which are vital for utility meters.

Automated meter reading: The aim of AMR is to avoid the routine process of visual inspection, which is labor intensive and prone to human errors or accessibility issues. Thus AMR is preferred for electronic meters. They read and communicate through such telecommunication technologies as infrared (light-emitting diode [LED]), radio frequency (short and long range), data modem via telephone, power-line carrier (PLC) (short to medium range), broadband, and Ethernet.

8.6.1 Advanced Billing

There are two billing technologies in current use:

- *Time of use*: This refers to the imposition of different tariffs by the utility for electricity consumption, depending on the time of day or the day of the week. Real-time clock (RTC) and calendar (RTCC) circuitry track the customer's usage in real time. This approach to billing promotes optimal use of the utility's daily capacity.
- *Prepaid*: In this approach, the customer can purchase finite amounts of service ahead of time and receive credits that are charged on smart cards.

8.6.2 Special Features and Benefits of AMR

Different manufacturers produce AMRs for utilities. In general, AMR increases efficiency by providing multiple meter reading and the most up-to-date information to office workers:

- Maintain data integrity
- Avoid system events and ensure utility standards for reliability and safety
- Improve productivity and customer satisfaction with accurate information
- Increase speed and accuracy of work creation

8.6.3 Advancement in AMR Technology

Since the early 1980s, the electronics industry has continued to facilitate the advancement of AMR beyond the meter technology evolution discussed previously. In recent years, communications technology has been used to retrieve and send data collected by advanced meters.

There are different communications options in use by the industry, including shared and dedicated phone lines and Ethernet. They allow:

1. LAN-based access to data
2. Wireless access to data
3. Shared telephone lines (from customer line)
4. Dedicated or Ethernet connections, which might be considered for all of the suggested communication options

Dependability and reliability are important concerns, and digital service may not be available in rural areas. In such cases, the utility might be forced to use mobile or cellular technology/networks, which can be limited in their reach.

8.6.4 Advances in Billing Technology

To promote automated meter reading (AMR), an information management strategy is employed to achieve reductions in supply prices while increasing the utility's competence in dealing with rising market supplies, spot markets, energy contracts, and prices. The following technologies are proposed:

1. Internet networks and high-performance private networks
2. Enterprise-integration platforms
3. Interenterprise or business-to-business integration platforms that can leverage the Internet as a key infrastructure network

8.7 Cost-Benefit Analysis (CBA) in Distribution Systems

Distribution automation has been a promising area of development in support of distribution systems. The potential benefits and costs associated with these functions are quantifiable. Several research works have been carried out to determine the costs and benefits in an effort to justify the feasibility of undertaking the automation of distribution system networks. The benefit to the customer may not be obvious or justifiable based on the existing technology and the present costs associated with it. However, we plan to evaluate the various benefits of the distribution functions discussed in the text to promote further work and acceptability of the new generation of distribution systems.

8.7.1 Cost-Benefit Analysis Methodology

The revenue-requirements model is the most rigorous methodology for electric utility cost-benefit analysis. This technique represents the complete financial environment of a utility, accounting for taxes, depreciation and the time-value cost of money, the actual capital investment, and operation and maintenance (O&M) expenses. Two or more alternatives, each of which may involve differing life cycles and cash flows, can be compared on a common basis. With the revenue-requirement methodology, several alternatives are defined, and the cash flows for capital investment and O&M expenses are determined.

8.7.2 Function/Payback Correlation

For many of the functions, the paybacks are highly dependent upon the existence of construction plans involving particular feeders or substations, and detailed analysis is required to calculate savings. Paybacks are functions of the utility's size, current development plans, cost of implementing a particular function, existing structure and work practices, and current costs.

The answers to the following questions lead to an algorithm for calculating the benefits that accrue when a DMS function allows a procedure to be modified:

- How is the task performed?
- What utility resources are used to perform the task?
- How could a DMS function modify the way the task is performed?
- What resources would be required for the modified procedure?

The algorithm, then, is the difference in required resources times the number of occurrences of the procedure during the study period. If the result is lower costs, the modified procedure has benefits that support the implementation of a DMS.

Functions of automation have been mentioned in Section 8.5.3 as layers of the substation automation (Figure 8.3). From this figure, the potential benefits of automating each of the functions are given in Table 8.2.

TABLE 8.2

Benefits of Distribution Automation

Substation Automation Benefits	Feeder Automation Benefits	Customer Interface Automation Benefits
Reduction in capital expenditure due to: · Deferment of additional substation facilities · Effective utilization of existing substation facilities Reduction in O&M costs of breaker or switching for: · Routine operations · Nonroutine operations Reduction in O&M costs in LTC operations for: · Routine LTC operations · Nonroutine LTC operations · Special LTC operations Reduction in O&M costs for: · Routine relay testing · Relay setting Reduction in O&M costs for: · Routine data collection · Nonroutine data collection · Data analysis · Testing of data logging devices · Repair of data logging devices Improved reliability Reduction in routine operations Reduction in gases Effective use of assets	Reduction in capital expenditure due to: · Deferment of additional feeders · Effective utilization of existing feeders Reduction in O&M costs of: · Fault location and isolation · Service restoration · Routine switching operations · Recloser setting · Recloser testing · Data collection · Data analysis · Feeder reconfiguration · Capacitor banks inspection Increased revenue due to: · Loss reduction (due to feeder reconfiguration) · Loss reduction (due to capacitor banks automation) · Faster service restoration	Reduction in O&M costs of: · Regular meter reading · Change-of-property meter reading · Special meter reads · Reprogramming of meters · Service center/disconnect for electric meters only · Service connect/disconnect for electric/gas meters · Processing of customer claims Increased revenue due to: · Manpower (O&M) · Service connect/disconnect with minimum delay · Process customer claims quickly · Reduce waste of manpower

8.8 Summary

The chapter described different functions of distribution management systems (DMS) in terms of the software and hardware support. The different functions, the enabling technologies, and the grand challenges/benefits of DMS are also discussed.

Problem Set 8

8.1 Explain fully, three differences between Distribution Management Sysems (DMS) and Energy Management Systems (EMS).

8.2 Draw and explain the general framework a Supervisory Control and Data Acquisition (SCADA) architecture.

a. Why do we need SCADA for distribution automation system work?

b. What are the overall functions of a SCADA system used in distribution automation?

8.3 Distribution automation functions and their benefits are numerous and the functions of a Distribution Management Systems (DMS) application are grouped into layers.

a. Provide examples of selected distribution automation functions for feeder automation substation automation.

b. Define and give examples of DMS layers.

8.4 Explain the term Automatic Meter Reading (AMR) and list its advantages.

8.5 What are the feataures allowed by the various communications options in use by the industry?

8.6 Accurate billing and efficient bill collection are important aspects of all distribution systems engineering and management.

a. Name the types of billing technologies in current use.

b. Provide detailed explanation of each types listen in (a).

8.7 Formulate steps for calculating the benefits that occur when a Distribution Mangement Systems (DMS) function allows a procedure to be modified.

8.8 Explain the term function/payback correlation. How is it useful to Distribution Management Systems (DMS)?

9

Communication Systems for Distribution Automation Systems

9.1 Introduction

As the energy enterprises are slowly restructured, utilities and customers are feeling the pressure to reduce costs, improve efficiency, and increase operating flexibility. This is accomplished through the introduction of communication options to support distribution systems. Technical devices associated with distribution automation functions include remote terminal units (RTU) and supervisory control and data acquisition (SCADA), as described in Chapter 8. Utilization of these devices provides the framework for the design and development of a distribution system.

To start with, we present the basic principles and concepts used in the development of power system communication. The background information here is readily available in texts, working papers of IEEE, and other relevant journals and papers.

9.1.1 What is Telecommunication?

Modern communication systems involve the integration of computers and telecommunication technology. Telecommunication is communication from afar using various forms of equipment, computers, networks, and different media over short to long distances. Early forms of communication from afar, including drums, mirrors, flags, and smoke, became extinct following the discovery of electricity by Edison. The value of both electricity and telecommunication has revolutionized our world and continues to penetrate our lives. Much progress has been made in research and development of telecommunication and its applications to power system automation. The telecommunication industry has facilitated several of the distribution automation functions, such as:

- Improved reliability
- Greater cost efficiency through automatic meter reading and billing

- Automatic outage analysis and maintenance
- Acceptance of various architectures and protocols for different data types and controls for efficient management
- Provision and handling of control strategies to improve the reconfigurability, restoration, and quality of supply

Significant developments in communication technology and economies of scale have made these devices available at affordable prices. For example, video recorders, remote terminal units (RTU), intelligent electronic devices (IED), and supervisory control and data acquisition (SCADA) are all part of the modern-day distribution automation system.

9.2 Telecommunication in Principle

Telecommunication is generally a transmission from a transmitter, which is a source, to another device (sink) called the receiver. Messages are coded in analog or digital encoder waveform and sent through a communication channel to a decoder or demodulator to an output signal to the message device. This communication line can be from one computer to another computer or from one device to another device with the capability to be configured as:

Simplex (one directional) where information flow can have any orientation, but it all flows in the same direction simultaneously

Half duplex, where information can flow in two directions, but only in one direction at a time

Duplex, also known as full duplex, where information can flow in two directions at the same time

Consider a possible information/data exchange for the integrated distribution system shown in Figure 9.1.

9.3 Data Communication in Power System Distribution Network

Telecommunication facilitates the transport of data and information among distribution agents for the purpose of system analysis and for remote use for storing, retrieving, and processing. The resulting enterprise is simply an information system with distribution management system (DMS) software that organizes data to produce information to benefit the distribution automation function. It provides:

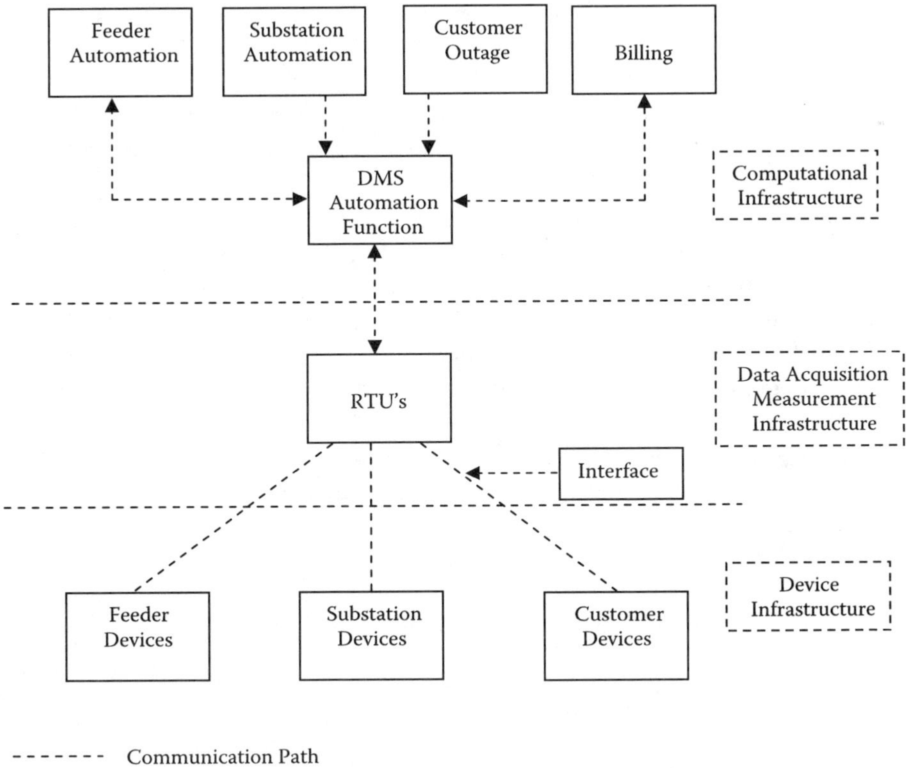

FIGURE 9.1
Integrated distribution systems.

- Opportunity to plan future activities and its supportive role
- Information that guides (controls) present activities
- Supportive information that is used to operate the enterprise

The telecommunication setup shown in Figure 9.1 involves data communication between RTU, DMS, and other automated distribution functions, which requires the use of data signals for automation and control.

9.4 Signal Representation

Information or messages are represented as electronic signals for specific performance by communication systems. It can be represented in data, text, voice, television, or musical formats. The classification of telecommunication signals are:

1. Analog signals, which vary continuously with respect to the information in time. They are the time arc of the natural form, with signals generated, transmitted, and coded.
2. Digital signal, which represents information in discrete forms. It assumes a limited set of values of one or zero (positive, zero, negative values).
3. Discrete signal, which represents information as a noncontinuous function whose value forms a discrete set and occurs at isolated points in the time domain.
4. Signals represented as functions of information that are divided on the basis of their certainty: deterministic, known at anytime; probabilistic, which is random and can be known in probabilistic/statistical terms only.
5. Noise signals are categorized as unwanted signals, and are different from recorded signals or corrupted signals that are attached to a message signal.

9.4.1 Communication Technology for Signal Description

Channel: Signals are typically represented in terms of magnitude and phase and are described in sinusoidal functions. They are sent through a medium or communication channel that represents a division in the transmission medium for sending streams of data at different frequencies.

Bandwidth: A channel typically made up of equipment between the transmitter and the receiver with sufficient bandwidth of frequency range to carry the signal. We define the bandwidth as the range of frequencies that encompass all the energy present in the given signal, i.e., we state that for a signal to be reconstructed at the receiver, it must be carried by the signal's bandwidth without distortion. For example if the power present in a signal is from 2f Hz to 5f Hz, then the bandwidth (BW) must be (5 – 2)f Hz = 3f Hz, which defines the bandwidth, otherwise called the passband (PB). This PB is the range of frequencies transmitted without distortion by a bearer and associated equipment, i.e., at no distortion, PB = BW. For bandwidth greater than passband, we have distortion present in the communication system. For effective communication, signals are transmitted at the passband of the bearer and are made large enough to accommodate the bandwidth of the signal to avoid distortion.

Signal-to-noise ratio (SNR): In communication systems, SNR is used to distinguish the ratio of power in a useful signal to power in a noise signal. It is measured in decibels. In general, good communication signals adhere to a SNR of about 50 dB. Given the power of the complex signals in amplitude and frequency, we decompose the signal into harmonic components, and hence it is necessary to extract or reproduce the actual symbols at the receiver as they existed in the original transmitter before decomposition.

Sampling: The process of determining the value of the amplitude of an analog signal instantaneously. When sampling of a band-limited signal occurs, we sample at a rate that is twice the highest frequency present. This is called the *Nyquist rate*. Therefore, sampling at 2 (BW) is given as 2W per second, and the Nyquist rate is 2W samples per second. If the sampling rate is equal to or greater than the Nyquist rate, the sampling yields a set of values that contains all of the information necessary to reconstruct the original so-called band-limited signal.

Quantizing: Using a process known as quantizing, sample values are expressed in octets to produce digital signals.

Aliasing: If the sampling rate is less than the Nyquist rate, the reconstructed signal will be a distorted version of the original. This effect is known as aliasing.

9.5 Types of Telecommunication Media

In recent years, new technology has been developed to facilitate communication via signal transmission. Such transmission can span long distances, between local phones and from computer to computer, hence providing the backbone for many communication networks. These networks are used in office buildings, industrial plants, and electric utility companies. All communications require a link between those originating the transmission and those receiving it. In electrical digital communications, the conversion of bit-stream signals can carry the information over the communication medium. Those signals are electromagnetic waves that are carried through a medium as radio waves or optical signals. Commonly used media are described in the following subsections.

9.5.1 Copper Circuit

The most widely used communications medium is still the copper circuit, consisting of a direct link using parallel conductors, twisted-pair conductors, or coaxial cable. Although copper is the best-known conductor, other metallic materials are used. One particularly interesting communications method available to the power supply industry is power-line carrier (PLC), where the conductors used to carry electric power are also used to carry communications signals using PLC techniques that have been extended from the use of EHV (extremely high voltage) transmission lines to include overhead lines and cable distribution systems and signaling data mains.

Propagation along a copper link is governed by the interaction of electrical and magnetic fields in the conductors. This leads to delay, distortion, attenuation , and reflection of the signals in the communications circuit, all of which complicate the transfer of information. The propagation speed in copper is about the speed of light (3×10^8 m/sec).

9.5.2 Twisted Pair

This is made from insulated copper wire and consists of a large number of pairs of copper wires of varying sizes in a cable. At high frequency, signals are able to leak out in twisted-pair cable. It is unsuitable for high-speed data transfer due to loading coils at the low-pass filter and bridge tap, which does not allow a direct path of electrical signal flows.

9.5.3 Coaxial Cable

Coaxial cable consists of a single-stranded iron wire core surrounded by shielding. It has a higher transmission speed than twisted pair.

9.5.4 Fiber Optics

A fiber-optics system is similar to a copper wire system. It uses light pulse signals instead of the electrical signals that are used to send information down copper wire systems. We provide here the characteristics of a fiber-optic cable.

A light-emitting diode (LED) is used to generate the light pulses, which move down the fiber-optic line. Fiber-optic cable is constructed from a fiber-optic strand or cable clad, which represents the strength of the material and is illustrated in Figure 9.2 below.

The optical receiver receives the light pulses and converts it to an electrical signal for further information processing. The electrical signal is then transmitted via a coaxial cable to the end user. It has a very wide application in power companies for monitoring and communication systems as well as in office buildings, universities, industries, etc.

The immediate advantage of optical fibers is that they have a high inherent immunity to external interference and do not generate interference. The signal used in this medium is light. Short bursts of light can be used to represent 1, and the absence of light can be used to represent 0. However, the propagation properties of the optical fibers can lead to delay, distortion, attenuation, and reflection of the transmitted signals.

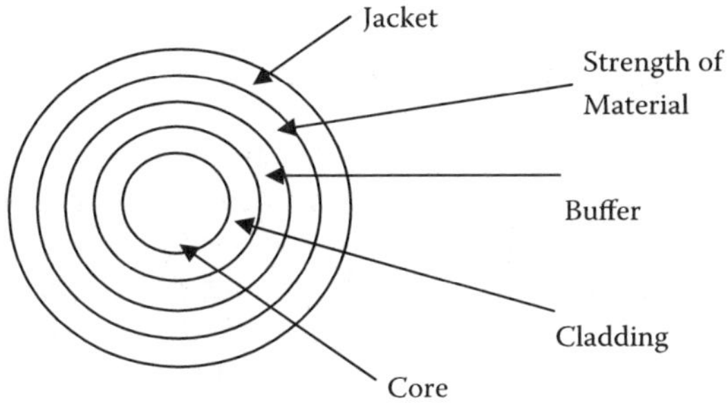

FIGURE 9.2
Cut view of a fiber-optic cable.

Because of its wide use, we enumerate both its merits and demerits. Fiber optics does not connect easily with current hardware, and so some amount of retrofitting has to take place. The speed gained is inhibited at the conversion points, and some malfunction can take place at the electronic interface hardware. However the greater bandwidth and speed outperform other media. It is important that a signal regenerator be used to boost the electronic pulse in a copper cable to keep the signal going in the fiber-optic system. An optical repeater is also used to transmit the pulse in a fiber-optic cable.

9.5.5 Microwave/Radio

These are transmission media above the Earth and the ionosphere. Microwave relay stations are built for line-of-sight-path communication to either another microwave relay station or a satellite communication site at about 22,000 miles above the Earth. The signals are then processed by another microwave relay station on Earth.

The immediate advantage of radio and microwave communications is that they do not require a physical link between the transmitter and receiver. The dependability of these systems relies on the capabilities of the base stations, since the medium is guaranteed and the range of transmission can be very great. Radio systems offer broadcast facilities by which a transmitted signal is received by several receivers, whereas microwave systems use directional capabilities so that the transmission is concentrated to a single receiver.

Radio and microwave transmissions are susceptible to delay, distortion, attenuation, and reflection. They are also susceptible to and generate interference.

9.5.6 Cellular Transmission

Cellular transmission is another type of medium used to transmit and receive communication over an integrated network.

9.6 Communication Modulation Techniques

Telecommunication systems are required to transmit data or signal over long distances efficiently and with minimum noise distortion at an economical power requirement. Modulation is a means of varying or changing a signal over a medium. It involves a signal-processing technique where one signal (the modulating signal) modifies another carrying signal, which enables the original signal to form a new composite signal (modulated signal = original signal + carrier signal). The two go together in the medium, and at the receiver, the modulating signal is recovered by an action called demodulation. To achieve this, the channel bandwidth must be equal to or greater than the base band frequency of the original signal.

The analog modulation is aimed at impressing an information-carrying analog waveform onto a carrier for transmission, while digital modulation is used to convert an information-bearing discrete time-symbol sequence into a continuous-time waveform impressed in a carrier waveform. In both cases, modulation is used to facilitate long-distance transmission over a high frequency range sending digital messages over a band-limited channel.

The three most commonly used modulation techniques are amplitude modulation, frequency modulation, and pulse-code modulation, which are discussed next.

9.6.1 Amplitude Modulation (AM)

Amplitude modulation refers to the instantaneous amplitude of the original (modulated) signal. It means that a carrier wave is modulated in proportion to the strength of the original signal. Figure 9.3 shows a typical AM waveform. AM is generally used for situations with low frequency. It is a simple, robust method to form a radio wave, but it suffers from static and high battery power requirements.

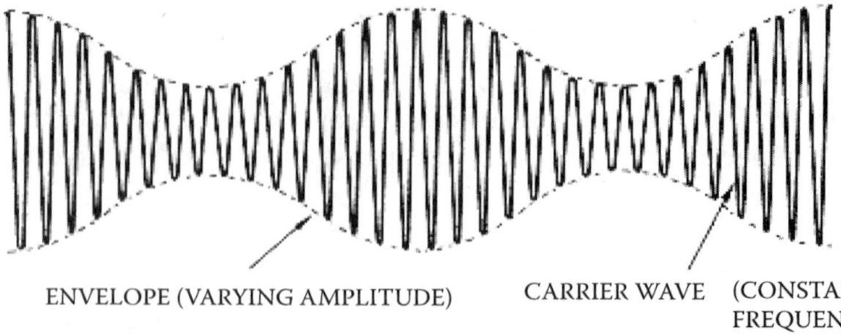

FIGURE 9.3
Amplitude modulation (AM) waveform.

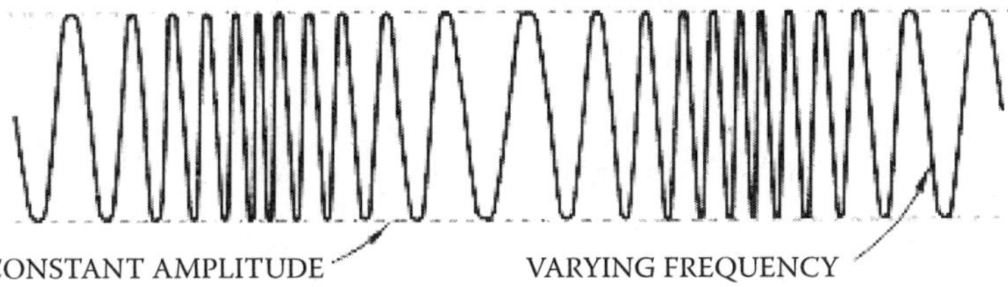

FIGURE 9.4
Frequency modulation (FM) waveform.

9.6.2 Frequency Modulation (FM)

Frequency modulation (Figure 9.4) is not frequency dependent; rather, it is a modulation technique to shape a radio wave, and not a service in itself. FM is the rate at which the signal varies a carrier wave, and not to any particular radio frequency it uses. It has little static, a good-quality signal, and is immune to electrical and atmospheric interference. FM requires less power for transmission when compared with AM, and it is used more often in TV audio and analog cellular.

9.6.2.1 Pulse Modulation (PM)

Suppose the signal is sent to be modulated with frequency W_m and Q_m. It is given as

$$M(t) = M\sin(W_m t + Q_m) \tag{9.1}$$

Let the carrier signal be represented as

$$C(t) = C\sin(W_c t + Q_c) \tag{9.2}$$

then

$$M(t) + C(t) = y(t) \tag{9.3}$$

$y(t)$ is the modulated signal:

$$y(t) = C\sin(W_c t + M(t) + Q_c) \tag{9.4}$$

This shows how $M(t)$ is modulated in phase. Thus PM is a special case of FM, where the frequency modulation is the time derivative of the modulating signal. This is shown here as follows.

9.6.2.2 Frequency Modulation

Angle modulation: the transmitted signal is

$$M(t) = M\cos(W_c t + \Phi(t)) = \Re\left\{Me^{j(W_c t+\Phi(t))}\right\} \tag{9.5}$$

with instantaneous phase

$$\theta_i(t) = W_c(t) + \Phi(t) \tag{9.6}$$

and instantaneous frequency

$$W_i(t) = \frac{d\theta_i(t)}{dt} = W_c + \frac{d\Phi(t)}{dt} \tag{9.7}$$

where $\Phi(t)$ is the instantaneous phase deviation, and $\frac{d\Phi(t)}{dt}$ is the instantaneous frequency deviation.

9.6.2.3 Amplitude Modulation

Consider the carrier wave of frequency

$$C(t) = C\sin(W_c t) \tag{9.8}$$

The equation for the simple sine wave of frequency W_m, the signal we wish to broadcast, is

$$M(t) = M\sin(W_m t + \Phi) \tag{9.9}$$

where Φ is the phase relative to $C(t)$

$$y(t) = [C + M\sin(W_m t + \Phi)\sin(W_c t)$$

$$\therefore\ y(t) = C\sin(W_c t) + \frac{M(\cos(\Phi - W_m - W_c)t)}{2} - \frac{M\cos((\theta + W_m + W_c)t)}{2} \tag{9.10}$$

The signal consists of carrier wave plus two sinusoidal waves at sideband frequency of $W_c \pm W_m$. As long as $W_c \pm W_m$ are spread out properly to avoid overlap, the station will not interfere.

9.6.3 Modulation Indices

Frequency modulation index (FMI): indicates how much the variables vary around their unmodulated level. For FM

$$h_{FM}(\text{index}) = \frac{\Delta f}{f_m} = \frac{f_\Delta |X_m(t)|}{f_m} \tag{9.11}$$

Amplitude modulation index (AMI): called the modulation depth, it represents how much the modulated variables vary around their original level. For AM

$$h_{AM}(\text{index}) = \frac{\text{peak value of } M(t)}{C} \tag{9.12}$$

9.6.4 Digital Modulation

Signals represented in digital form are also transmitted in amplitude and phase representation, analogous to the discussion in the previous section. We develop here the three commonly used schemes:

- Frequency-shift keying, an FM variation
- Phase-shift keying (PSK)
- Amplitude-shift keying (ASK)

Frequency-Shift Keying: sends data by slightly shifting frequencies (keying means forming or creating a signal). Figure 9.5 shows a frequency-shift-keying waveform. This method gives you two states to send information: 0 or 1 in two different frequency ranges. The instantaneous frequency is shifted between two discrete values termed the mark frequency and the space frequency.

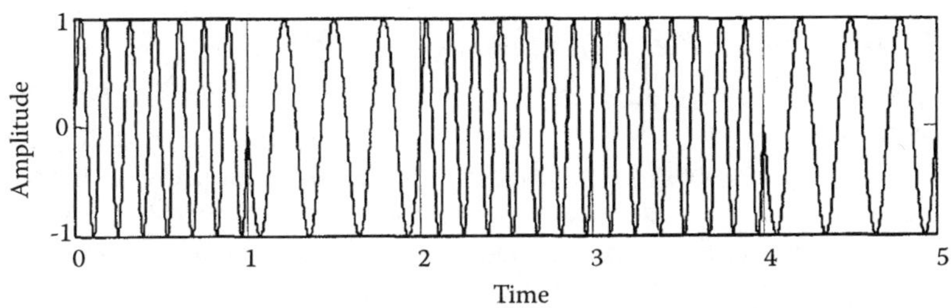

FIGURE 9.5
Frequency-shift-keying waveform.

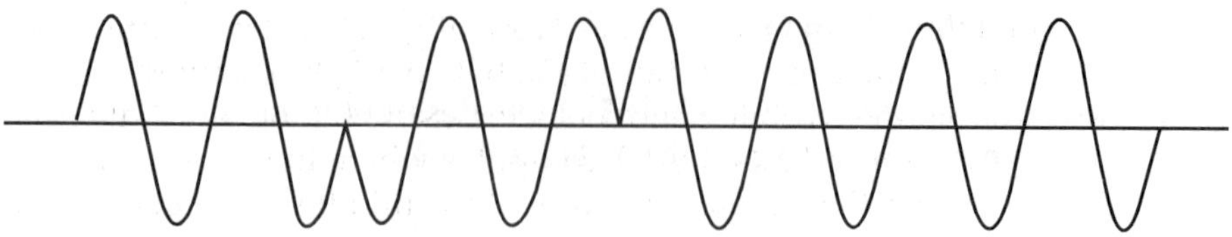

FIGURE 9.6
Phase-shift-keying waveform.

Amplitude-Shift Keying (ASK): a form of modulation that represents digital data as a variation in the amplitude of a carrier wave. The amplitude of the carrier signal is in the on or off position, which is 0 or 1. The modulated signal 0 is represented by the absence of a carrier, thus giving an off/on operation. It is like AM: not sensitive to noise and distortions but requires excessive BW and hence more power.

Phase-Shift Keying (PSK): The simplest version is called quadrature phase-shift keying or QPSK. Figure 9.6 shows a phase-shift-keying waveform. It uses two phases that are separated by 180°, called 2-PSK. It changes a sine wave's normal pattern. It shifts or alters a wave natural fall to rest or 0°. By frequency exchange in a sine wave, you can convey information.

These techniques are used in various telecommunications applications such as GSM (global system for mobile communication) mobile phones and, in most cases, over long-distance transmission. There are many more variants of modulation techniques that can be reviewed in advanced textbooks.

9.6.4.1 Asynchronous/Synchronous Communications

Serial data communication involves the conversion of bytes of data into time sequences of electrical signals. The time that each bit spends in a particular state is equal to the inverse of the bit rate. For example, a relay that sends data 9600 bits per second (bps) sets an electrical signal level for each bit of data for 1/9600 of a second, or 104 μsec per bit.

9.6.4.1.1 Asynchronous Data Transmission

Asynchronous communication is clearly the most common form of communications in today's IEDs. This method of clocking the bits of data depends only on the clock sources at each end. In particular, decoding of each bit in a message depends on detecting a start bit in the transmitted message.

The device that encodes and decodes the serial data is known as a universal asynchronous receiver transmitter (UART). On transmission, UART accepts one or more bytes of data from the processor, loads the data into a shift register, adds the start bit, parity bit (if required), and stop bit, and proceeds to shift out the data one bit at a time. Normally, data is transmitted starting with the least significant bit (LSB) of the byte being transmitted.

The UART on the receiving end looks for a start bit, on which it will time the rest of the received bits. UART computes where the middle of the start bit is and samples the analog waveform at the middle of the period of each successive bit. As the asynchronous name implies, the start bit of each byte of data can be sensed independently of any other start bit of any other byte of data. As a result, clock accuracy is critical to proper decoding.

9.6.4.1.2 Synchronous Data Transmission

In asynchronous communications, the detection of the bit positions is determined by the detection of the start bit in the data stream. To guarantee that the transmission line is in a suitable state to start, a stop bit must be appended to the transmitted data. As a result, it takes 10 bits on the wire to transmit 8 bits of information. Otherwise stated, 20% of the bandwidth of the communications medium is used for timing purposes.

Synchronous communications takes a different approach to bit timing in that the bit detection is based on a clock signal that is included with the data. The clock can either be transmitted via a separate wire or embedded in the modulation technique. As there is still a need to determine the start of a transmission, a sync character is usually placed at the beginning of a message. The universal synchronous/asynchronous receiver transmitter (USART) decoding device is placed in a search sync mode and continually searches for the sync character. Once detected, each eight bits of data that are received are interpreted as the next byte of data, thereby achieving almost 20% improvement in bandwidth.

9.6.4.2 Intelligent Electronic Devices (IEDs)

Since the introduction of communicating IEDs in the electric utility environment, there has been an increasing demand for corporate access to field device data and the capability to automatically control system equipment. Utilities are aware of the great advantages in utilizing communicating IEDs to minimize integration and automation costs, and to improve system operation and customer service. To realize the full potential of communicating IEDs, information exchange with field devices is not merely data retrieval and limited control, but an advanced level of data integration and processing for exchange with enterprisewide information systems.

IED is a broad term for communicating devices used in transmission and distribution systems, and includes substation host computers, remote terminal units (RTUs), programmable logic controllers (PLCs), communication processors, digital protective relays, sequence of events and fault recorders, and automatic system controllers, e.g., automatic VAr controllers.

IEDs are microprocessor-based and have the ability to exchange digital data. IEDs perform multiple functions, and in some restricted cases, such as relay coordination between substations, they can communicate with each other at a simple level. IEDs communicate with a certain language, the language depending on the type of application, which may call for proprietary utility protocol implementation. Some IEDs have built-in protocol support for a number of protocols, e.g., DNP 3.0, Modbus, Modbus Plus, etc.

The primary application of IEDs is in the digital monitoring and protection of electric system equipment (lines, switch gear, buses, transformers, and feeders), with the transfer of basic, raw data and control commands between IEDs and external systems, such as SCADA systems. Another application is in the acquisition and processing of protection, control, and operating data for exchange with system applications and enterprisewide users, such as large-scale distribution management systems. The advanced application of IEDs will provide true integration and sharing of data through networking and distributed processing.

9.7 Communication Networking

Information is distributed over a variety of connections:

1. One-to-one connection of one location to another, e.g., telephone
2. One-to-many connection of one location to many other locations, e.g., cable TV
3. Many-to-many connection of many locations to many locations, such as a conference arrangement or the so-called local area network (LAN)

This combination of connection types has led to new configurations, referred to as local area networks (LAN), wide area networks (WAN), and metropolitan area networks (MAN).

9.7.1 Local Area Network

A local area network (LAN) consists of two or more personal computers, printers, and high-capacity disk storage (file servers) that allow each computer in the network to access a common set of rules. A LAN has operating system software that interprets input, instructs network devices, and allows users to communicate with each other. Each hardware device on a LAN, such as a computer or printer, is called a node. A LAN can operate or integrate up to several hundreds of computers. LANs provide high-speed data communication over a geographical spread of 1 to 10 km. A LAN can also access other LANs or tap into wide area networks (WAN). LANs with similar architectures are called "bridges," which acts as transfer points, while

LANs with different architectures are linked by "gateways" that convert data as it passes between systems.

LAN is a shared-access technology. This means that all of the devices attached to the LAN share a common medium of communication, such as coaxial, twisted pair, or fiber-optic cable. A physical connection device called a network interface card (NIC) connects stations to the network. The network software manages communication between stations on the system.

Special attributes of a LAN include:

Resource sharing: this is the greatest advantage of LANs. It allows intelligent devices such as storage devices, programs, and data files to share resources. LAN users can use the same printer on the network. The database and the software installed on the network can also be shared by multiple users in the network.

Area covered: LANs are normally restricted to a small geographical area such as an office building, utility, or a university campus.

Low cost: connection to LANs is low cost. The application software and interface devices have become more affordable, making LANs more commonplace.

High channel speed: LANs possess high channel speed, with the ability to transfer data at rates between 1 million to 10 million bits per second.

Flexibility: LANs have the flexibility to grow with low probability of error, and they are easy to maintain and operate.

9.7.1.1 Method of Transmission in LAN

Data transmission in a LAN falls under three categories, namely:

1. Unicast transmission: involves transmission of single-packet data from a source to a destination on a network. Data packets are sent from the source node through an address in the distribution.
2. Multicast: consists of a single data packet that is copied and sent to a specific subset of nodes on the network. The source node addresses the packet by using the multicast addresses.
3. Broadcast transmission: consists of a single data packet that is copied and sent to all nodes on the network. Again, the source node addresses the packet by using the broadcast address.

The following terminology used in LAN/WAN is useful in design and application fields:

Bridge: consists of two or more networks that use the same protocol at the media control sublayer of the data-link layer

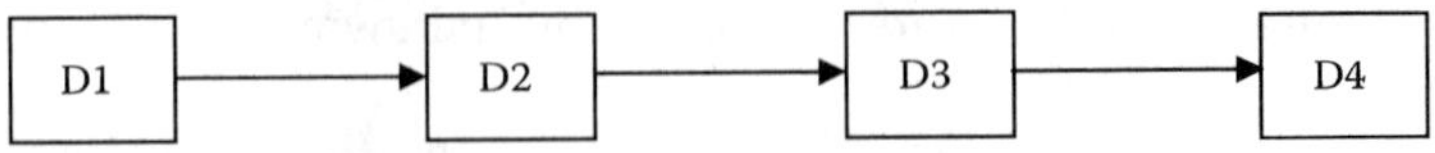

FIGURE 9.7
LAN–bus topology.

Router: operates at the network level of the OSI model with more sophisticated addressing software than a bridge

Gateway: operates at or above the OSI transport layer and links LANs or networks that employ different architectures and use dissimilar protocols

Switch: switches data to its destination by a point-to-point connection

9.7.1.2 LAN Topologies

LAN topologies define the manner in which the network devices are organized. Four common organizational structures are in common use: bus, ring, star, tree.

Bus topology: a linear LAN architecture (Figure 9.7) in which transmission from a network station propagates the length of the medium and is received by all other stations connected to it

Ring-bus topology: a ring LAN architecture (Figure 9.8) that consists of a series of devices connected to one another by unidirectional transmission links to form a single closed loop

Star topology: a LAN architecture (Figure 9.9) in which the endpoints on a network are connected to a common central hub or switch by dedicated links

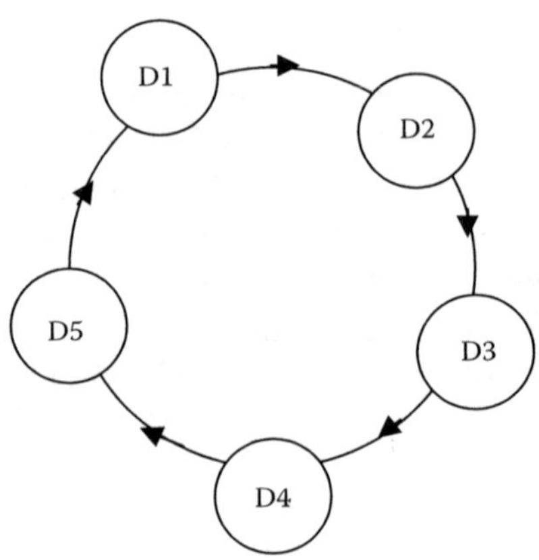

FIGURE 9.8
LAN–ring bus topology.

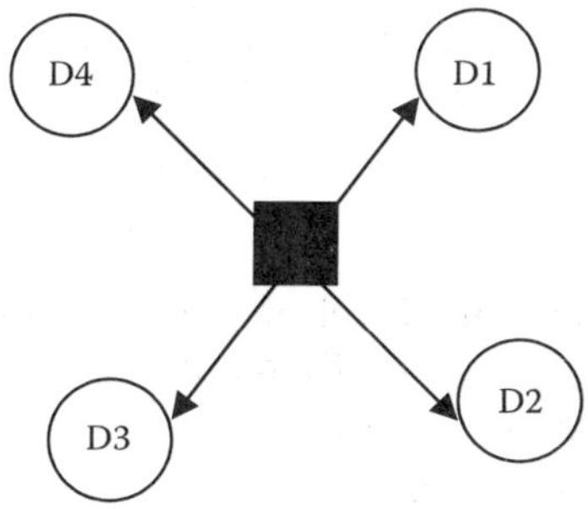

FIGURE 9.9
LAN–star topology.

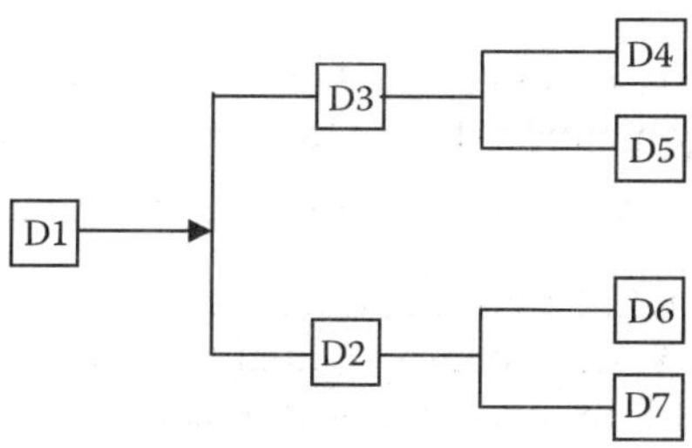

FIGURE 9.10
LAN–tree topology.

Tree topology: a LAN architecture (Figure 9.10) that is identical to the bus topology except that branches with multiple nodes are also possible

The various devices and software used in LANs utilize a standard protocol such as Ethernet/IEEE 802.3 or Token Ring/IEEE 802.5 or 880.2, which are easily available through IEEE Press. They are mapped into both physical and data layers in the OSI reference model to be discussed in Section 9.10.

9.7.2 Metropolitan Area Network (MAN)

A metropolitan area network (Figure 9.11) is a system of LANs connected throughout a city or metropolitan area. The main connections between LANs are done through a local exchange carrier, and they follow required protocols and interface standards defined by RS-232, frame relay and ISDN/dedicated T1 line, and asynchronous transfer mode (ATM).

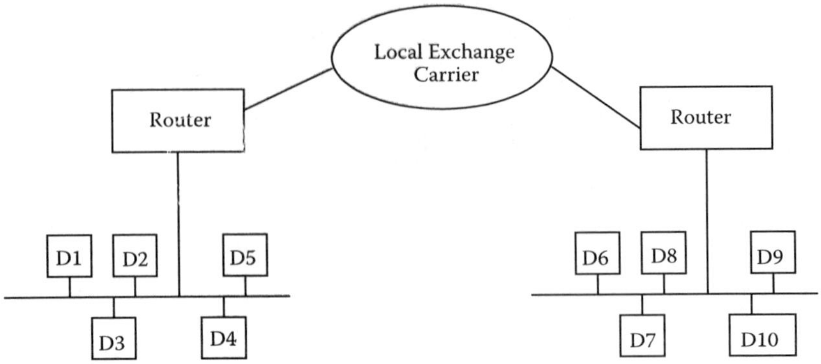

FIGURE 9.11
MAN configuration.

9.7.3 Wide Area Network (WAN)

A wide area network (WAN) is a network system connecting cities, countries, and continents together. WANs are connected using one of the communication media. WANs use long-distance carriers and are linked by cable, optical fibers, or satellites, but their users commonly access the network via a modem.

The largest wide area network is the Internet, which is a collection of networks and gateways linking million of computer users all over the globe. The industry practice based on TCP/IP (transmission control protocol/Internet protocol) is generally used.

9.7.3.1 Types of WAN Connection

There are three main WAN connection services, namely:

1. *Connection services* (*X.25*): This uses the OSI model layer network and packet for packet switching.
2. *ISDN (integrated services digital network)*: This service is based on digital physical connection. It uses data-link voice and a video network with an X.25 upgrade topology with a point-to-point connection between sender and receiver with asynchronous clocking.
3. *ATM (asynchronous transfer mode)*: It is based on Least Line Service (LLS)/network and status rooting.

In general, WAN ties together large geographic regions using microwave and satellite transmission or telephones.

9.7.4 Types of Computing Connectivity

Several types of computing connectivity exist, such as:

1. Terminal to host: involves the use of a dumb terminal to access applications and databases that reside on the host mainframe computer
2. File server: transfers data and programs to PCs on the network, where the PCs perform computer processing tasks
3. Client/server: uses applications and databases that reside on specialized host computers (servers), with processing being shared between the host server and the client; in most cases, client and server may be different types of computers

9.8 Frame-Relay Communications

Frame relay is a high-performance WAN communication protocol that transports data in a "packet" format that maximizes bandwidth on communication circuits. Frame relay was designed for use across integrated services digital network (ISDN) interfaces and uses a packet-switching technology that other communication protocols cannot handle. It was developed to accommodate the following needs:

1. Increased need for high speeds
2. Increased need for large bandwidth efficiency, particularly for clumping and busting traffic
3. Increase in intelligent network devices that lower protocol processing
4. Need to connect LANs and WANs

Advantages of frame relay:

- In many scenarios involving long-haul, high-speed connection, it is cheaper than dedicated lines.
- It is a cheap solution to incorporate redundancy in the network.
- Mixed speeds can be converted; traffic bursts can be buffered.
- Less hardware is needed for the same number of connections.
- Frame relay is protocol independent; it accepts data from many different protocols (IP, Internet-work Packet Exchange (IPX), systems network architecture [SNA], etc.).

Disadvantages of frame relay:

- There may be jams; no guaranteed bandwidth.
- In a point-to-point scenario, it is not economically feasible.
- In short hauls it is not economically feasible.
- Frame only supports byte-oriented data types.

Frame relay is often described as a streamlined version of the X.25 protocol for point-to-point connection. As a packet-switching relay, it allows streams of data broken into discrete blocks of data called frames or packets. Frame relay is simply a way of sending information over a wide area network (WAN) that divides the information into frames or packets. Each frame contains information necessary to route it to the correct destination. It can carry multiple network layer protocols, including internet protocol (IP).

Frame relay is strictly a layer-2 protocol suite, whereas X.25 provides services at layer 3 (network layer) as well. Thus frame relay offers high performance and greater transmission efficiency than X.25, making it suitable for WAN applications. Its capability for a connection-oriented approach makes frame-relay label or DLCI (data-link connection identifier) a simple reference to a virtual connection.

9.8.1 Frame-Relay Standardization

The standardization of frame relay was developed by a specification in CCITT and the protocol was extended with features that provide capabilities for a complex internetworking environment. The frame extension is referred to as the local management interface (LMI). Currently, the International Telecommunication Union–Telecommunication Standard Section (ITU-T) standardizes the frame relay internationally. In the U.S., the frame relay is standardized by an American National Standards Institute (ANSI) standard.

Frame relay provides connection-oriented data-link-layer communication. This implies that each pair of devices (DTE [data terminal equipment] and DCE [data circuit-terminating equipment]) and these connections are associated with a connection identifier. The service is offered by a permanent virtual circuit (PVC) dedicated connection through the shared frame-relay network that replaces a dedicated end-to-end line. PVC circuits provide a bidirectional communication path from one DTE device to another and are uniquely identified by a data-link connection identifier (DLCI). The DLC identifier is used as the logical address for frame-relay multiplexing. A frame-relay virtual circuit can be divided into two categories, namely:

- Switched virtual circuits (SVC)
- Permanent virtual circuits (PVC)

9.8.2 Switched Virtual Circuits

These are temporary connections used in situations where data transfer from DTE devices across frame-relay networks is sporadic. The communication setup in this case consists of four operational states.

1. Call setup: the virtual circuit between two frame-relay DTE devices
2. Data transfer: data transmitted between DTE devices over the established virtual circuit
3. Idle: active connection between DTE devices, but no data transfer
4. Call termination: termination of the virtual circuit between DTE devices

9.8.3 Permanent Virtual Circuits

These are permanent connections that are used for frequent and consistent data transfers between DTE devices across a frame-relay network. It has two operational states.

1. Data transfer: occurs between DTE devices over the virtual circuit
2. Idle: active connection between DTE devices, but no data transfer

Frame-relay virtual circuits are identified by data-link connection identifiers (DLCI). DLCI values are typically assigned by the frame-relay service provider, for example the telephone company.

9.8.4 Frame-Relay Handling of Congestion Error

1. *Congestion control mechanism*: Frame relay is equipped with a congestion notification mechanism rather than explicit virtual flow control. The two mechanisms for congestion control are forward-explicit congestion notification (FECN) and backward-explicit congestion notification (BECN). They are controlled by a single bit contained in the frame-relay header.
2. *Discard eligibility (DE)*: The discard eligibility bit is used to indicate that a frame has lower importance than other frames. A bit of 1 in the header frame means the frame has lower importance than other frames.
3. *Frame error checking (CRC)*: By using a cyclic redundancy check (CRC) scheme, the frame relay is able to determine errors in transmission from source to destination.

A local management interface (LMI) provides enhancement to frame-relay specifications to be able to handle global addressing, virtual circuit status messages, and multicasting.

9.8.5 Frame-Relay Network Implementation

Frame-relay network implementation is proposed to consist of a number of DTE devices such as routers, bridges, and frame relay access devices (FRADs).

Frame router: translates existing data communication protocol for transmission over a frame-relay network, then routes the data across the network to another frame router or other frame-relay-compatible devices. Each router is able to support physical data interfaces and serve multiple user ports. Routers can handle traffic from other WAN protocols with a congestion-control scheme.

Bridges: easy to configure and maintain, these are used to connect a branch office to a hub location. Bridges and routers are able to aggregate and convert data into frame-relay products.

FRAD: formatted outgoing data required by a frame-relay network and can also function as a router. FRADs work well in applications where a site is already equipped with bridges and routers or when sending mainframe traffic over a frame-relay network.

A typical frame-relay network consists of a number of DTE devices, routers, and remote ports via T1, fractional T1, or 50-kB circuits. An example of a frame-relay network is presented in Figure 9.12.

9.8.5.1 Public-Carrier-Provided Networks

In public-carrier-provided frame-relay networks, the frame-relay-switching equipment is isolated in the central offices of a telecommunications carrier. Subscribers are charged based on their network use but are relieved from administering and maintaining the frame-relay network equipment and service.

Generally, the DCE equipment is also owned by the telecommunications provider. DTE equipment either will be customer-owned or perhaps will be owned by the telecommunications provider as a service to the customer. The majority of today's frame-relay networks are public-carrier-provided networks.

9.8.5.2 Private Enterprise Networks

More frequently, organizations worldwide are deploying private frame-relay networks. In private frame-relay networks, the administration and maintenance of the network are the responsibilities of the enterprise (a private

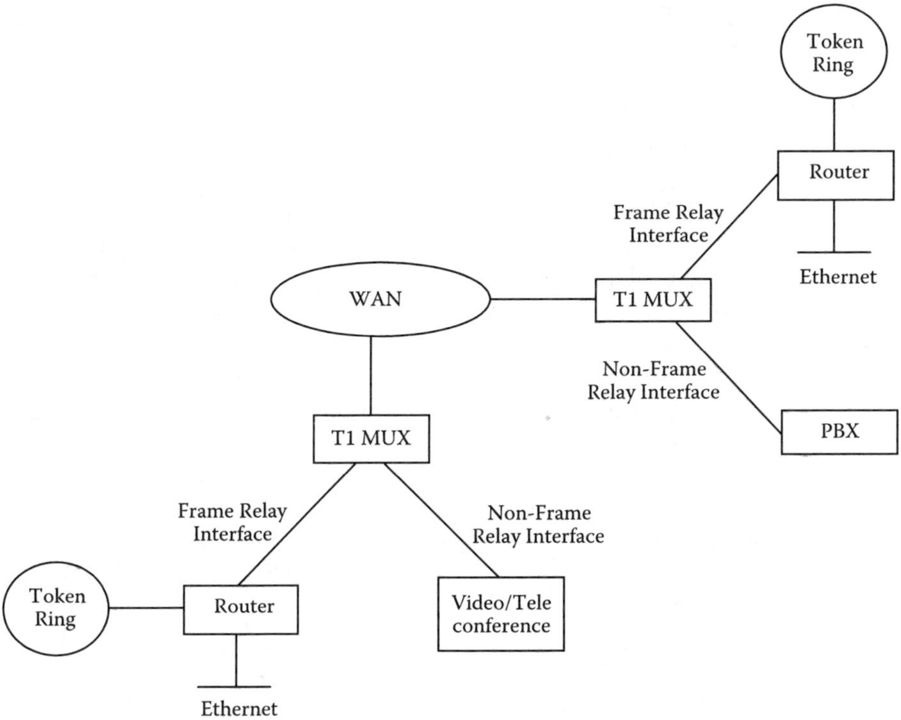

FIGURE 9.12
Simple frame-relay network.

company). The customer owns all the equipment, including the switching equipment, in a private enterprise network.

9.8.6 Frame-Relay Frame Formats

Standard frame-relay frames consist of the fields illustrated in Figure 9.13. The following descriptions summarize the basic frame-relay frame fields illustrated in Figure 9.13.

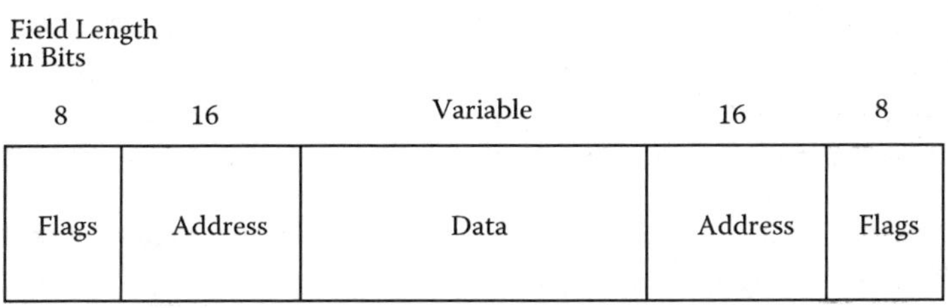

FIGURE 9.13
Frame-relay frame.

Flag: delimits the beginning and end of the frame. The value of this field is always the same and is represented either as the hexadecimal number 7E or as the binary number 01111110.

Address: contains the following information:

- DLCI: the 10-bit DLCI is the essence of the frame-relay header. This value represents the virtual connection between the DTE device and the switch. Each virtual connection that is multiplexed onto the physical channel will be represented by a unique DLCI. The DLCI values have local significance only, which means that they are unique only to the physical channel on which they reside. Therefore, devices at opposite ends of a connection can use different DLCI values to refer to the same virtual connection.
- Extended address (EA): the EA is used to indicate whether the byte in which the EA value is 1 is the last addressing field. If the value is 1, then the current byte is determined to be the last DLCI octet. Although current frame-relay implementations all use a two-octet DLCI, this capability does not allow longer DLCIs to be used in the future. The eighth bit of each byte of the address field is used to indicate EA.
- C/R: the C/R is the bit that follows the most significant DLCI byte in the address field. The C/R bit is not currently defined.
- Congestion control: this consists of the three bits that control the frame-relay congestion-notification mechanisms. These are the FECN, BECN, and DE bits, which are the last three bits in the address field.

Data: contains encapsulated upper-layer data. Each frame in this variable-length field includes a user data or payload field that will vary in length up to 16,000 octets. This field serves to transport the higher-layer protocol packet (PDU) through a frame-relay network.

Frame-check sequence: ensures the integrity of transmitted data. This value is computed by the source device and verified by the receiver to ensure integrity of transmission.

Frame relay is a networking protocol that works at the bottom two levels of the OSI reference model: the physical- and data-link layers. It is an example of packet-switching technology, which enables end stations to dynamically share network resources.

Frame-relay devices fall into the following two general categories:

1. Data terminal equipment (DTE), which includes terminals, personal computers, routers, and bridges
2. Data circuit-terminating equipment (DCE), which transmit the data through the network and are often carrier-owned devices (although,

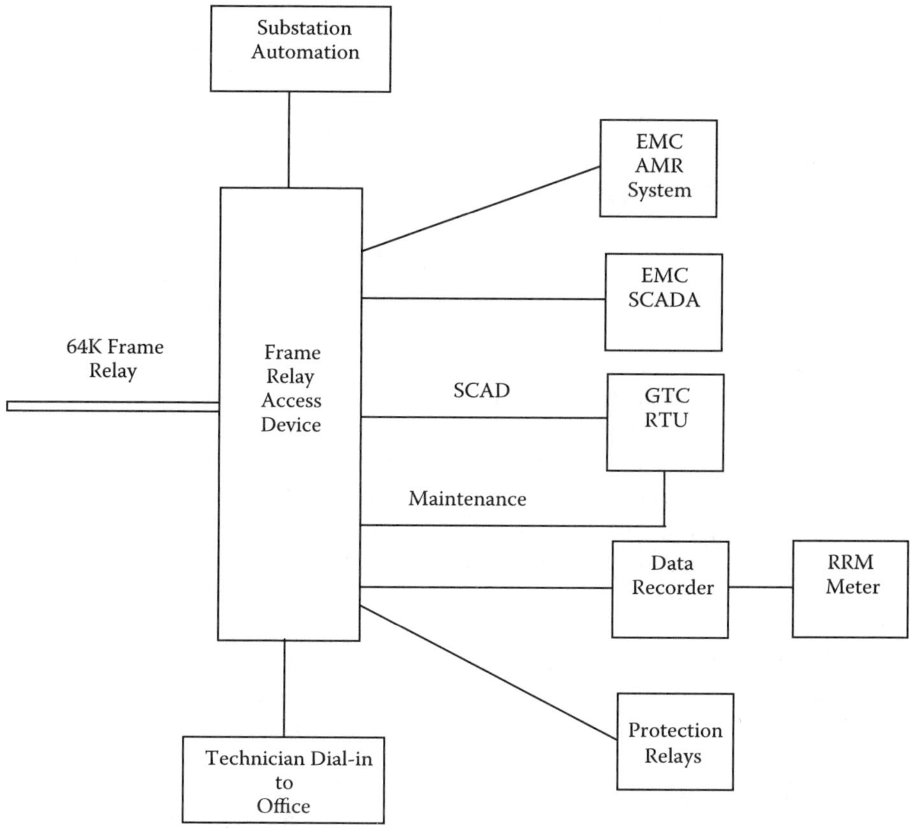

FIGURE 9.14
Typical use of a frame relay in distribution automation.

increasingly, enterprises are buying their own DCEs and implementing them in their networks)

Figure 9.14 illustrates a typical use of a frame relay in distribution automation.

9.9 Communication Standards Overview

The communication practices in the utility industry have been specialized and are standardized to meet the best practices to ensure efficiency, reliability, and cost effectiveness. There are many standards defined by the IEEE working group on standards. We review the standardization process and bodies involved in the development.

9.9.1 Standards Bodies

The groups responsible for the standard are:

1. *IEEE Standards Coordinating Committee*: This is a professional society that develops national (North America) standards for communicating with electric, gas, and water meters. The institute pioneered the standards for local area networks.
2. *International Standards Organization (ISO)*: This organization is concerned with improving international collaboration and communication.
3. *International Electrotechnical Commission (IEC)*: This commission develops standards for the utility industry to promote safety, compatibility, interchangeability, and acceptability of electrical standards. In particular, it is responsible for UCA (utility communication architecture)-compliant intercontrol center communication protocol.
4. *ITU (International Telecommunication Union)*: This group is concerned with the creation of standards that facilitate international telecommunication.

To avoid conflicting requirements, harmonization of standards by different organizations is done at joint international working groups from the national bodies on standards until an international standards is acceptable. The two most common bodies responsible for most of the utility communication are ISO and IEC. Other industry standards committees such as ANSI, NIST, and IEEE are well established for promoting communication for utility standards. ITU — the International Telecommunication Union, telecommunication standardization sector (formerly called CCITT) — is an organization that provide standards for data telecommunication.

9.9.2 Suite of Standards

Different communication enterprises, especially in the utility industry, require different standards. Among these, the most popular ones applicable to distribution system automation are:

1. Transport standards: apply to the local area network and wide area network (LAN/WAN). They require X.25 network and fiber optics and a router to interconnect the networks. They obey the OSI standards, which will be presented in Section 9.10.
2. Standards for speed, reliability, and simplicity: standards for proprietary protocols for different communication devices, intelligent relays, and other simple terminal devices are designed to meet special requirements on speed, reliability, and simplicity. Examples are IEEE and IEC standards on device performance.

3. TCP/IP standards: an industry standard for Internet activity developed by open system foundation (OSF). It provides a comprehensive solution to most of the communication standards. It is currently being migrated to an international ISO standard.
4. Other international standards: FTAM, MMS, and MITS are migrating into the ISO organization and are generally accepted by the industry.
5. EPRI (Electric Power Research Institute) utility communication architecture (UCA): a special architecture of communication protocol that is in use by utility communication interface. Again, this protocol is being planned for migration to ISO standards protocol. Under the current arrangement, the UCA specification has specified the use of seven-layer and three-layer protocol states.
6. Utilities standards: ongoing efforts to develop protocol standards that will facilitate support for all data exchange between utilities and between control centers within a utility.
7. Intelligent electronic device (IED): protocols being added by a number of working groups to make a UCA IED standard protocol.
8. DNP (3.0) (distribution network protocol): a simple important standard protocol with asynchronous capability; handles data priority levels in classes with TCP/IP transport for handling intelligent electronic devices. It is designed for low- to medium-speed functional applications. It is based on a three-layer architecture called the enhanced performance architecture layer, the application layer, and the physical layer.
9. Master-to-remote protocol (MRP): based on MMS, was developed by EPRI. It is being considered for an IEEE/IEC standard. Other remote-access protocols have to choose from several standards, which could be national or international, for example, MMS directory service and remote database access that satisfies, in principle, the OSI layer.
10. EIA RS-232: Electronic Industry Alliance standard, represented as RS-232, was originally specified for the connection of an electromechanical teletypewriter to a modem. Currently, the incorporation of personal computers and other devices has contributed to a rename of the standard. The current revision is TIA 232-F, interface between data terminal equipment and data equipment employing serial binary data interchange. The revision has helped to improve its harmonization with the CCITT standard. The EIA 232 standard specifies connections for several features that involve 25-pin connectors and cables, the required voltage levels, and the connector-compliant interface with DCE.

9.9.3 Interconnection Standards and Regulations

The penetration of distribution generation in the transmission grid has led to some technical requirements for the safe and reliable operation of the power systems to which they are connected. The lack of uniform interconnection standards and tests for operational safety and maintenance of the interconnections has become a concern for federal and state agencies. In response to these concerns, bodies under IEEE 1547 have developed connection standards for fuel cells, photovoltaics, and other distributed energy generation and storage systems to grids. The IEEE 1547 establishes criteria and requirements for interconnection. It is, however, not an application guideline; rather, it provides the minimum functional technical requirements universally needed for a sound technical interconnection.

The IEEE 1547 service of interconnection standards for DG are given in different versions to account for the following:

1. [1547-2003]: establishes criteria and requirements for interconnection of distributed resources (DR) within the electric power system
2. [1547.1]: specifies the type of production and commissioning tests that should be performed to demonstrate the interconnection functions and equipment that connect DG
3. [1547.2–1547.3]: provides technical background and application for monitoring, information exchange, and control of DR and application details to support 1547 series interconnections

For a detailed description of standard 1547 series-consist IEEE interconnection standard, see http://standards.ieee.org/reading/ieee/std_public/description/powergen/1547-3003_desc.html for a discussion on the scope and purpose of these standards. Further work in the international arena to harmonize IEEE standards 1547 and IEC standards is ongoing in an effort to ensure consistency in scope and purpose of the applicable standards for DG interconnection.

The summary of standards and protocols leads to one of the most important international standards models, called the open system international (OSI) protocol.

9.10 OSI Model

Advances in automation techniques for power systems require the use of information system technology. These advances enable the user and utility to exchange just-in-time information at the right speed and accuracy. The information obtained is processed in milliseconds for protection relay tripping, in seconds for breaker tripping, within minutes for billing reports, within hours for maintenance scheduling, and in months or years for reliability-data collection. The information in these time scales is needed to support the automation function in distribution power systems. The use of information technology (IT) to ensure the smooth flow of data and just-in-time decision making requires a standards protocol process to achieve a reliable, healthy, expandable, secure, flexible, and integrated system. The standard OSI model is discussed here as a background for designing computational devices and measuring infrastructure requirements and specifications.

9.10.1 Description of OSI Model

The OSI model provides a generalized description of the functions needed to perform reliable data communication. The model is organized in seven layers, which form the basis of this discussion. These are generally classified as:

Lower layer: consists of the physical layer, data-link layer, network layer, and transport layer of the OSI reference model

Upper layer: consists of session layer, presentation layer, and application layer of the OSI reference model

9.10.1.1 Transport Layers or Lower Layers

The aim of the lower layer is to provide data transmission services of increasing reliability and scope.

9.10.1.1.1 Physical Layer (Level 1)

This layer defines the "physical, electrical, functional and procedural characteristics to establish, maintain, and disconnect the physical link." It is concerned with interfaces to media, such as how the 0 and 1 bits are modulated, what bit rates are used, and what pin connections are plugs. The most common implementation standard associated with the physical layer is RS-232.

9.10.1.1.2 Data-Link Layer (Level 2)

This layer defines the protocols required to send blocks (packets) of data over a single physical link, e.g., between two nodes in a network. It is concerned with how a node recognizes which bits signal the start of a block of data and where the block ends. It detects and recovers from transmission errors (typically by resending any block with an error). It also regulates data flows between two nodes to avoid traffic jams or buffer overflows. For example, the data-link layer in LANs is made of two sublayers:

Medium access control (MAC): offers multiplexed access to the physical-layer transmission facilities of the LAN

Logical link control (LLC): based on the MAC sublayer service, offers a traditional connectionless- or connection-mode data-link service between arbitrary systems attached to a LAN.

Common link-layer protocols that have been implemented include IBM's venerable Bisync protocol, HDLC, Ethernet, IBM's Token Ring, DEC's DDCMP, and IEC 870-5 FT1.2.

9.10.1.1.3 Network Layer (Level 3)

This layer defines the logical or virtual circuits through a network of nodes. Specifically, it is concerned with how a block of data must be routed from node to node as it makes its way through the communications network on its way from host A to host B. It is also concerned with rerouting data if failures or traffic jams prevent the data from going to the node normally next on its path. This layer offers a high degree of reliability.

The CCITT X.25 packet-switching protocol has become an internationally accepted standard associated with this layer. The most widespread network protocol is probably the IP protocol of TCP/IP. Newer network layer protocols include IS-IS ISO/IEC 10589 and ES-IS routing ISO/IEC 9542, as well as connectionless ISO/IEC 8348, 8648, and 8473.

The data transfer phase uses normal data, interruption data, and two way exchange protocol for data transfer, with the data independent of flow control.

9.10.1.1.4 Transport Layer (Level 4)

This layer defines the end-to-end control of complete messages. In particular, it can handle the segmentation of long messages into short packets for transmission through the network and then reorder and reassemble those packets into complete messages at the far end. It also can route messages through gateways between different networks, even when these networks use totally different protocols at the physical, link, and network layers.

Common transport-layer protocols include the TCP layer of TCP/IP and the connection-oriented ISO/IEC 8072 and 8073. The transport layer has classes of service based on quality of service (QOS) described in terms of speed, accuracy, protection, and priority.

9.10.1.2 Application Layers or Upper Layers

The upper OSI layer consists of session, presentation, and application layers of the OSI reference model. These rely on the support service to give suitable, reliable, end-to-end data transfer, where the degree of reliability is specified by QOS of the transport-service user.

9.10.1.2.1 Session Layer (Level 5)

This layer defines how to initiate a session (e.g., dial up the other party) and how to terminate a session (e.g., hanging up). Dial-up control within a group of two or more application processes includes setting of synchronization marks in the data streams and rolling back to these marks. It is concerned with handshaking and security, checking to ensure that a valid and authorized connection is made.

The most commonly used session protocol layer is telephone dialing and ringing. For data, the ISO/IEC protocols are the connection-oriented 8326 and 8327. For connectionless data channels, the session layer is null by definition.

9.10.1.2.2 Presentation Layer (Level 6)

This layer converts data structures into representative structures acceptable to individual processes. The layer defines the data formats to be utilized by the users. It is concerned with what types of messages can be transmitted, how many bits are used for each kind of data, the meaning of flags, and the type of encryption (if any) that is used.

Some common standards are ASCII, integers, floating-point representations, etc. The ISO/IEC protocols are 8822 and 8823. Abstract Syntax Notation #1(ASN 1) is the new method for defining data formats.

9.10.1.2.3 Application Layer (Level 7)

This layer defines the user's interface to the communications system. It is concerned with what data should be sent where, when, and how frequently. It also determines the validity of the data from a user's standpoint. It is typically required to retrieve and store data from/to databases in the host computer. This layer is also the area requiring the greatest work in creating standards, since users in different industries and with different requirements must develop different types of interfaces to the communication system. Simply put, this layer directs support for various types of distribution applications.

Some common application-layer protocols include the top portions of IBM's systems network architecture (SNA), DEC's DECNet, the remote procedure call (RPC) and file transfer protocol (FTP) in the TCP/IP suite, ISO FTAM, ISO directory services, and MMS ISO 9506.

Figure 9.15 shows OSI model layers communicating with other layers.

9.10.2 Message Handling

Within ISO, the message-handling system is known as MOTIS (message-oriented text interchange system) and within ITU-T as MHS (message-handling system). Both are based on message-handling agents called:

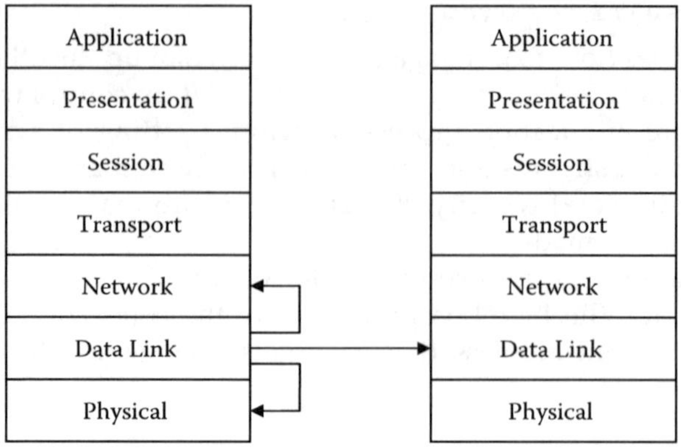

FIGURE 9.15
OSI model layers communicating with other layers.

1. User agents (UA): mediate the interaction between the users of the message system and the subsystem that actually transfer the message
2. Message-transfer agents (MTA): cooperate to provide a message-transfer service offered by a message-transfer system (MTS); in relation to the OSI model, these agents are all applications entitled as invocations of MOTIS

9.11 Distribution Network Protocol (DNP3)

The DNP3 is designated for low- to medium-speed control applications. It is a simpler version of the OSI model, consisting of an application layer, a data-link layer, and a physical layer.

1. The application layer determines data characteristics and passes data to the user through an interface.
2. The data-link layer checks and extracts user data from the frames and passes it to the application layer.
3. Data arrives at the physical layer and is passed to data link.

It is a simple implementation of Information Embedded Power System (IEPS) capable of multiple operating modes for pooling response end reports by exception, unsolicited response, and peer-to-peer communication. This open-system protocol stack is recommended by IEEE for RTU-to-IED messages. It is a three-layer version of the seven-layer OSI model.

9.11.1 DNP3 Protocol Three-Layer Structure Description

The three layers used in DNP3 are briefly described here:

1. *Application layer*: This layer is user supported and uses application functions such as initialization, clock synchronization, and file transfer. It uses its application functions to communicate with the data-link layer.
2. *Data-link layer*: This layer assembles or disassembles frames and detects errors and provides procedures for recovery. The class of services provided includes:
   - Send/no reply
   - Send/confirm and request/respond
   - Confirmation services
3. *Physical layer*: This layer is directly connected to the communication media. It is made reliable by the data-link layer. It is done in synchronized bits. It uses an interface such as RS-232, RS-483, etc. It is implemented in octal data bits over a twisted-pair cable, fiber-optic cable, or radio waves.

9.12 Utility Communication Architecture (UCA)

9.12.1 Overview and Application

The current utility computing environment consists of major networks of networks, which include business functions, accounting and engineering applications, and EMS functions for real-time applications such as dispatching and operation. For example, computers are connected on LAN or WAN network arrangements. The different operating networks overlap and are specified for special stand-alone operations. In the past, the connections between them were unable to communicate across business, plant operations, and real-time operation of a typical EMS. Now with the advent of distribution automation and control through EMS and DMS, much work is needed to support:

- New data exchange processing
- Different protocols and standards in use in the industry

These new standards are derived from ISO, frame relay, and other variations of standards from the professional bodies. The new standard in UCA allows for the interchange of information between the control system and business and other application programs.

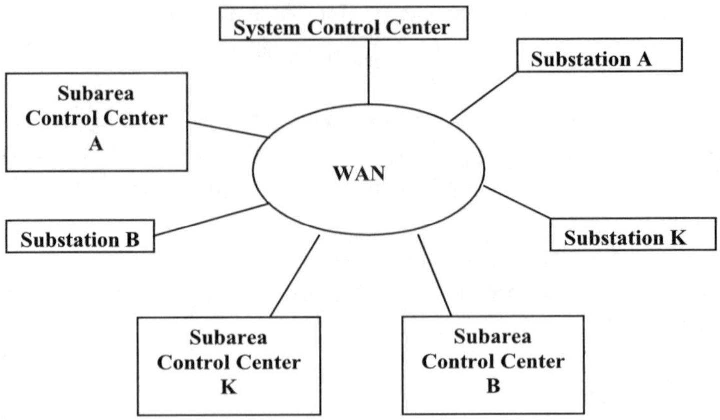

FIGURE 9.16
Integrated utility communications architecture system.

A brief introduction to the history of UCA is presented. The legacy system is to provide utility process control and a business information system for an integrated enterprise and feasibility level. It is an EPRI product under contract no. EPRI RP 2949. The vision is aimed at designing an architecture design to accommodate different software applications, facilitating the interaction of mainframes, PCs, EMS, and support for real-time monitoring of substation distribution automation. Figure 9.16 shows the integrated UCA.

The following networks are connected through a WAN via communication processor for each of the subnetworks: power plant network, corporate network, distribution automation/DMS network, transmission network, and control center network. This architecture has the potential to:

1. Serve as a communication highway to distribution automation, communicating with substations and power plants at a reduced telecommunication cost
2. Facilitate module interface with vendor equipment
3. Provide quick response to contain changes in the power system through an open-access environment

UCA is properly documented for utility usage with dedicated report volumes covering:

1. Functional descriptions
2. Communication requirements
3. Standards assessments
4. User guides and specifications for interactions and standards

OSI features: UCA has the OSI specification requirements meeting all of the layer requirements. In particular, UCA provides support to LAN/WAN technology to include other standards such as X.25, token ring, and Fixed Scheduled Dynamic Network (FSDN).

Security of UCA: UCAs are designed to include the best security standards. They are designed to handle data retrieval and reporting with a strong security model based on best practices. A decentralized communication structure is used in UCA to enhance security and encourage fast reconfiguration and peer-to-peer communication.

9.13 Power-Line Carrier Communication

9.13.1 Introduction

Power-line communication (PLC), also called mains communication or power-line telecoms (PLT) or power band, is a term describing several different systems for using power distribution wires for simultaneous distribution of data. The carrier can communicate voice and data by superimposing an analog signal over the standard 50- or 60-Hz alternating current (AC). It includes broadband over power lines (BPL) with data rates sometimes above 1 Mbit/sec and narrowband over power line with much lower data rates. For increased speed over that of the Internet and fiber optics, a conventional power-line carrier is widely used to provide real-time communications for protection of high-voltage transmission lines. Therefore PLC is often the most economical and reliable high-speed dedicated channel available for protective relaying.

Traditionally, electric utilities used low-speed power-line carrier circuits for control of substations, voice communication, and protection of high-voltage transmission lines. High-speed data transmission has been developed using the lower-voltage transmission lines used for power distribution. A short-range form of power-line carrier is used for home automation and intercoms.

9.13.2 PLC Architecture

A power-line carrier system includes three basic elements: a transmission line, presenting a channel for the transmission of carrier energy; tuning, blocking, and coupling equipment, providing connection to the high-voltage transmission line; and transmitters, receivers, and relays. The simplified functional diagram of a power-line carrier system is shown in Figure 9.17.

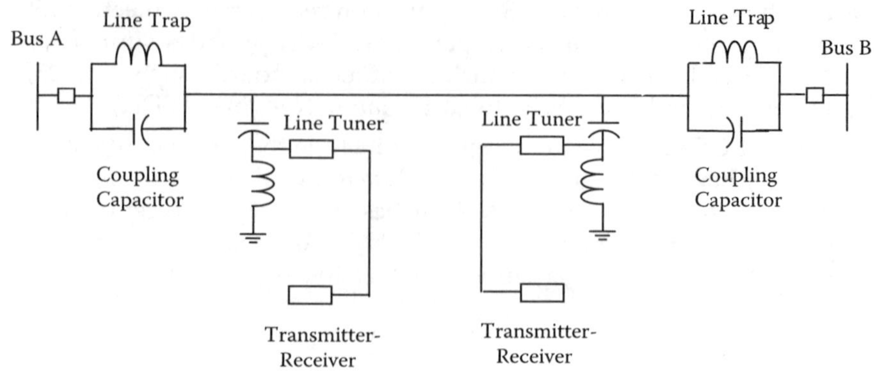

FIGURE 9.17
Power-line carrier communication system.

9.13.2.1 Line Traps

Line traps provide blocking of the carrier signal, preventing it from continuing into other transmission line sections. Single- and two-frequency line traps are parallel L-C circuits with parameters of variable inductances and capacitances selected so as to resonate at a specific frequency or at two frequencies, thus blocking the carrier frequency. Line traps are available in various inductance ratings and continuous power frequency ranges. Figure 9.18 illustrates an equivalent circuit diagram of line traps.

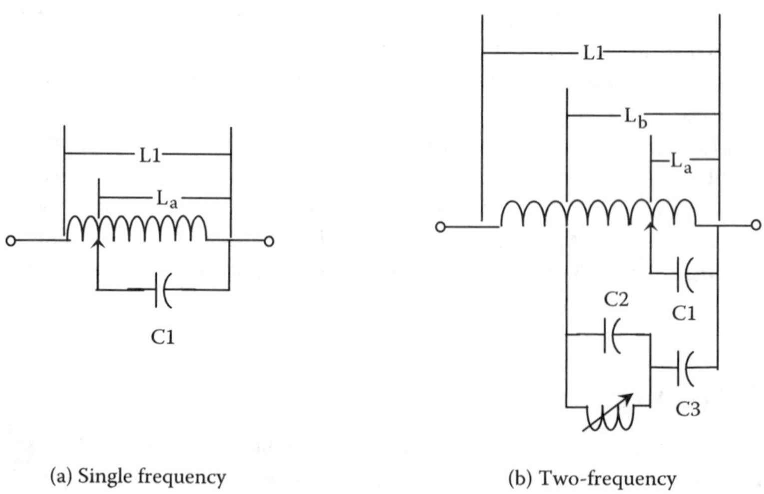

FIGURE 9.18
Equivalent circuit diagram of line traps.

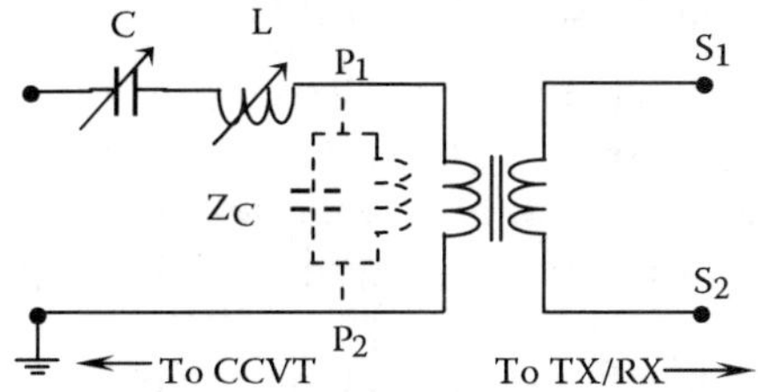

FIGURE 9.19
Equivalent circuit diagram of line-tuning unit.

9.13.2.2 Line-Tuning Units

Line-tuning units (LTUs) or line tuners are used to tune to the carrier frequency and provide impedance matching between the power line and the transmitter/receiver. The LTU includes an impedance-matching transformer; a series-resonant L-C circuit tunes to the carrier frequency and also serves as a protective device. Figure 9.19 illustrates an equivalent circuit diagram of a line-tuning unit.

9.13.2.3 Hybrids

Auxiliary coupling devices can be defined as any component of a PLC coupling scheme used to mix or separate transmit/receive frequencies on the 50-∫ side of the LTU. Hybrids and filters are passive auxiliary coupling devices, as opposed to active devices that combine PLC functions using unidirectional amplifiers. The hybrids can work in both directions (bilateral), and therefore can be applied for cases of two inputs and a single output or one input and two outputs, as shown in Figure 9.20 for the resistive hybrid.

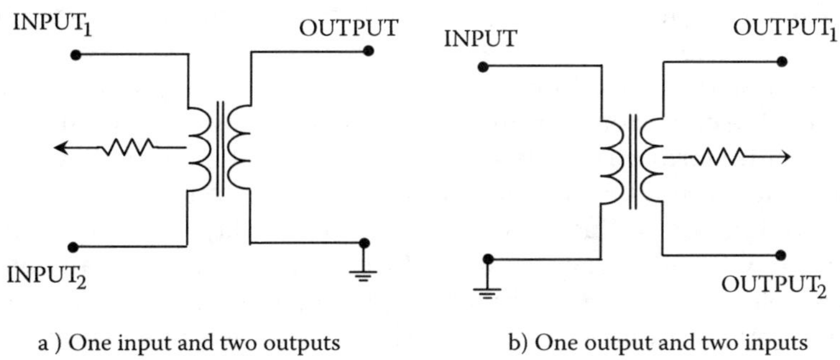

FIGURE 9.20
Equivalent circuit diagram of balanced resistive hybrid.

9.13.3 Broadband over Power Lines (BPL)

Broadband over power lines (BPL), also known as power-line Internet, is the use of PLC technology to provide broadband Internet access through ordinary power lines. A computer (or any other device) would need only to plug a BPL "modem" into any outlet in an equipped building to have high-speed Internet access.

BPL offers obvious benefits over regular cable or DSL connections: the extensive infrastructure already available would appear to allow more people in more locations to have access to the Internet. Also, such ubiquitous availability would make it much easier for other electronics, such as televisions or sound systems, to hook up. However, variations in the physical characteristics of the electricity network and the current lack of IEEE standards mean that provisioning of the service is far from being a standard, repeatable process, and the amount of bandwidth a BPL system can provide compared with cable and wireless is in question. High-speed data transmission, or broadband over power line, uses the electric circuit between the electric substations and home networks.

PLC modems transmit in medium and high frequency (1.6- to 30-MHz electric carrier). The asymmetric speed in the modem is generally from 256 to 2.7 Mbit/sec. In the repeater situated in the meter room, the speed is up to 45 Mbit/sec and can be connected to 256 PLC modems. In the medium-voltage stations, the speed from the head ends to the Internet is up to 135 Mbit/sec. To connect to the Internet, utilities can use an optical fiber backbone or a wireless link.

9.13.4 Standards

Several competing standards are evolving, including the Home Plug Powerline Alliance, Universal Powerline Association, and ETSI, and the IEEE X.10 is a de facto standard. The standards will be developed in sufficient detail to allow interoperability between equipment from different manufacturers and the coexistence of multiple power-line systems within the same environment. Harmonized standards will be developed to allow presumption of conformity with the relevant EU/EC directives.

The proliferation of this technology has been delayed due to its potential to interfere with radio transmissions. As power lines are typically untwisted and unshielded, they are essentially large antennas and thus will broadcast large amounts of radio energy. Because of their lack of shielding, the BPL systems are also at risk of being interfered with by outside radio signals.

9.13.5 Current Trends and Applications

BPL does bridge the digital divide in the Third World, bringing broadband to isolated village and farms. It is acclaimed because the infrastructure is already there, so that there is ostensibly no need to deploy fiber, satellite, WiMAX, or other new communication infrastructure. BPL has a range of benefits, including:

- Potential to provide every home with a fast Internet connection, given that almost all homes are on an electricity supply grid
- Wiring within buildings transforms every power socket into an Internet access point
- Provision of an always-on service with the same characteristics as DSL and cable modem connectivity
- Surveillance from any Internet connection, e.g., monitoring your children or people who are in need of regular help
- Improved safety and efficiency of the power network, with remote control and monitoring of appliances via power lines

Applications of mains communications vary enormously, as would be expected of such a widely available medium. One natural application of narrowband power-line communication is the control and telemetry of electrical equipment such as meters, switches, heaters, and domestic appliances. There are a number of active developments that are considering such applications from a systems point of view, such as "demand-side management." In this, domestic appliances would intelligently coordinate their use of resources, for example limiting peak loads.

Control and telemetry applications include both utility-side applications, which involve equipment belonging to the utility (i.e., between the supply transformer substation up to the domestic meter), and consumer-side applications, which involves equipment on the consumer's premises. Possible utility-side applications include automatic meter reading, dynamic tariff control, load management, load profile recording, credit control, prepayment, remote connection, fraud detection, and network management, and such applications could be extended to include gas and water.

A project of EDF, France, includes demand-side management, control of street lighting, remote metering and billing, customer-specific tariff optimization, contract management, expense estimation, and gas applications safety.

There are also many specialized niche applications that use the mains supply within the home as a convenient data link for telemetry. For example, in the U.K. and Europe, a TV-audience-monitoring system uses power-line communications as a convenient data path between devices that monitor TV viewing activity in different rooms in a home and a data concentrator, which is connected to a telephone modem.

9.14 Security in Telecommunications and Information Technology

The communication infrastructure has to be efficient to be able to handle increasing demands and the business enterprise. The utility protocol and communication infrastructure have the responsibility of providing security for the gathered information. The security provisions range from details in protocol to management of networks. The security standard is aimed at safeguarding information about the data, device, and computational or application software against all forms of vulnerability, threats, and risks. Security policy is a function of organizational structure, from the chief officer to low-rank officers, partners/customers, and the general public. This section provides an overview of security in telecommunications and information technologies, describes practical issues, and indicates how different aspects of security in today's applications are addressed by ITU-T and its relevance to the power system utility.

The security architecture is defined in terms of two major concepts: layers and planes. Security layers address requirements that are applicable to the network elements and systems that constitute the end-to-end network. A hierarchical approach is taken in dividing the requirements across the layers so that the end-to-end security is achieved by building on each layer. The three layers are infrastructure layer, services layer, and applications layer. The vulnerabilities at each layer are different, and thus countermeasures are to be defined to meet the needs of each layer.

Infrastructure layer: consists of the network transmission facilities as well as individual network elements. Examples of components that belong to the infrastructure layer are individual routers, switches, and servers as well as the communication links between them.

Services layer: addresses security of network services that are offered to customers. These services range from basic connectivity offerings such as leased-line services to value-added services such as instant messaging.

Application layer: addresses requirements of the network-based applications used by the customers. These applications can be as simple as e-mail or as sophisticated as collaborative visualization, where very-high-end video transfers are used in oil exploration, designing automobiles, etc.

9.14.1 Vulnerabilities, Threats, and Risks

A security vulnerability is a flaw or weakness in a system's design, implementation, or operation that could be exploited to violate the system's security (RFC 2828). A security vulnerability is not a risk, a threat, or an attack.

Vulnerabilities can be of four types:

1. Threat model vulnerabilities originate from the difficulty to foresee future threats.
2. Design and specification vulnerabilities come from errors or oversights in the design of the protocol that make it inherently vulnerable.
3. Implementation vulnerabilities are vulnerabilities that are introduced by errors in a protocol implementation.
4. Operation and configuration vulnerabilities originate from improper usage of options in implementations or weak deployment policies (e.g., not enforcing use of encryption in a WiFi network, or selection of a weak stream cipher by the network administrator).

According to X.800, a security threat is a potential violation of security, which can be active (when the state of a system can be changed) or passive (unauthorized disclosure of information without changing the state of the system). Masquerading as an authorized entity and denial of service are examples of active threats, and eavesdropping to steal a clear password is an example of a passive threat. Agents of threats can be hackers, terrorists, vandals, organized crime, or state sponsored, but in a significant number of cases, threats come from the insiders of an organization.

A security risk originates when a security vulnerability is combined with a security threat. For example, an overflow bug in an operating system application (i.e., a vulnerability) associated with a hacker's knowledge, appropriate tools, and access (i.e., a threat) can develop the risk of a Web server attack. Consequences of security risks are data loss, data corruption, privacy loss, fraud, downtime, and loss of public confidence.

While threats change, security vulnerabilities exist throughout the life of a protocol. With standardized protocols, protocol-based security risks can be very large and global in scale. Hence it is important to understand and identify vulnerabilities in protocols.

9.14.2 Security Architecture Elements in ITU-T X.805

It consists of three layers, as follows:

- Vulnerabilities
  Security layers
  Application security
  Services security
  Infrastructure security
  End-user plane
  Control plane

Management plane

- Eight security dimensions

  Access control

  Authentication

  Nonrepudiation

  Data confidentiality

  Communication security

  Data integrity

  Availability

  Privacy

- Threats/attacks

  Destruction

  Corruption

  Removal

  Disclosure

  Interruption

9.14.3 Privacy and Data Confidentiality

The concept of privacy is a fundamental motivator for security. Privacy is commonly understood as the right of individuals to control or influence what information related to them may be collected and stored and by whom and to whom that information may be disclosed. By extension, privacy is also associated with certain technical means (e.g., cryptography) to ensure that this information is not disclosed to any one other than the intended parties, so that only the explicitly authorized parties can interpret the content exchanged among them.

9.14.4 Authentication

Authentication is the provision of proof that the claimed identity of an entity is true. Entities here include not only human users, but also devices, services, and applications. Authentication also provides for assurance that an entity is not attempting a masquerade or an unauthorized replay of a previous communication. There are two kinds of authentication: data origin authentication (i.e., authentication requested in a connection-oriented association) and peer entity authentication (i.e., authentication in a connectionless association).

9.14.5 Data Integrity

Data integrity is the property that data have not been altered in an unauthorized manner. By extension, data integrity also ensures that information is protected against unauthorized modification, deletion, creation, and replication and provides an indication of these unauthorized activities.

9.14.6 Nonrepudiation

Nonrepudiation is the ability to prevent users from denying later that they performed an action. These actions include content creation, origination, receipt, and delivery, such as sending or receiving messages, establishing or receiving calls, participating in audio and video conferences, etc. The term "nonrepudiation" is referenced in several ITU-T recommendations, including F.400, F.435, F.440, J.160, J.93, J.95, M.60, T.411, X.400, X.805, X.813, and X.843.

9.14.7 Other Dimensions Defined in X.805

In addition to privacy and data confidentiality, authentication, integrity, and nonrepudiation, ITU-T X.805 defines the three other security dimensions: access control, communication, and availability.

The *access control* security dimension protects against unauthorized use of network resources. Access control ensures that only authorized personnel or devices are allowed access to network elements, stored information, information flows, services, and applications. Access control is defined in ITU-T X.810 section 6.3 and in X.812. It is related but beyond the scope of authentication.

The *communication* security dimension is a new dimension defined in X.805 that ensures that information flows only between authorized endpoints. This dimension deals with measures to control network traffic flows for prevention of traffic diversion and interception.

The *availability* security dimension ensures that there is no denial of authorized access to network elements, stored information, information flows, services, and applications due to network interruption. Network restoration and disaster recovery solutions are included in this category.

9.14.8 Security Framework Requirements

The requirements for a generic network security framework have been derived from different sources:

- Customers/subscribers need confidence in the network and the services offered, including availability of services (especially emergency services) in case of major catastrophes (including terrorist actions).
- Public authorities demand security by directives and legislation to ensure availability of services, fair competition, and privacy protection.
- Network operators and service providers themselves need security to safeguard their operations and business interests and to meet their obligations to the customers and the public.

Security requirements for telecommunication networks and services should preferably be based upon internationally agreed security standards, as it increases interoperability as well as avoids duplication of efforts and reinventing the wheel. The provisioning and usage of security services and mechanisms can be quite expensive relative to the value of the transactions being protected. There is a balance to consider between the cost of security measures and the potential financial effects of security breaches. It is therefore important to have the ability to customize the security provided in relation to the services being protected. The security services and mechanisms that are used should be provided in a way that allows such customization. Due to the large number of possible combinations of security features, it is desirable to have security profiles that cover a broad range of telecommunication network services.

9.14.9 Information Security Goals

Confidentiality: ensuring that the information is not disclosed to unauthorized persons

Integrity: ensuring that the information held in the system is a proper representation of the information intended and that it has not been modified, created, or deleted by an unauthorized person

Availability: ensuring that the information processing resources are not made unavailable by malicious action

Accountability: ensuring that actions of an individual can be uniquely traced to that individual

Nonrepudiation: ensuring that agreements made electronically can be proven to have been made

9.15 Illustrative Examples

9.15.1 Example 1

Compute the Nyquist channel capacity

$$C = \mathrm{BW}\log_2 L$$

where L is the number of signaling levels, e.g.,

$$L = 2 \quad \{0\,\mathrm{or}\,1, \mathrm{or}\,(0, +5)\}$$

$$\mathrm{BW} = 3000\ \mathrm{Hz} \quad \log_2 2 = 1$$

$$C = 2 \times 3000 \times \log_2 2 = 2 \times 3000 \times 1$$

$$= 6000\ \mathrm{bps}$$

Shannon capacity

$$C = \mathrm{BW}\log_2\left(1 + \frac{s}{w}\right)$$

$$C = \mathrm{BW}\left[\frac{\log_{10}\left(1 + \frac{s}{w}\right)}{\log_{10} 2}\right]$$

$$C = \mathrm{BW}\left[\frac{\log_{10}\left(1 + \frac{s}{w}\right)}{0.301}\right]$$

If BW = 3 kHz, we have the following:

$$\frac{s}{w} = 10$$

$$\therefore C = 3 \times \left(\frac{\log_{10}(1+10)}{0.301} \right) = 10.37\ \text{kbps}$$

9.16 Summary

This chapter deals with basic concepts of communication and technology trends and their applications to distribution automation functions (DAF). The special technology used in communications applications for distribution systems — such as remote terminal unit (RTU) and supervisory control and data acquisition (SCADA), and enabling technologies such as Ethernet, broadband, and wireless/sensor technology — are briefly discussed. The use of automatic meter reading (AMR) and billing based on advances in communication systems using communications or intelligent systems are also briefly discussed.

Problem Set 9

9.1 List the telecommunications standards organizations of the U.S. and the international community.

9.2 Briefly discuss the purpose of using a network and what network types apply to a given situation.

9.3 Identify and briefly describe the following:

a. Four lower layers of the OSI model
b. Three upper layers of the OSI model

9.4 Describe the following terms and briefly discuss their functions with regards to communication.

a. RS-232 and RS-422A
b. Physical level TP Protocol standardized by EIA

(RS = recorded standard.)

9.5 What are five reasons to use a local area network? Construct the LAN topologies and compare with WAN.

9.6 In communication systems, what are the functions of routers and bridges? Explain.

   a. What are the methods of multiplexing? Explain FDM and TDM.
   b. Design an analog-to-digital and digital-to-analog converter.

9.7 Explain the terms amplitude modulation, frequency modulation, and phase modulation.

   a. What is differential-phase-shift keying?
   b. State the general means of modulation and give the purpose of modulation.

9.8 What is one major advantage of frame relay over IC-25?

   a. What error checking techniques can be used to detect the quality of the signal being transmitted to the relay?
   b. Distinguish frame relay from earlier protocols.

9.9 Explain the terms simplex, half duplex, and full duplex as they pertain to communication.

9.10 Design an automated enterprise in power system services telecommunication as an enabling technology of choice.

9.11 Explain the concepts of Signal to Noise Ratio (SNR) and provide descriptive examples of sources of noise in communication systems.

9.12 Sampling, Nyquist Rate, and Aliasing are common concepts in telecommunication science. Carefully explain these three concepts and state why they ae important in communication theory.

9.13 Describe the process for converting an analogue signal from a digital signal. (Draw the necessary block diagram to supplement your answer).

9.14 Design a communication layer between two different machines (two smart transducer/computers) in a typical substation environment were data for fault, power quality, harmonics, etc. are being transferred back and forth in order to control the power system. Model the layers 1 to 7 and explain the communication between the computers and design and discuss the special features of an appopriate communication scheme.

9.15 Why is security of data important in telecommunication? What are the criteria to guarantee information security?

10

Epilogue

10.1 Challenges to Distribution Systems for a Competitive Power Utility Environment

This book treats the broad issues of distribution automation, control, and its various functions as well as the penetration of distributed generation with various options, renewable energy, and performance assessment. Government agencies and utility companies have proffered a variety of roadmaps identifying the characteristics and features of future distribution systems, the required technologies, and the scope of coverage for research and educational purposes. We have labored to present some of these timely topics in this book.

This chapter briefly discusses the areas of future work that will improve the distribution so that it can become flexible, reliable, and smart. Several government organizations and utility companies have proposed some of the challenges in building the so-called smart-connection technology platforms for distribution systems. Simply stated, smart-connect is an attempt to develop communication and technology controls that enable a distribution system with distributed generation (DG) to be upgraded with smart, reconfigurable, self-healing, restorative, and reliable systems.

We summarize here some of the urgent research work needed to further develop the topics covered in this book:

1. Development of technology for handling two-way electrical flows as well as communication that permits distributed generation to be dispatched, monitored, and controlled from a central source.
2. Design of enabling technology for distribution substations to improve monitoring, equipment diagnosis, fault recording, and multimedia support with audio and visual cues for signature analysis.
3. Development of low-cost sensors, energy storage, and power electronics to ensure greater interaction to achieve improved reliability and efficiency.

4. Use of multiagent schemes to design reconfiguration, restoration, and load balancing of distribution systems under fault.
5. Integration of distribution systems with new communication and electronics technology to improve capability for self-diagnosis, self-healing, self-reconfiguration, and the ability to handle congestion and instability.
6. Development of a brokerage system for pricing and marketing of distribution generation along with utility generation in a competitive power market.

In the near term, the research road map using the concepts discussed in the book will increase the capability of future distribution systems. In this chapter, we have classified these as near-term and long-term research works, addressed as grand challenges/problems.

10.2 Protection

The current protection schemes have limitations in a distributed-generation-based distribution system. We need to develop a new generation of protection schemes capable of detecting faults and restoring the system in minimum time for a two-way power flow.

10.3 Demand Response

Using voltage sag, frequency, power factors, or harmonics changes, a demand-response strategy is needed to control distribution system contingency impacts.

10.4 Communication Advances

Advances in low-cost communication and Ethernet technology can easily be the option for handling the features of distribution management systems. For example, the use of phasor measurement units (PMU) and state estimation could enhance real-time management and control using advances in global positioning systems (GPS) and Internet technology.

10.5 Microgrid

Increased penetration of DG units in electrical proximity to the loads for autonomous generation operation from the grid has led to the development of the microgrid concept. These developments of DG units are derived from different renewable options such as wind, solar, geothermal, etc. The ability to control and communicate with the microgrid requires computational tools that are capable of handling system dynamics, uncertainties, and interconnection issues. To this end, advances in wireless communication and robust dynamic optimization schemes, such as adaptive dynamic programming (ADP), will be useful for real-time operation.

10.6 Standards and Institutional Barriers

Much work and documentation have been done to establish standards for the interconnection of DG to future distribution to achieve reliability and safety objectives. There is further work to be done in standardization of the software tools needed for distribution automation. For example, a common format and benchmark test system is needed for researchers to customize the research products, discussed in Section 10.1 as grand challenges.

To overcome institutional barriers to the development and deployment of the new features of distribution automation, research products that integrate the demand-side management (DSM) function using communication and intelligent systems need to be available for adaptation by the utility.

Finally, plug-and-play technology that will facilitate deployment of control measures with embodied intelligence is needed to achieve a self-healing, safe, reliable, and cost-effective distribution system.

10.7 Pricing and Billing

The distribution system is the business endpoint for obtaining a return on investment in a power system. The ability to collect bills on time and guarantee power at an affordable price is one of the overall missions of automating the distribution system. Smart meters and good pricing structures are needed to justify the investment, using cost-benefit analysis tools. Further research is needed to promote enabling technology to achieve accurate and correct billing and competitive pricing for future and current owners of distribution systems.

Finally, an enabling environment for billing and sustaining distribution automation and control for a competitive power market requires the use of innovative tools and dedicated effort. Funded research work and prototype products currently available require further testing and customization by the power industry.

Glossary

Chapter 1

Circuit breaker: a high-current device that automatically disconnects faulted equipment

Distribution automation: a term used to (a) define the application of communication, optimization, and intelligent systems to improve the performance and functions of a distribution system during normal and abnormal operation and (b) facilitate efficiency, quality of service, and security of the power system

Fuse: a circuit-breaking device that melts when an overload current passes through it

Recloser: a device that serves as a special-purpose light-duty circuit breaker that interrupts overloads but not faults

Relay: a device designed to protect against excessive voltage, frequency, or current

Sectionalizer: a device that automatically isolates faults on a line segment from a disturbance

Chapter 2

Automotive voltage regulation: designed to provide a boost of voltage magnitude along a line or change in phase to control flows of power between systems

Branch: electrical wires connecting nodes to nodes

Distribution transformer: a device that provides an electric link to the customer; it operates at a voltage level safe to use on the customer side of the premises

Lateral branch: the branch emanating from the main feeder

Leaf node: represents the top of the highway from substation to the far end of the service station

Main feeder: designed as the branch-connecting substation to the outside world

Phase shifter: a regulating transformer aimed at controlling power flows and losses within a distribution system

Power factor: the phase angle between voltage and current, given as $\cos(\theta_V - \theta_I)$

Voltage sag: a sudden reduction in the supply of voltage followed by a voltage recovery after a short period of time

Power transformer: a device that reliably and efficiently changes voltage and current at high/low power levels

Chapter 3

Auxiliary relay: a device that provides miscellaneous functions with other relaying systems, e.g., timers are examples of auxiliary relays

Fuse: a one-time, nonreusable device for interrupting a fault current; the metallic conductor within the fuse melts in the presence of an overload current, thereby opening the circuit

Monitoring relay: a device that monitors conditions within the power system and sends an alarm when conditions are unstable; used for returning signals and system voltage levels

Programming relay: a device that detects sequences of events; used to control and monitor synchronization

Recloser: special-purpose automatic circuit recloser that protects distribution circuits from temporary disturbances; a self-controlled device that automatically interrupts overloads but not severe faults

Regulatory relay: a device used to determine whether a parameter such as voltage, current, or impulse has exceeded its allowable limit; sends an alarm when parameter exceeds its limit

Relay: an electromechanically or microprocessor-controlled electronic system that senses faulty or abnormal conditions in a distribution system (such as overcurrent, overvoltage, overfrequency or undercurrent, undervoltage, underfrequency); an excessive value generates a trip signal to a current breaker

Sectionalizer: a device that automatically isolates faulted line segments from a distribution system

Chapter 4

ASAI: average service availability index; a measure of the average annual outage time or the availability of the power supply, defined as

$$ASAI = \frac{\text{Sum of hours of available service to customers}}{\text{customer hours service demand}}$$

Breakdown: condition requiring repair or corrective maintenance to restore the system to an acceptable status

CAIDI: customer average interruption duration index; represents the average time taken to restore service to the customer, defined as

$$CAIDI = \frac{\text{Sum of customer-interruption durations}}{\text{total number of customer interruptions}}$$

CAIFI: customer average interruption frequency index; used to calculate failure rate of distribution systems or the interruption rate to which customers are subjected, defined as

$$CAIFI = \frac{\text{total number of customer interruptions}}{\text{total number of customers affected}}$$

Connected load: includes the connection of a transformer, a peak-load metered demand on the circuit, or a portion of the interrupted circuit

Corrective maintenance: maintenance based on restoring equipment to an operable condition after failure or some other malfunction has occurred

Customer interruption: supply of power interrupted by component outages, system instability, thermal overloads, or undervoltages

EACI: expected annual cost of interruption; an alternative index for measuring adequacy of customer service; designed to provide an economic value to reliability or a cost of unreliability

Event-tree method: an analysis method used to provide a detailed examination of possible scenarios initiated by a faulty event or component within a distribution system

Inspection check: careful scrutiny of an item carried out without dismantling and using all senses to detect the cause of an item's failure to operate

Interrupting device: a device that disconnects or restores service by automatic or manual control; such devices include transmission circuit breakers, feeder breakers, line reclosers and fuses, sectionalizers, and switches

Interruption: where customer experiences an outage due to a problem in the distribution system

Interruption duration: the time period from initiation of an interruption of service until it has been restored

Loss of service: a complete loss of voltage on at least one normally energized conductor to one or more customers

Momentary interruption: a single separation of an interrupting device that results in a zero voltage

Momentary interruption event: an interruption of duration limited to a period required to restore service by an interrupting device switching operation; this operation must be completed within a specified time of 5 min or less

Monitor: inspection with partial dismantling of parts, measurement, and nondestructive tests for unsatisfactory performance of an item

Outage: the state of a component when it is not available to perform its intended function; an outage may or may not cause interruption of service, depending on the configuration of the system

Overhaul: a minor overhauls is limited to lubrication and replacement of consumable points; a major overhaul involves major dismantling and replacement of items

Planned interruption: loss of power when a component is deliberately taken out of service or for construction/maintenance

Planned outage: state of a component when it is not available to perform its intended function due to a planned event directly associated with that component

Postfault management: inspection and diagnostic tests to establish whether equipment is in acceptable condition and, if needed, corrective action to restore service

Prevention: planned maintenance carried out as a result of an inspection or report, but not the result of a breakdown

Reliability: ability of the power network to deliver uninterrupted power at prescribed levels of quality and security to its customers

Reliability indices: used to assess past performance or to predict future performance

Routine: maintenance carried out in accordance with a predetermined policy or plan to prevent breakdown or reduce the likelihood of an item of the plant failing to meet an acceptable condition; also includes operational checks and diagnostic testing for acceptable positions

SAIDI: system average interruption duration index; indicates the average duration of service interruptions for the system, defined as

$$\text{SAIDI} = \frac{\text{Sum of customer-interruption durations}}{\text{total number of customers}}$$

SAIFI: system average interruption frequency index; indicates how often the average customer experiences a sustained interruption over a period of time for the area

$$\text{SAIFI} = \frac{\text{total number of customer interruptions}}{\text{total number of customers served}}$$

State space diagram: represents a system by defining all possible states of interest that the system can adopt

Sustained interruption: any unplanned interruption not classified as part of a momentary testing lasting more than 5 min

Unplanned interruption: interruption caused by an unplanned event/outage

Visual check: eyeball check to detection of anything that might cause an item to fail due to an unacceptable position

Chapter 5

Demand-side management: an effective means of modifying the consumer demand to cut operating expenses from expensive generators and defer capacity addition in the long run

Harmonics: nonfundamental components of a distorted 60-Hz waveform; they have frequencies that are integral multiples of the fundamental frequency of 60 Hz

Outage: a complete loss of voltage, usually covering a time period varying from 30 cycles up to several hours or even days

Restoration: provides an ample amount of power to nonfaulty out-of-service areas for as many customers as possible while guaranteeing the safety and optimum reliability of the distribution systems

Surge: important anomaly caused by transient voltage or current that can have extremely short duration and high magnitude

Undervoltage: anomaly experienced when voltage is less than the proper (or contractual) nominal voltage

Chapter 6

Artificial intelligence: a subfield of computer science that investigates how the thought and action of human beings can be modeled or mimicked by machine

Artificial neutral network: differs from an expert system in that it does not need a knowledge base to work; instead, it must be trained with numerous actual cases

Expert system: also referred to as a knowledge-based system; embodies human expertise in a narrow field or domain in a machine-implementation form

Fault analysis: involves consideration of what happens after a fault occurs, identifying the location of the fault, and assessing the nature of the damage caused by the fault

Fuzzy set: a function that maps a value that might be a member of the set to a number between zero and one, indicating the actual degree of membership

Inference engine: a data-driven or goal-driven function that uses facts and rules to deduce new facts, which allows the firing of other rules

Knowledge base: a collection of domain-specific knowledge that is processed by a logic component (inference engine) to solve a problem

Network reconfiguration: refers to balancing the load distribution in a power system during or after a disturbance while accounting for power-loss-minimization voltage, thermal-generation constraints, and power-outage costs

Power quality: refers to a large number of anomalies related to voltage, current, and frequency deviation that result in failure or abnormal operation of customer/utility equipment

Reconfiguration: principal aim of reconfiguration is to minimize distribution losses, optimize voltage profiles and relieve overload requirements while maintaining the radial structure of the network

Restoration: provides an ample amount of power to nonfaulty, out-of-serve areas for as many customers as possible while guaranteeing the safety and optimum reliability of the distribution systems

Chapter 7

Bioenergy: the energy derived from biomass organic matter such as corn, wheat, soybeans, wood, and residues that can produce chemicals and materials

Gibbs energy: the energy to do external work, neglecting any work done by changes in pressure and or volume; Gibbs energy represents the external work involved in moving electrons around an external circuit

Insulation: a term used in PV system to describe the available solar energy for conversion to electricity

Chapter 8

Customer information system: developed to solve the customer-accounting function and the trouble-call analysis function

Geographic information system: links automated digital maps of utility infrastructure to databases containing nonspatial facility-management data

Man–machine interface: accesses data from the process database and presents it in the form of single-line-diagram tabular displays and reports

Remote terminal units: installed in distribution substations at various feeders to facilitate automation of the distribution network; also used as a digital communication interface with computer-based substation control systems

SCADA: a platform with basic functionality to classify or handle events, alarm processing, monitoring, and the limits of measurable power qualities; it consists of a process database, a man–machine interface, and application software

Chapter 9

AMI: also called the modulation depth; it represents how much the modulated variable varies around its original level

Amplitude modulation: the instantaneous amplitude of the original (modulated) signal; a carrier wave is modulated in proportion to the strength of the original signal

Analog modulation: aimed at impressing an information-carrying analog waveform onto a carrier for transmission

Bridge: consists of two or more networks that use the same protocol at the media control sublayer of the data-link layer

Broadband over power lines (BPL): also known as power-line Internet; involves the use of PLC technology to provide broadband Internet access through ordinary power lines

Broadcast transmission: consists of a single data packet that is copied and sent to all nodes on the network

Channel: a division in the transmission medium for sending streams of data at different frequencies

Data integrity: the property that data have not been altered in an unauthorized manner

Digital modulation: used to convert an information-bearing discrete time-symbol sequence into a continuous time waveform impressed in a carrier waveform

Duplex: known as full duplex, where information can flow in two directions at the same time

FMI: indicates how much the variables vary around their unmodulated level

Frame relay: a high-performance WAN communication protocol that transports data in a "packet" format that maximizes bandwidth on communication circuits

Gateway: operates at or above the OSI transport layer and links LAN or networks that employ different architectures and use dissimilar protocols

Half duplex: where information can flow in two directions, but only in one direction at a time

IEEE Standards Coordinating Committee: a professional society that develops national (North America) standards for communicating with electric, gas and water meters; the institute pioneered the standards for local area networks

International Electrotechnical Commission (IEC): a commission that develops standards for the utility industry to promote safety, compatibility, interchangeability, and acceptability of electrical standards

International Standards Organization (ISO): an organization that is concerned with improving international collaboration and communication

International Telecommunication Union (ITU): an organization that is concerned with the creation of standards that facilitate international telecommunication

Line-tuning units (LTU): line tuners that are used to tune to the carrier frequency and provide impedance matching between the power line and the transmitter/receiver

Local area network (LAN): consists of two or more personal computers, printers, and high-capacity disk storage (file servers) that allow each computer in the network to access a common set of rules

Logical link control (LLC): based on the MAC sublayer service, offers a traditional connectionless- or connection-mode data-link service between arbitrary systems attached to LAN

Medium access control (MAC): offers multiplexed access to the physical-layer transmission facilities of the LAN

Metropolitan area network (MAN): a system of LANs connected throughout a city or metropolitan area

Modulation: a means of varying or changing a signal over a medium; it involves a signal-processing technique where one signal (the modulating signal) modifies another carrying signal, which enables the original signal to form a new composite signal (modulated signal = original signal + carrier signal)

Multicast: consists of a single data packet that is copied and sent to a specific subset of nodes on the network

Permanent virtual circuits: permanent connections that are used for frequent and consistent data transfers between DTE devices across a frame-relay network

Power-line communication (PLC): also called main communication, power-line telecoms (PLT), or power band; a term describing several different systems for using power distribution wires for simultaneous distribution of data

Router: operates at the network level of the OSI model with more sophisticated addressing software than a bridge

Signal-to-noise ratio: used in communication systems to distinguish the ratio of power in a useful signal to power in a noise signal; it is measured in decibels

Simplex: (one directional) one way, where information flow can have any orientation, but it all flows in the same direction simultaneously

Switch: switches data to its destination by a point-to-point connection

Switched virtual circuits: temporary connections used in situations where data transfer from DTE devices across frame-relay networks is sporadic

Telecommunication: communication from afar using various forms of equipment, computer, networks, and different media over short to long distances

Unicast transmission: involves transmission of single-packet data from a source to a destination on a network

Wide area network (WAN): a network system connecting cities, countries, and continents together

References

Major Textbooks

Bergen, A. and Vittal, V., *Power Systems Analysis*, Prentice Hall, Englewood Cliffs, NJ, 2000.

Coffer, W. and Faulkenberry, L., *Electrical Power Distribution and Transmission*, Prentice Hall, Englewood Cliffs, NJ, 1996.

El-Hawary, M., *Electrical Power Systems Design and Analysis*, John Wiley and Sons, New York, 2003.

Kirschen, D. and Strbac, G., *Fundamentals of Power System Economics*, John Wiley and Sons, London, 2004.

Momoh, J., *Electric Power Systems Applications of Optimization*, Marcel Dekker, New York, 2005.

Fault Analysis

Aucoin, B.M. and Russell, B.D., Distribution of high impedance fault detection utilizing high frequency current component, *IEEE Trans. Power Appar. Syst.*, 101, 1596–1606, 1982.

Balser, S.J., Clements, K.A., and Lawrence, D.J., A microprocessor based technique for detection of high impedance faults, *IEEE Trans. Power Delivery*, 1, 252–258, 1986.

Benner, C.L., Carswell, P.W., and Russell, B.D., Improved Algorithm for Detecting Arching Faults Using Random Fault Behavior, paper presented at Southern Electric Industry Application Symposium, New Orleans, Nov. 15–16, 1988.

Bernard, J.P. and Durocher, D., An Expert System for Fault Diagnosis Integrated in Existing SCADA System, paper presented at IEEE 1993 PICA, 1993, pp. 313–319.

Butler, K.L., Momoh, J., and Dias, L.G., Expert System Assisted Identification of Line Faults on Delta–Delta Distribution Systems, paper presented at IEEE NAPS Conference, 1992.

Ebron, S., A Neural Network Approach to the Detection of Incipient Faults in Power Distribution System Feeders, paper presented at IEEE and Distribution Conference, April 2–7, 1989.

Eickhoff, F., Handschin, E., and Hoffman, W., Knowledge-based alarm handling and fault location in distribution networks, *IEEE Trans. Power Syst.*, 6, 358–364, 1991.

Electric Power Research Institute, High Impedance Fault Detection Using Third Harmonic Current, EPRI Report EL 2430, prepared by Hughes Aircraft Co., June 1982.

Fauquembergue, P., Kezunovic, M., Gonzalez-Sabato, M.V., and Sanz, A., Intelligent System Applications to Protections, Control and Monitoring within Substations, paper presented at CIGRE Symposium on Integrated Control and Communication Systems, Helsinki, Finland, August 1995.

Fukui, C. and Kawakami, J., An expert system for fault section estimation using information from protective relays and circuit breakers, *IEEE Trans. Power Delivery*, 1, 83–90, 1986.

Girgis, A. and Johns, M.B., A hybrid expert system for faulted section identification, fault type classification and fault location algorithms, *IEEE Trans. Power Delivery*, 4, 978–985, 1989.

Huang, C.L., Chu, H.Y., and Chen, M.T., Algorithm comparison for high impedance fault detection based on stage fault tests, *IEEE Trans. Power Delivery*, 3, 1427–1435, 1988.

Jeerings, D.I. and Linders, J.R., Unique Aspects of Distribution System Harmony due to High Impedance Ground Faults, paper presented at the IEEE/PES Conference and Exposition on Transmission and Distribution, New Orleans, April 1989.

Kandil, N., Sood, V.K., Khorasani, K., and Patel, R.V., Fault identification in an AC-DC transmission system using neural networks, *IEEE Trans. Power Syst.*, 6, 285–292, 1991.

Kezunovic, M., Fault Analysis Using Intelligent Systems, paper presented at IEEE T&D Conference, Los Angeles, September 1996.

Kezunovic, M., Rikalo, I., Fromen, C.W., and Sevcik, D.R., New Automated Fault Analysis Approaches Using Intelligent System Technologies, paper presented at International Conference on Power System Technology, Beijing, October 1994.

Kezunovic, M., Rikalo, I., Sobajic, D.J., Fromen, C.W., and Sevcik, D.R., Automated Fault Analysis Using Neural Network, paper presented at Fault Disturbance Conference, College Station, TX, March 1994.

Kim, C.J. and Don Russel, B., Harmonic behavior during arcing faults on distribution feeders, *Electric Power Syst. Res.*, 14, 219–225, 1988.

Kohonen, T., Self-organized formation of topologically correct feature maps, *Biol. Cybern.*, 43, 59–69, 1982.

Minakawa, T. and Kunugi, K., Development and implementation of a power system fault diagnosis expert system, *IEEE Trans. Power Syst.*, 10, 932–940, 1995.

Minsky, M.L. and Papert, P., *Perceptions*, MIT Press, Cambridge, MA, 1969.

Momoh, J.A., Integrated Detection and Protection Schemes for High Impedance Faults on Distribution Systems, report prepared for the Office of Research Administration, Howard University, Washington, D.C., October, 1990.

Momoh, J., Dias, L., and Butler, K., Selection of Artificial Neural Networks for Distribution System Fault Diagnosis, in Proceedings of the International Conference on Intelligent Systems Application to Power Systems (ISAP), Montpellier, France, September 1994.

Momoh, J., Laird, D., Dias, L.G., and Thor, T., Rule-based decision support system for single line fault detection in a delta-delta connected distribution system, *IEEE Trans. Power Syst.*, 9, 782–788, 1994.

Momoh, J., Oliver, W. Jr., and Shaw, A., Application of Wavelet Theory to Terrestrial and Nonterrestrial Power Distribution Systems for Fault Detection, in Proceedings of the 1995 North American Power Symposium, Canada, 1995.

Momoh, J., Shaw, A.D., and Butler, K.L., A Sensitivity Study Using a Clustering Based ANN for Fault Diagnosis, in Proceedings of the 1994 North American Power Symposium, Manhattan, Kansas, October, 1994, pp. 413–419.

Momoh, J., Sobajic, D., and Dolce, J., An Evaluation of Intelligent Systems for Fault Diagnosis, in Proceedings of the IEEE International Conference on Systems, Man and Cybernetics, San Antonio, Texas, October 1994.

Niebur, D.A. and Germond, J., Power System Static Security Assessment Using the Kohonen Neural Network Classifier, IEEE, 1991.

Oliver, W.E. Jr., Momoh, J., and Dolce, J., Fault Analysis of Space Station DC Power Systems Using Two-Stage ANN, in Proceedings of the 26th North American Power Symposium, Manhattan, Kansas, September 1994, pp. 721–731.

Pao, Y. and Sobajie, J., Combined Use of Unsupervised and Supervised Learning for Dynamic Security Assessment, IEEE, 1991.

Park, Y.M., Kim, G.W., and Sohn, J.M., A Logic Based Expert System (LBES) for Fault Diagnosis of Power System, Paper 96-WM-298-0 PWRS, presented at IEEE/PES Winter Meeting, 1996.

Power Technologies, Detection of High Impedance Faults, EPRI Report EL 2413, prepared by Power Technologies, Schenectady, NY, June 1982.

Riechelt, D. and Glavitsch, H., Features of a Hybrid Expert System for Security Enhancement, IEEE, 1991.

Russell, B.D. and Aucoin, M., Detection of distribution high impedance faults using burst noise signals near 60 Hz, *IEEE Trans. Power Delivery,* 2, 342–348, 1987.

Russell, B.D., Chinchali, R.P., and Kim, C.J., Behavior of low frequency spectra during arching fault and switching events, *IEEE Trans. Power Delivery,* 3, 1485–1492, 1988.

Russell, B.D., Mehta, K., and Chinchali, R.P., An arcing fault detection technique using low frequency current components: performance evaluation using recorded field data, *IEEE Trans. Power Delivery,* 3, 1493–1500, 1988.

Sekine, Y., Yokoyama, A., and Okamoto, H., A Real-Time Expert System for Fault Section Estimation Using Cause–Effect Network, in Proceedings of 10th PSCC, Graz, 1990.

Sobajic, D.J. and Pao, Y.-H., Artificial neural net based dynamic security assessment for electric power systems, *IEEE Trans. Power Syst.,* 4, 220–228, 1989.

Von Der Malsburg, C., Self-organization of orientation sensitive cells in the striate cortex, *Kybernetik,* 14, 85–100, 1973.

Wang, S.M. et al., A negotiation methodology and its application to cogeneration planning, *IEEE Trans. Power Syst.,* 9, 202–208, 1994.

Wook, H.K., Gi, W.L., and Young, P.M., High impedance fault detection utilizing incremental variance of normalized even order harmonic power, *IEEE Trans. Power Delivery,* 6, 557–564, 1991.

Yang, H., Huang, Y., and Huang, C., A New Intelligent Hierarchical Fault Diagnosis System, Paper 96-WM-296-4 PWRS, presented at IEEE/PES Winter Meeting, 1996.

Power Quality References

Arseneau, R. and Oulette, M., The effect of supply harmonics on the performance of compact fluorescent lamps, *IEEE Trans. Power Delivery,* 8, 473–479, 1993.

Bohman, L. and Plante, C., A Harmonic Survey of Switched Modes Power Supply Loads and Their Buildings, in Proceedings of NAPS Conference, Reno, NV, October 1992, pp. 180–188.

Day, A. and Mahmoud, A., Methods of evaluation of harmonic levels in industrial plant distribution systems, *IEEE Trans. Power Delivery,* 3, 498–503, 1988.

Farach, J., Gardy, W., and Arapostathis, A., An Optimal Procedure for Placing Sensors and Estimating the Locations of Harmonic Sources in Power Systems, Paper 92-SM497-8 PWRD, presented at IEEE/PES Summer Meeting, 1992.

Fuchs, E. and Roesler, J., Sensitivity of electrical appliances to harmonics and fractional harmonics of power system's voltage: parts I and II, *IEEE Trans. Power Delivery,* 2, 437–453, 1987.

Fuller, J., Fuchs, E., and Roesler, D., Influence of harmonics on power distribution system protection, *IEEE Trans. Power Delivery,* 3, 549–557, 1988.

Gentil, T.J., Pileggi, D.J., and Emanuel, A.E., The effect of modern compact fluorescent lights on voltage distortion, *IEEE Trans. Power Delivery,* 7, 1451–1459, 1992.

George, T.A. and Bones, D., Harmonic power flow determination using the fast Fourier transform, *IEEE Trans. Power Delivery,* 6, 530–535, 1991.

Hegazy, Y.G. and Salama, M.M.A., Important Issues in the Evaluation of Distribution System Harmonics, in Proceedings of 26th NAPS, Part I, 1994, pp. 281–285.

Heydt, G.T., The Identification of Harmonic Sources by a State Estimation Technique, IEEE Winter Power Meeting, New York, January 1988.

IEEE Task Force on the Effects of Harmonics on Equipment, Effects of harmonics on equipment, *IEEE Trans. Power Delivery,* 2, 672–680, 1993.

IEEE Task Force on the Effects of Harmonics on Equipment, Effects of harmonics on equipment and loads, *IEEE Trans. Power Appar. Syst.,* 104, 672–680, 1985.

Meliopoulous, A.P. and Cokkinedes, G.J., Effects of Transmission Line Model Accuracy on the Computation of Harmonic Resonance Parameters, in Proceedings of the International Conference on Power System Harmonics, pp. 8–14.

Mendis, S.R. and Bishop, M.T., Utility Interface Concerns with In-Plant Generation in a Harmonic Environment, IEEE IAS Annual Meeting, Houston, October 1992.

Najjar, M. and Heydt, G.T., Computational Enhancements to the Power System State Estimator at Harmonic Frequencies, in Proceedings of 22nd NAPS, October 1990, pp. 44–53.

Olejniczak, K. and Heydt, G., Basic mechanisms of generation and flow of harmonic signals in balanced and unbalanced three phase power system, *IEEE Trans. Power Delivery,* 4, 2162–2170, 1989.

Pileggi, D., Harish Chandra, N., and Emanuel, A., Prediction of harmonic voltages in distribution systems, *IEEE Trans. Power Appar. Syst.,* 100, 1307–1315, 1981.

Powers, E., Grady, W.M., and Hofh, P., Power Quality Assessment via Wavelet Transform Analysis, Paper 95-SM-371-5 PWRD, presented at IEEE/PES Summer Meeting, 1995.

Radmer, D.T., Montgomery, R., and Bala, J., Harmonic and Loss Reduction in Electric Distribution Systems, in Proceedings of the 26th NAPS, September 1994.

Rizy, D.T., Gunther, G.W., and McGranaghan, M.F., Transient and harmonic voltage associated with automated capacitor switching on distribution systems, *IEEE Trans. Power Syst.*, 2, 713–723, 1987.

Robertson, D., Camps, O., Mayer, J., and Gish, B., Wavelets and Electromagnetic Power S265-0 PWR System Transients, Paper 95-SM-391-3 PWRD, presented at IEEE/PES Summer Meeting, 1995.

Tang, Y. and Mahmoud, A.A., Evaluation and reduction of harmonic distortion in power systems, *Electric Power Res.*, 17, 1989.

Valcarcel, M. and Mauyordomo, J.G., Harmonic Power Flow for Unbalanced Systems, Paper 93-WJ\'I-061-2 PWRD, presented at IEEE/PES Winter Meeting, 1993.

Wagner, V.E., Effects of harmonics on equipment, *IEEE Trans. Power Delivery*, 8, 672–680, 1993.

Watson, N.R. and Arrilaga, J., Frequency dependent AC: system equivalents for harmonic studies and transient converter simulation, *IEEE Trans. Power Delivery*, 3, 1190–1203, 1988.

General References

Bunch, J.B. et al., A distribution automation evaluation using digital techniques, *IEEE Trans. Power Appar. Syst.*, 104, 169–175, 1985.

Barruncho, L. and Vidigal, A., GIS and Distribution Management System Design and Integration Issues, in Proceedings of IEEE, Stockholm Power Tech Conference, June 1995.

Blair, W.E. et al., A methodology for economic evaluation of distribution automation, *IEEE Trans. Power Appar. Syst.*, 104, 2954–2960, 1985.

Gordon, M.E. and Redmon, J.R., Electric Cooperatives and Distribution Automation: a Survey, in Proceedings of 1991 Rural Electric Power Conference, Dearborn, MI, April 1991, pp. A1/1–A1/6.

Venkata, S.S. et al., Applying AT system in the T&D arena, *IEEE Comput. Applic. Power*, 6, 29–34, 1993.

Wada, M. et al., Development of Remote Meter Reading System for Distribution Automation, in Proceedings of Seventh International Conference on Metering Apparatus and Tariffs for Electric Supply, Glasgow, November 1992.

Demand-Side Management

Bhatnagar, R. and Rahman, S., Dispatch of direct load control for fuel cost minimization, *IEEE Trans. Power Syst.*, 1, 1986.

Bohlin, P. et al., Successful implementation of a nation-wide load management system, *IEEE Trans. Power Syst.*, 1, 90–95, 1986.

Chan, M.L. et al., Integrating load management into energy management system's normal operations: primary factors, *IEEE Trans. Power Syst.*, 1, 152–157, 1986.

Chen, J., Lee, F.N., Briehpohl, A.M., and Adapa, R., Scheduling Direct Load Control To Minimize System Operational Cost, Paper 95 WIVI 196-6 PWRS, presented at Winter IEEE/PES Meeting, January 31, New York.

Chen, C.S. and Leu, J.T., Interruptible load control for Taiwan Power Company, *IEEE Trans. Power Syst.*, 5, 460–465, 1990.

DeAlmeida, A.T. and Vine, E.L., Advanced monitoring technologies for the evaluation of demand-side management programs, *IEEE Trans. Power Syst.*, 9, 1691–1697, 1994.

Effler, L. et al., Optimization of energy procurement and load management, *IEEE Trans. Power Syst.*, 7, 327–333, 1992.

Gellings, C.W., The concept of demand side management alternatives, *Proc. IEEE*, 73 (10), 1468–1470, 1985.

Geiger, D.L. and Samaneigo, G.M., Evaluation of load management as an electric system resource, *IEEE Trans. Power Syst.*, 1, 137–143, 1986.

Gustafson, M.W. et al., Direct water heater load control: estimating program effectiveness using an engineering model, *IEEE Trans. Power Syst.*, 8, 41–47, 1993.

Joskow, P.L. and Marron, D.B., What does a megawatt really cost? Evidence from utility conservation programs, *Energy J.*, 14 (3), 715–720, 1992.

Majumdar, S., Chattopadhyay, D., and Parikh, J., Interruptible Load Management Using Optimal Power Flow Analysis, Paper 95 SM 501-7 PWRS, IEEE/PES Summer Meeting, Portland, OR, 1995.

McRae, M.R., Scheer, R.M., and Smith, B.A., Integrating load management programs into utility operations and planning with a load reduction forecasting system, *IEEE Trans. Power Appar. Syst.*, 104, 1031–1043, 1985.

Nelson, S.K. and Hobbs, B.F., Screening DSM programs with a value-based test, *IEEE Trans. Power Syst.*, 7, 1–1043, 1992.

Runnels, J.E. and Whyte, M.D., Evaluation of demand side management alternatives, *Proc. IEEE*, 73, 1489–1495, 1985.

Shekel, J.S., Hardware/firmware considerations for integrating load management into system operations, *IEEE Trans. Power Syst.*, 1, 132–136, 1986.

White, K., The economics of conservation, *IEEE Trans. Power Appar. Syst.*, 100, 4546–4552, 1981.

Voltage/VAr Control

Abdul-Rahman, K.H. and Shahidehpour, S.M., Application of fuzzy sets to optimal reactive power planning with security constraints, *IEEE Trans. Power Syst.*, 9, 589–597, 1994.

Aoki, K. et al., An Efficient Algorithm for Load Balancing of Transformer and Feeders by Switch Operation in Large Scale Distribution Systems, Paper 87-SM 543-2, IEEE/PES Summer Meeting, 1987.

Aoki, K. et al., Voltage Drop Constrained Restoration of Supply by Switch Operation in Distribution Systems, Paper 87-SM 544-0, JEEE/PES Summer Meeting, 1987.

Aoki, K. et al., Optimal VAr planning by approximation method for recursive mixed-integer linear programming, *IEEE Trans. Power Syst.*, 3, 1741–1747, 1988.

Borozan, V. et al., Improved method for loss minimization in distribution networks, *IEEE Trans. Power Syst.*, 10, 1420–1425, 1995.

Borozan, V., Minimum Loss Reconfiguration of Unbalanced Distribution Networks, Paper 96WM 343-4 PWRD, presented at 1996 IEEE/PES Winter Meeting, Baltimore, 1996.

Chen, T. et al., Distribution system power flow analysis: a rigid approach, *IEEE Trans. Power Delivery*, 6, 1146–1153 1991.

Cheng, C.S. and Shirmohammadi, D., A three phase power flow method for real-time distribution system analysis, *IEEE Trans. Power Syst.*, 10, 671–679, 1995.

Chiang, H.D. and Jean-Jumeau, R., Optimal Network Reconfigurations in Distribution Systems: Part 2, Solution Algorithms and Numerical Results, Paper 90 WJVI 165-1, presented at 1990 IEEE/PES Winter Meeting, Atlanta, 1990.

Chiang, H.D. and Jean-Jumeau, R., Optimal network reconfigurations in the distribution systems: part 1, a new formulation and a solution methodology, *IEEE Trans. Power Delivery*, 5, 1902–1909, 1990.

Cova, B. et al., Contingency constrained optimal power flow procedure for voltage control in planning and operation, *IEEE Trans. Power Syst.*, 10, 602–608, 1995.

Deeb, N. and Shahidehpour, S.M., Decomposition approach for minimizing real power losses in power system, *Proc. lEEE C*, 138, 27–38, 1991.

Deeb, N. and Shahidehpour, S.M., Cross decomposition for multi-area optimal reactive power planning, *IEEE Trans. Power Syst.*, 8, 1539–1544, 1993.

Dwyer, A. et al., Load to voltage dependency tests at B.C. Hydro, *IEEE Trans. Power Syst.*, 10, 709–715, 1995.

El-Kady, M.A. et al., Assessment of real-time optimal voltage control, *IEEE Trans. Power Syst.*, 1, 98–107, 1986.

Fan, J. et al., Distribution Network Reconfiguration: Single Loop Optimization, Paper 96WM 168-5 PWRS, presented at 1996 IEEE/PES Winter Meeting, Baltimore, 1996.

Goldberg, D.E., *Genetic Algorithm in Search*, Addison-Wesley, Reading, MA, 1989.

Grainger, J.J. and Civanlar, S.C., Voltage/VAr control on distribution systems with lateral branches using shunt capacitors and voltage regulators: parts I, II, III, *IEEE Trans. Power Appar. Syst.*, 104, 3169–3175, 1985.

Gu, Z. and Rizy, D.T., Neural Networks for Combined Control of Capacitor Banks and Voltage Regulators in Distribution Systems, paper presented at IEEE/PES Winter Meeting, Baltimore, 1996.

Hsu, Y.Y. et al., Voltage control using a combined integer linear programming and rule-based approach, *IEEE Trans. Power Syst.*, 7, 744–752, 1992.

Hsu, Y.Y. and Yang, C.C., A hybrid artificial neural network-dynamic programming approach for feeder capacitor scheduling, *IEEE Trans. Power Syst.*, 9, 1069–1075, 1994.

Huneault, M. et al., A study of knowledge engineering tools in power engineering applications, *IEEE Trans. Power Syst.*, 9, 1825–1832, 1994.

Iba, K., Reactive power optimization by genetic algorithm, *IEEE Trans. Power Syst.*, 9, 685–692, 1994.

Kim, H. et al., Artificial neural network based feeder reconfiguration for loss reduction in distribution systems, *IEEE Trans. Power Delivery*, 8, 1991.

Kim, H. et al., Network reconfiguration algorithm for automated distribution systems based on artificial intelligence approach, *IEEE Trans. Power Delivery*, 8, 1933–1941, 1993.

Liu, C. and Tomsovic, K., An expert system assisting decision making of reactive power and voltage control, *IEEE Trans. Power Syst.*, 1, 195–201, 1986.

Markushevich, N.S. et al., Functional Requirements and Cost-Benefit Study for Distribution Automation at B.C. Hydro, paper presented at 1993 IEEE PICA Conference, Scottsland, AZ, May 1993, pp. 169–178.

Markushevich, N.S., Voltage and VAr Control in Automated Distribution Systems, in Proceedings of 3rd International Symposium on Distribution Automation and Demand Side Management, Palm Springs, California, 1993, pp. 478–485.

Merlin, A. and Back, H., Search for a Minimal Loss Operating Spanning Tree Configuration for an Urban Power Distribution, in Proceedings of PSOC, Cambridge, 1975.

Mondon, E. et al., MARS: an aid for network restoration after local disturbance, *IEEE Trans. Power Delivery*, 6, 850–855, 1991.

Naga Raj, B. and Rao, K.S.P., A new fuzzy reasoning approach for load balancing in distribution system, *IEEE Trans. Power Syst.*, 10, 1420–1432, 1995.

Nara, K. et al., Implementation of Genetic Algorithms for Distribution System Loss Minimization Reconfiguration, Paper 91 SM 467-1 PWRS, presented at IEEE/PES Summer Meeting.

Qiu, J. and Shahidehpour, S.M., A new approach for minimizing power losses and improving voltage profile, *IEEE Trans. Power Syst.*, 2, 1044–1051, 1987.

Rama Iyer, S. et al., Optimal reactive power allocation for improved system performance, *IEEE Trans. Power Appar. Syst.*, 103, 287–295, 1984.

Ramos, J.L.M. et al., A Hybrid Tool To Assist the Operator in Reactive Power/Voltage Control and Optimization, Paper 94 SM 537-1 PWRS, presented at IEEE/PES Summer Meeting, July 1994, San Francisco.

Santoso, N.I. and Tan, O.T., Neural-net based real-time control of capacitors installed on distribution systems, *IEEE Trans. Power Delivery*, 5, 266–272, 1990.

Shirmohammadi, D. et al., A compensation based power flow method for weakly meshed distribution and transmission networks, *IEEE Trans. Power Syst.*, 3, 753–776, 1988.

Shirmohammadi, D. and Hong, H., Reconfiguration of electric distribution networks for resistive line losses reduction, *IEEE Trans. Power Delivery*, 4, 1492–1498, 1989.

Taleski, R. and Rajicic, D., Distribution Network Reconfiguration for Energy Loss Reduction, Paper 96W1V1 305-3 PWRS, presented at 1996 IEEE/PES Winter Meeting, Baltimore, 1996.

Wagner, W.R. et al., A rule-based approach to decentralized voltage control, *IEEE Trans. Power Syst.*, 5, 643–651, 1990.

Distributed Generation and Storage Dispatch

Baughman, M.L. et al., Optimizing combined cogeneration and thermal storage systems: an engineering economic approach, *IEEE Trans. Power Syst.*, 4, 974–980, 1989.

Billinton, R. and Chowdhury, A., Generation adequacy impacts of cogenerator sources, *IEE Proc. C*, 137, 1990.

Billinton, R. and Ghedy, F., Effects of Non-Utility Generators on Composite System Adequacy Evaluation, paper presented at IEEE/PES Summer Meeting, Seattle, 1992.

Ghoudjehbaklou, H.G. and Puttgen, H.B., Optimization topics related small power producing facilities under spot pricing policies, *IEEE Trans. Power Syst.*, 2, 296–302, 1987.

Hamound, G.A. et al., Reliance on non-utility generation in planning customer delivery systems, *IEEE Trans. Power Syst.*, 9, 1795–1802, 1994.

Kim, J.C. and Ann, B.H., On the economics of cogeneration: pricing and efficiency in government owned utilities, *Energy J.*, 11 (1), 87–99, 1990.
Kuwahata, A. and Asano, H., Utility-cogenerator game for pricing power sales and wheeling fees, *IEEE Trans. Power Syst.*, 9, 1875–1879, 1994.
MacGregor, P.R. and Puttgen, H.B., A spot price control mechanism for electric utility systems with small power producing facilities, *IEEE Trans. Power Syst.*, 6, 683–690, 1991.
MacGregor, P.R. and Puttgen, H.B., The integration of non-utility generation and spot prices within utility generation scheduling, *IEEE Trans. Power Syst.*, 9, 1302–1308, 1994.
Maeda, A. and Kaya, Y., Game-theory approach to use of non-commercial power plants under time-of-use pricing, *IEEE Trans. Power Syst.*, 7, 1052–1059, 1992.
Mukerji, R. et al., Evaluation of wheeling and non-utility generation (NTJG) options using optimal power flow, *IEEE Trans. Power Syst.*, 7, 201–207, 1992.
Ponrajah, R.A. and Galiana, F.D., Derivation and applications of optimum bus incremental costs in power system operation and planning, *IEEE Trans. Power Appar. Syst.*, 104, 3416–3422, 1985.
Prince, W. et al., Current operational problems associated with non-utility generation, *IEEE Trans. Power Syst.*, 4, 1534–1541, 1989.
Puttgen, H.B. and MacGregor, P.R., Optimum scheduling procedures for cogeneration of small power producing facilities, *IEEE Trans. Power Syst.*, 4, 957–964, 1989.
Ramanathan, R., Real-time wheeling losses computation techniques for energy management systems, *IEEE Trans. Power Syst.*, 1, 314–320, 1986.
Rooijers, F.R. and Amerongen, R.A.M.V., Static Economic Dispatch for Cogeneration Systems, Paper 93 SM 468-9 PWRS, presented at IEEE/PES Summer Meeting, July 1993, Vancouver.
Whipple, D.P. and Trefny, F.J., Current electric system problems from a cogenerator's viewpoint, *IEEE Trans. Power Syst.*, 4, 1037–1042, 1989.

Communication

Celik, M.K., Integration of Advanced Applications for Distribution Automation, IEEE, 0-7803-4403-0, 1998, pp. 366–369.
Chao, Y., Lee, S., and Chang, H., Application of Automated Mapping System to Distribution Transformer Load Management, IEEE, 0-7803-7525-4, 2002, pp. 1179–1184.

Automatic Meter Reading

Aberson, M., Smart Meter's Protocol: Ewit Base, IEEE, 0-7803-3879-0, pp. 293–297.
Albuyeh, F. and Alaywan, Z., California ISO Formation and Implementation, *IEEE Computer Applications in Power*, ISSN 11895-0156, October 1999, pp. 30–34.
Ando, N., Takashima, M. et. al., Automatic Meter Reading System Adopting Automatic Routing Technology, IEEE, 0-7803-755-4, 2002, pp. 2305–2309.

Cavdar, I.H., A solution to remote detection of illegal electricity usage via power line communications, *IEEE Trans. Power Delivery*, 19, 1663–1667, 2004.
Cho, M.Y. and Huang, C.W., Development of PC Based Energy Management System for Electrical Energy Saving of High Voltage Customer, IEEE, 0-7803-7055-4, 2001, pp. 7–12.
Cooper, D., Low-data-rate narrow-band power-line communications on the European domestic mains: symbol timing estimation, *IEEE Trans. Power Delivery*, 20, 664–667, 2005.
De, S., Anand, R. et al., E-Metering Solution for Checking Energy Thefts and Streamlining Revenue Collection in India, IEEE, 0-7803-8110-6, 2003, pp. 654–658.
Duncan, B.K. and Bailey, B.G., Protection, metering, monitoring, and control of medium-voltage power systems, *IEEE Trans. Ind. Applic.*, 40, 33–40, 2004.
Fischer, R.A., Laakonen, A.S., and Schulz, N.N., A general polling algorithm using a wireless AMR system for restoration confirmation, *IEEE Trans. Power Syst.*, 16, 312–316, 2001.
Kearney, S., The Age of Advanced Metering Arrives, January 2005, pp. C6-1–C6-4.
Lim, T.Y. and Chan, T., Experimenting remote kilowatt hour meter reading through low-voltage power lines at dense housing estates, *IEEE Trans. Power Delivery*, 17, 708–711, 2002.
Newbury, J. and Miller, W., Potential metering communication services using the public Internet, *IEEE Trans. Power Delivery*, 14, 1202–1207, 1999.
Newbury, J. and Miller, W., Multi-protocol routing for automatic remote meter reading using power line carrier systems, *IEEE Trans. Power Delivery*, 16, 1–5, 2001.
Oya, H., Hase, S., and Shimada, T., Multi-Service Network for Advanced AMR, IEEE, 0-7803-5935-6, 2000, pp. 1680–1684.
Wallin, F., Bartusch, C. et. al., The Use of Automatic Meter Readings for a Demand-Based Tariff, Paper 0-7803-9114-4, presented at 2005 IEEE/PES Transmission and Distribution Conference and Exhibition: Asia and Pacific, Dalian, China, 2005, pp. 1–6.
Wang, H. and Schulz, N.N., A revised branch current-based distribution system state estimation algorithm and meter placement impact, *IEEE Trans. Power Syst.*, 19, 207–213, 2004.
Wu, C., Chang, S., and Huang, Y., Design of a Wireless ARM-Based Automatic Meter Reading and Control System, pp. 1–6.
Zhao, S., Li, B. et al., Research on Remote Meter Automatic Reading Based on Computer Vision, Paper 0-7803-9114-4, presented at 2005 IEEE/PES Transmission and Distribution Conference and Exhibition: Asia and Pacific, Dalian, China, 2005, pp. 1–4.

Communication Media

Chaffanjon, D. and Duval, G., Differential and common mode propagation in PLC low voltage networks, *IEEE Trans. Power Delivery*, 14, 327–334, 1999.
Matsuo, T. and Maekawa, M., Field Test of the World First 200 Mbps PLC Modems, IEEE, 0-7803-8834-8, 2005, pp. 5330–5332.
Nissen, T. and Peterchuck, D., Substation integration pilot project, *IEEE Power Energy*, 2, 42–49, 2003.

Ostertag, M. and Imboden, Ch., High Data Rate, Medium Voltage Powerline Communications for Hybrid DA/DSM, IEEE, 0-7803-5515-6, 1999, pp. 240–245.

Patrick, A., Newbury, J., and Gargan, S., Two-way communications systems in the electricity supply industry, *IEEE Trans. Power Delivery,* 13, 53–58, 1998.

Sidhu, T.S., Demeter, E., and Faried, S.O., Power System Protection and Control Integration over Ethernet-Based Communication Channels, IEEE, 0-7803-8253-6, 2004, pp. 225–228.

Distribution Automation and Decision Analysis

Moon, Y. and Cho, B., Fault Restoration Algorithm Using Fast Tracing Technique Based on the Tree-Structured Database for the Distribution Automation System, IEEE, 0-7803-6420-1, 2000, pp. 411–415.

General Distribution Automation Papers

Ackerman, W.J., Substation Automation and the EMS, IEEE, 0-7803-5515-6, 1999, pp. 275–279.

Ault, G.W., Foote, C.E.T., and McDonald, J.R., U.K. Research Activities on Advanced Distribution Automation, IEEE, 0-7803-9156-X, 2005, pp. 1–4.

Baran, M.E., Data Requirements for Real-Time Monitoring and Control of Feeders, IEEE, 0-7803-4403-0, 1998, pp. 374–376.

Booth, C., McDonald, J.R., and Verster, P., Dynamic Network Reconfiguration for Medium Voltage System Automation, IEEE, 0-7803-5515-6, 1999, pp. 746–752.

Chan, F.C., Distribution Automation System: from Design to Implementation, IEEE, 0-7803-5935-6, 2000, pp. 2368–2373.

Fanning, R. and Huber, R., Distribution Vision 2010: Planning for Automation, IEEE, 0-7803-9156-X, 2005, pp. 1–2.

Fujisawa, A. and Kurokawa, N., Overseas Distribution System Based on Japanese Experience, IEEE, 00-7803-7525-4, 2002, pp. 1164–1169.

Geisler, K.I., Nielsen, T.D. et al., The Rise of Energy Delivery Management Systems, IEEE, 01TD191, March 2001, pp. 895–900.

Gupta, R.P., Tiwari, S., and Varma, R.K., Novel Software Architecture for Power Distribution Automation, IEEE, 0-7803-7989-6, 2003, pp. 1598–1603.

He, Y., Andersson, G., and Allan, R.N., Modeling the Impact of Automation and Control on the Reliability of Distribution Systems, IEEE, 0-7803-6420-1, 2000, pp. 79–84.

Hoffman, R. and Aouad, P., Communications for a Large Distribution Automation Project in Thailand, IEEE Power Engineering Society General Meeting, Vol. 2, July 2003, pp. 1598–1603.

Hunt, R.K. and Proudfoot, D., Improving the Operation of Distribution Substations, IEEE, 0-7803-7285-9, 2001, pp. 511–515.

Khedkar, M.K. and Gohokar, V.N., An Integrated Approach for Automation of Distribution System, IEEE, 0-7803-7525-4, 2002, pp. 2106–2110.

Kim, M.S. and Hyun, D.H., The Development of an Intelligent and Integrated Gateway System for the Automation Systems in Power Utilities, IEEE, 0-7803-7525-4, 2002, pp. 2–5.
Kusano, N., New Trends in Protection Relays and Substation Automation Systems in Japan, IEEE, 0-7803-7525-4, 2002, pp. 624–628.
Ockwell, G.L., Implementation of Network Reconfiguration for Taiwan Power Company, IEEE, 0-7803-7989-6, 2003, pp. 2430–2434.
Pahwa, A., Planning and Analysis Tools to Evaluate Distribution Automation Implementation and Benefits, IEEE, 0-7803-9156, 2005, pp. 1–2.
Proudfoot, D., Innovative Substation Design: the Bay Controller Concept, IEEE, 0-7803-5589-X, 1999, pp. 953–959.
Rudolph, D.L., An Integrated Solution to Substation Automation, IEEE Rural Electric Power Conference, April 1998, pp. C2-1–C2-9.
Seol, J., Ha, B. et al., Microprocessor and Integrated Electronic Technology, IEEE, 0-7803-9114-4, presented at 2005 IEEE/PES Transmission and Distribution Conference and Exhibition: Asia and Pacific, Dalian, China, 2005, pp. 1–5.
Soma, O., Valet, Z.A. et al., Object-Oriented Agents in Power Distribution Automation, paper presented at 10th IEEE Mediterranean Electrotechnical Conference, MEleCon, vol. III, 2000, pp. 891–894.
Staszesky, D. and Meisinger, M., Use of Distributed Intelligence for Reliability Improvement Using Minimum Available Distribution Assets, IEE, 0-7803-9114-4, presented at 2005 IEEE/PES Transmission and Distribution Conference and Exhibition: Asia and Pacific, Dalian, China, 2005, pp. 1-5.
Su, C.L., Lu, C.N., and Lin, M.C., Wide Area Network Performance Study of a Distribution Management System, IEEE, 0-7803-55I5-6, 1999, pp. 136–141.

Distribution Automation System Graphical User Interface

Fan, J., Zhao, H. et al., A New Design of Modern Power Automation Platform, IEEE, 0-7803-9114-4, presented at 2005 IEEE/PES Transmission and Distribution Conference and Exhibition: Asia and Pacific, Dalian, China, 2005, pp. 1–5.
Lee, S.T., The EPRl Common Information Model for Operation and Planning, IEEE, 0-7803-5569-5, 1999, pp. 866–871.
Lee, S., Park, J. et al., Visual Power Distribution Load Flow Simulator for Insertion of Distributed Generations, IEEE, 0-7803-9156-X, 2005, pp. 1–6.

Distribution Papers

Ackerman, W.J., Substation Automation and EMS, IEEE, 0-7803-5515-6, 1999, pp. 274–279.
Adams, R.C., Moeller, K., and Rockway, J.W., The Joint Tactical Radio and the Navy RF Distribution System Distribution, pp. 359–362.
Borlase, S.H., Advancing to true station and distribution system integration in electric utilities, *IEEE Trans. Power Delivery,* 13, 129–134, 1998.

Cho, I.-K. and Meyn, S.P., Dynamics of Ancillary Service Prices In Power Distribution Systems, proc. 42nd IEEE Conf. Decision and Control, Maui, December 2003, pp. 2094–2099.

Fujisaki, H., On Distributions of Multiple Access Interference for Spread Spectrum Communication Systems Using M-Phase Spreading Sequences of Markov Chains, IEEE, 0-7802-8251-X, pp. IV-609–IV-602.

Hayashi, H., Oka, M. et al., Rapidly Increasing Application of Intranet Technologies for SCADA (Supervisory Control and Data Acquisition System), IEEE, 0-7803-7525, 2002, pp. 22–25.

Khodr, H.M., Molea, J. et al., Standard levels of energy losses in primary distribution circuits for SCADA application, *IEEE Trans. Power Syst.*, 17, 615–620, 2002.

Singh, N., Kiissel, R. et al., Power system modeling and analysis in a mixed energy management and distribution management system, *IEEE Trans. Power Syst.*, 13, 1143–1149, 1998.

Michelena, E.D. and Gutman, S.I., An Automatic Meteorological Data Collection System that Is Installed at Global Positioning System Monitoring Stations, IEEE, 0-7803-7534-3, 2002, pp. 1930–1934.

Ostertag, M. and Imboden, Ch., High Data Rate, Medium Voltage Powerline Communications for Hybrid DA/DSM, IEEE, 0-7803-5515-6, 1999, pp. 240–245.

Pimpa, C. and Premrudeepreechacharn, S., Voltage Control in Power System Using Expert System Based on SCADA System, IEEE, 0-7803-7322-7, 2002, pp. 1282–1286.

Su, C.L., Lu, C.N., and Lin, M.C., Wide Area Network Performance Study of a Distribution Management System, IEEE, 0-7803-55I5-6, 1999, pp. 136–141.

Wang, Q. and Qian, Q., Design and Analysis of Communication Network for Distributed SCADA System, IEEE, 0-7803-5935-6, 2000, pp. 2062–2065.

Wang, H. and Schulz, N.N., A revised branch current-based distribution system state estimation algorithm and meter placement impact, *IEEE Trans. Power Syst.*, 19, 207–213, 2004.

Demand-Side Management

Celik, M.K., Integration of Advanced Applications for Distribution Automation, IEEE, 0-7803-4403-0, 1998, pp. 366–369.

He, Y. and Deng, Y., A Novel Architecture of Distribution Management System, IEEE, 0-7803-6420-1, 2000, pp. 67–72.

Jaaksoo, U. and Utkin, V.I., Automatic Control, Proc. 11th Triennial World Congress Int. Fed. Automatic Control, Tallinn, Estonia, USSR, vol. VI, August 1990, pp. 24–30.

Jackson, C.E. and Evans, J.W., A Network-less Automation Implementation: Case Study, IEEE, 0-7803-6420-1, 2000, pp. 579–582.

Kezunovic, M., Integrating Data and Sharing Information from Various IEDs To Improve Monitoring, Condition-Based Diagnostic, Maintenance, Asset Management and Operation Tasks, EPRI Substation Equipment Disturbance Conference, New Orleans, Louisiana, February 2004, pp. 1–11.

Kontogiannis, C.C. and Safacas, A.N., An Expert System for Power Plants, Department of Electrical and Computer Engineering, University of Patras, 2002, pp. 1–7.

Miranda, V. and Matos, M., Intelligent Tools in a Real World DMS Environment, IEEE, 0-7803-6420-1, 2000, pp. 163–168.
Moore, M.S., Monemi, S., and Wang, J., Integrating Information Systems in Electric Utilities, Volume 1, October 2000, pp. 399–404.
Moser, A., Ejebe, G.C., and Frame, J.G., Network and Power Applications for EMS within a Competitive Environment, IEEE, 0-7803-5515-6, 1999, pp. 280–285.
Oman, P.W., Roberts, J., and Schweitzer, E.O. III, Tools for Protecting Electric Power Systems from Electronic Intrusions, Proceedings of the 3rd Annual Western Power Delivery Automation Conferences, Spokane, WA, April 2001, pp. 1–21.
Serizawa, Y., Ohba, E. et al., Conceptual Design for Distributed Real-Time Computer Network Architecture, IEEE, 0-7803-7535-4, 2002, pp. 26–31.
Silva, M.P., Saraiva, J.T., and Sousa, A.V., A Web Browser Based DMS Distribution Management System, IEEE, 0-7803-6420-1, 2000, pp. 2338–2343.
Singh, N., Kiissel, R. et al., Power system modeling and analysis in a mixed energy management and distribution management system, *IEEE Trans. Power Syst.*, 13, 1143–1149, 1998.
Su, R. and Yurcik, W., A Survey and Comparison of Human Monitoring of Complex Networks, 10th International Command and Control Research and Technology Symposium (CCRTS), Mclean, Virginia, 2005, pp. 1–10.
Tram, H., The ASP Model for Energy Delivery Information Systems, IEEE, 0-7803-7285-9, 2001, pp. 754–758.

Intelligent Systems References

Holland, J.H., Genetic Algorithms and the Optimal Allocation of Trials, *SIAM Journal on Computing*, Vol. 2, No. 2, pp. 88–105, 1973.
Zadeh, L.A., Fuzzy Sets, *Information and Control*, 8(3), pp. 338–353, 1965.
Zadeh, L.A., A New Direction in AT, Toward a Computational Theory of Perceptions, *American Association for Artificial Intelligence*, pp. 73–84, 2001.

Fault Analysis

Mori, H., Yamanaka, T. et al., A Fault Detection Technique with Preconditioned ANN in Power Systems, IEEE, 0-7803-7525-4, 2002, pp. 758–763.

GIS

Arai, C., Matsuda, N., and Shikada, M., Management of Mapping in Local Government Using Remote Sensing and the Real Time GIS, IEEE, 0-7803-7536-X, 2002, pp. 3145–3147.
Choi, H., Kim, K., and Lee, J., Design and Implementation of Open GIS Component Software, IEEE, 0-7803-6359-0, 2000, pp. 2105–2107.

Han, M., Tian, X., and Xu, X., Research on Data Collection and Database Update of GIS Based on GPS Technology, IEEE, 0-7803-9050-4, 2005, pp. 920–923.
Kim, D., Kim, K. et al., The Design and Implementation of Open GIS Service Component, IEEE, 0-7803-7031-7, 2001, pp. 1922–1924.
Lav, C.T., Staley, D.B., and Olsen, T.W., Practical Design Considerations for Application of GIS MV Switchgear, IEEE, 0-7803-7956-X, 2003, pp. 93–100.
Lav, C.T., Staley, D.B., and Olsen, T.W., Practical design considerations for application of GIS MV Switchgear, *IEEE Trans. Ind. Appl.*, 40, 1427–1434, 2004.
Mulaku, G.C., Accurate mapping: the first step to better spatial information management by African utilities, *AJST*, 5 (1), 29–33, 2004.
McCoy, J., GIS and Joint Use Management: a Productive Combination, Rural Electric Power Conference, May 2005, pp. C2/1–C2/4.
Okuno, A. and Shikada, M., Application of Real Time GIS Using Remote Sensing and RTK-GPS for Local Government, IEEE, 0-7803-8742-2, 2004, pp. 4790–4792.
Sun, Q., Chi, T. et al., Design of Middleware Based Grid GIS, IEEE, 0-7803-9050-4, 2005, pp. 854–857.
Teng, W., Pollack, N. et al., GIS and Data Interoperability at the NASA Goddard DAAC, IEEE, 0-7803-7031-7, 2001, pp. 1953–1955.
Yanfeng, S., Zhuo, C. et al., A Compensation Mechanism in GIS Web Service Composition, IEEE, 0-7803-9050-4, 2005, pp. 940–943.
Yang, J. and Yang, C., Research on High-Performance Web GIS System for Map Symbol Dynamic Editing and Network Publishing, IEEE, 0-7803-8742-2, 2004, pp. 2971–2974.

General Papers

Aarts, E.H.L. and Korst, J., *Simulated Annealing and Boltzmann Machines*, John Wiley and Sons, New York, 1989.
Arrillaga, J., Arnold, C.P., and Harker, B.J., *Computer Modelling of Electrical Power Systems*, John Wiley and Sons, New York, 1983.
Baran, M.E. and Wu, F.F., Network reconfiguration in distribution systems for loss reduction and load balancing, *IEEE Trans. Power Delivery*, 4, 1401–1407, 1989.
Bergen, A.R., *Power Systems Analysis*, Prentice-Hall, Englewood Cliffs, NJ, 1986.
Broadwater, R.P., Khan, A.H., Shaalan, H.E., and Lee, R.F., Time Varying Load Analysis to Reduce Distribution Losses through Reconfiguration, Paper 92 WM 269-1, presented at IEEE/PES 1992 Winter Meeting, New York, 1992.
Chen, T.H., Generalized Distribution Analysis System, Ph.D. dissertation, The University of Texas at Arlington, May 1990.
Chen, T.H., Chen, M.S. et al., Distribution system power flow analysis: a rigid approach, *IEEE Trans. Power Delivery*, 6, 1146–1152, 1991.
Chiang, H.D., A decoupled load flow method for distribution power networks: algorithms, analysis and convergence study, *Electr. Power Energy Syst.*, 13 (3), 130–138, 1991.
Chiang, H.D. and Baran, M.E., On the existence and uniqueness of load flow solution for radial distribution power networks, *IEEE Trans. Circuits Syst.*, 37, 410–416, 1990.

Chiang, H.D. and Jean-Jumeau, R.M., Optimal network reconfigurations in distribution systems: part 1, a new formulation and a solution methodology, *IEEE Trans. Power Delivery*, 5, 634–642, 1990.

Chiang, H.D. and Jean-Jumeau, R.M., Optimal network reconfigurations in distribution systems: part 2, a solution algorithm and numerical results, *IEEE Trans. Power Delivery*, 5, 643–649, 1990.

Civanlar, S., Grainger, J.J. et al., Distribution feeder reconfiguration for loss reduction, *IEEE Trans. Power Delivery*, 3, 1217–1223, 1988.

Erny, V., Thermodynamic approach to the traveling salesman problem: an efficient simulation algorithm, *J. Optimization Theory Appl.*, 45, 41–51, 1985.

Kendrew, T.J. and Marks, J.A., Automated distribution comes of age, *IEEE Comput. Appl. Power*, Volume 2, 7–10, 1989.

Kirkpatrick, S., Gelatt, C.D. Jr., and Vecchi, M.P., Optimization by Simulated Annealing, IBM Research Report RC 9355, 1982.

Kirkpatrick, S., Gelatt, C.D. Jr., and Vecchi, M.P., Optimization by simulated annealing, *Science*, 220, 671–680, 1983.

Kojovic, L.A. and Witte, J.F., Improved Relay Coordination and Relay Response Time by Integrating the Relay Functions, IEEE, 0-7803-6420-1, 2000, pp. 1202–1207.

Metropolis, N., Rosenbluth, A. et al., Equation of state calculations by fast computing machines, *J. Chem. Phys.*, 21, 1087–1092, 1953.

Nara, K., Shiose, A. et al., Implementation of Genetic Algorithm for Distribution Systems Loss Minimum Reconfiguration, Paper 91 SM 467-1, presented at IEEE/PES 1991 Summer Meeting, San Diego, July 1991.

Steve, K., How Outage Management Systems Can Improve Customer Service, IEEE, 0-7803-4883-4, 1998, pp. 172–178.

Taylor, T. and Lubkeman, D., Implementation of heuristic search strategies for distribution feeder reconfiguration, *IEEE Trans. Power Delivery*, 5, 239–246, 1990.

Tsai, S., Wu, S. et al., Integrated Home Service Network on Intelligent Intranet, IEEE, 0-7803-6301-9, 2000, pp. 104–105.

Van Laarhoven, P.J.M. and Aarts, F.H.L., *Simulated Annealing: Theory and Applications*, Reidel, Dordrecht, 1987.

ACKNOWLEDGMENT Please note that I acknowledge the use of work which contributed to this text but was overlooked in the noted references above. Those contributions are greatly appreciated, and I hope that the reader's knowledge will be improved through the combinations of all of these works.

Index

D

E

F

N

O

P

Q

R

NEW PERSPECTIVES ON

Adobe® Photoshop® CS6

INTRODUCTORY

Jane Hosie-Bounar

COURSE TECHNOLOGY
CENGAGE Learning®

Australia • Brazil • Japan • Korea • Mexico • Singapore • Spain • United Kingdom • United States

COURSE TECHNOLOGY
CENGAGE Learning

New Perspectives on Adobe Photoshop CS6, Introductory, International Edition

Vice President, Career and Computing: David Garza

Editor in Chief: Marie L. Lee

Director of Development: Marah Bellegarde

Executive Editor: Donna Gridley

Associate Acquisitions Editor: Amanda Lyons

Senior Product Manager: Kathy Finnegan

Product Managers: Leigh Hefferon, Julia Leroux-Lindsey

Senior Brand Manager: Elinor Gregory

Market Development Managers: Gretchen Swann, Kristie Clark

Developmental Editor: Mary Pat Shaffer

Senior Content Project Manager: Jennifer Goguen McGrail

Composition: GEX Publishing Services

Art Director: Marissa Falco

Text Designer: Althea Chen

Cover Designer: Roycroft Design

Cover Art: © Deb Casso/OJO Images/Getty Images

Copyeditor: Karen Annett

Proofreader: Brandy Lilly

Indexer: Rich Carlson

Library of Congress Control Number: 2012945797

International Edition:

ISBN-13: 978-1-133-56268-9

ISBN-10: 1-133-56268-X

Cengage Learning International Offices

Asia
www.cengageasia.com
tel: (65) 6410 1200

Australia/New Zealand
www.cengage.com.au
tel: (61) 3 9685 4111

Brazil
www.cengage.com.br
tel: (55) 11 3665 9900

India
www.cengage.co.in
tel: (91) 11 4364 1111

Latin America
www.cengage.com.mx
tel: (52) 55 1500 6000

UK/Europe/Middle East/Africa
www.cengage.co.uk
tel: (44) 0 1264 332 424

Represented in Canada by
Nelson Education, Ltd.
tel: (416) 752 9100 / (800) 668 0671
www.nelson.com

Cengage Learning is a leading provider of customized learning solutions with office locations around the globe, including Singapore, the United Kingdom, Australia, Mexico, Brazil, and Japan. Locate your local office at: **www.cengage.com/global**

For product information: **www.cengage.com/international**
Visit your local office: **www.cengage.com/global**
Visit our corporate website: **www.cengage.com**

Printed in the United States of America
1 2 3 4 5 6 7 16 15 14 13 12